The Electric City

THE ELECTRIC CITY

*Energy and the Growth of the
Chicago Area, 1880–1930*

Harold L. Platt

The University of Chicago Press
Chicago and London

Harold L. Platt is professor of history at Loyola University of Chicago.

The University of Chicago Press, Chicago 60637
The University of Chicago Press, Ltd., London
© 1991 by The University of Chicago
All rights reserved. Published 1991
Printed in the United States of America

99 98 97 96 95 94 93 92 91   54321
Library of Congress Cataloging-in-Publication Data

Platt, Harold L.
    The electric city : energy and the growth of the Chicago area,
1880–1930 / Harold L. Platt.
        p.   cm.
    Includes index.
    ISBN 0-226-67075-9 (alk. paper) :
        1. Electric utilities—Illinois—Chicago Metropolitan Area—
History. 2. Electrification—Illinois—Chicago Metropolitan Area—
History. 3. Chicago Metropolitan Area (Ill.)—History.
4. Commonwealth Edison Company—History. 5. Insull, Samuel,
1859–1938. I. Title.

To Harold and Fern Hyman

Contents

Illustrations

Plates

Maps

Charts

Tables

Preface

This book is about the city and about the ways urban technology has affected everyday life. Today we are completely dependent on intensive energy use to sustain daily life in our homes, shops, factories, and public spaces, yet most of us are unconscious of the "invisible" world of energy that envelops us except on the rare occasions when a power blackout plunges us into darkness and everything comes to a sudden stop. Only then do we pause to contemplate the technological environment we live in. By studying the historical process that led to the building of this energy-intensive world, we can better understand our society and culture.

As industrialization fueled the rapid growth of cities in the nineteenth century, the need for public utilities and services became painfully evident. Terrifying epidemics, great fires, impassable streets, and other disorders led to the construction of a complex infrastructure of pipes and pavements, wires and railways to alleviate these problems. The rise of the "networked city" progressively transformed densely packed centers of population from death traps to places of convenience and luxury unparalleled in the rural heartland.[1] In the years following the Civil War, running water, indoor plumbing, gaslights, horse-drawn streetcars, and steam commuter trains—not to mention services such as home delivery of ice and coal—became generally available to the relatively few city dwellers who could afford them. Gradually, however, the luxuries of the elite became common necessities. Although public utilities and services lifted urban life to unprecedented levels of comfort, inherent flaws in existing technology and new problems arising from the city's continuing growth spurred the search for better solutions. In the 1870s, a brighter system of lighting and a mechanical substitute for the horsecar ranked high on the list of pressing needs for improving urban life.

Efforts to harness the powerful yet mysterious force of electricity engaged a wide range of scientists, inventors, and equipment manufacturers in Europe and the United States. Thomas Edison, of course, is the best known of these pioneers, but he was certainly not alone in the drive to open new frontiers of electrical technology. In fact Edison was a member of a fairly large community of electrical men who mostly worked in the telegraph business. A general awareness of demands for better urban services among this fraternity helps explain why several inventors here and abroad would lay claim almost simultaneously to being first to perfect a telephone in the mid-1870s and a commercially practical lighting device by the end of the decade. These widely heralded marvels of modern life set off a wild scramble for "ground floor" opportunities as entrepreneurs attempted to meet the city's insatiable pent-up demand for better communication and more light. Less than ten years later, the achievement of an electric motor for the streetcar would trigger a similar race to upgrade the city's transportation services.

Yet the rapid diffusion of electric light and power systems across the urban landscape of America was neither inevitable nor driven by some inherent technological imperative. On the contrary, contemporary social values and cultural orientations determined the pace and direction of technological change. In the decade following World War I, monopolistic central station utility companies and their regional networks of power emerged throughout the United States. This outcome, however, was the result of human choice in the form of business decisions, public policies, and consumer preferences, not the inexorable logic of the machine.[2] At the beginning stages of the electrical revolution, from the 1880s to the 1910s, sharp competition among technological systems offered several alternative routes to the future. Small as well as large-scale self-contained systems coexisted alongside the central station company, which also faced serious economic challenges from other utility operators in the same locality. The reasons the development of the supply of electricity narrowed to a single path are to be found in a study of American society and politics, and uncovering them is one of the chief purposes of this book. Equally important, a cultural ethos of progress through technology stimulated an increasingly popular desire for things electrical. An investigation into the interplay of technology and culture also figures predominantly in the story of the social construction of an energy-intensive style of life.

To shed light on the historical roots of our modern, highly technological environment requires that one focus on a single locality. Chicago, Illinois, offers an ideal test case for two related reasons. First, when Edison's light bulb was introduced Chicago was the fastest-growing big city in the world—the horror and wonder of the age. With 500,000 inhabitants in 1880, the urban center of the Midwest would gain more than a million more people over the next twenty years and surpass 2 million by the time of the 1910 census. Certainly no place in the United States offers a better opportunity to study the social and political processes of city building. Second, this incredible growth meant opportunity, attracting some of the most innovative electrical men and utility entrepreneurs of the day.

Among those seeking their fortunes in the preeminent "shock city" of the late nineteenth century was a young man named Samuel Insull. After serving an apprenticeship under Edison, the English immigrant arrived in Chicago at the time of the World's Fair of 1893—the city's coming-of-age as a cosmopolitan center of art, culture, and high society. He soon emerged as not only the largest central station operator in the Chicago area but also the most important national leader of the electric utility industry. Insull was a fascinating and controversial figure who still remains a subject of debate. He became the master salesman of central station service and the director of a vast empire of utility holding companies. Rising to become one of the most influential men in America during the twenties, he fell from power during the Great Depression of the thirties. Insull and his gospel of consumption epitomized contemporary aspirations to create a life-style of comfort, luxury, and leisure, a ubiquitous world of energy.

A large-scale research project like this one would not have been possible without the generous support and encouragement of a large number of people. It is impossible to acknowledge everyone who contributed, but, the list must include the individuals listed below. Perhaps the best place to begin is at home, Loyola University of Chicago. I am grateful for the support of the history department and especially its chairman, Joseph Gagliano. The university provided considerable financial aid through the help of the Office of Research Services, directed by Thomas Bennett and Timothy Austin. Abigail Byman supplied legal advice on copyright matters. I also received invaluable assistance from Brother Michael Grace and Valerie Gerrard Browne of the University Archives, which contain the Samuel Insull Papers.

As readers will soon notice, most of the primary source materials used in this book come from the historical archives of the Commonwealth Edison Company. I thank its chairman, James J.

O'Connor, for granting me permission to use these records. Many other company employees helped me gain access to the documents, including John Hogan, Ed Mantel, John Baker, and Mike Duba. In addition, company librarians Barbara Kelly and Grace Pertel deserve praise for their cooperation with an outside researcher.

Many colleagues helped make this a better book. I am especially indebted to Lewis Erenberg, who proved not only an untiring source of encouragement but a wellspring of creative ideas as well. I owe to him an appreciation of popular culture and its crucial importance to the social history of technology. Joel Tarr deserves special mention because he too served double duty. He has opened up the field of the history of urban infrastructure, giving it credibility and importance. He has also been a great supporter of my work, especially during its early stages. Many, many others have helped along the way, including Mark Rose, Paul Barrett, Ann Keating, Marty Melosi, Gabriel Dupuy, Michael Ebner, Cathy Conzen, Sue Hirsch, Alan Teller, Eric Monkkonen, and Roger Lochin. Finally, I wish to thank Harold Hyman, whose loyalty has never wavered in spite of my wandering far from the field of constitutional and legal history.

I appreciate the help of several people in making my study into a book. Janice Petit was a wonderful copy editor whose patience and quiet encouragement kept the project on track. Jerry Zbiral of the Collected Image helped bring the photographic and graphic reproductions used here up to the highest possible standards. Members of the University of Chicago Press staff have also extended a warm and generous spirit.

Most important, members of my family deserve my deepest gratitude for enduring the hardships of living with a writer of books. Words completely fail to express my feelings of love and appreciation for my wife, Carol Woodworth. My son, Dylan, deserves a thank you for his patience and understanding. My daughter, Abbey, also kept the faith over the long haul required to complete this project.

Part I

The Energy Revolution in the City, 1880–1898

1

Introduction: The Search for Better Lighting, 1848–1880

On the evening of 25 April 1878, a crowd gathered at the Water Tower on Chicago's North Side to see a "long talked of experiment." The superintendent of the city's telegraph fire alarm system, "Professor" John P. Barrett, and his assistants hurriedly made final preparations for the test as twilight faded to night. Sputtering and flashing, an unsteady but "very powerful" light turned the darkness back into day. Although "the light produced flickered a good deal," the crowd was awed and amazed, because just two of the newfangled devices generated more light than 650 of the standard gas-powered streetlamps then in use. By wiring the arc lights to a storage battery, Barrett opened a new era of electrical energy in Chicago.[1]

Like the dawn of most technological revolutions, the early experiments with electric lights in Chicago yielded mixed results. Attracted by newspaper reports of the first test, a much larger audience gathered at the Water Tower the following night to see the brilliant new lights for themselves. But the crowd—swelled by curious boys and nightlife "hoodlums," according to press reports—was disappointed. A long delay, a brief glow, and then darkness. Promising to make it work, Barrett affirmed his faith in the invention in spite of the failure.[2]

The newspapers were also eager to anticipate the benefits of the new energy form. The *Chicago Times*, for example, declared the coming substitution of electricity for both gas lighting and horse-drawn streetcars a heavenly "blessing to mankind." To the newspaper, the promise of technology went far beyond solving two of the chronic problems of urban life and entered the loftier realms of politics and business morality. Proclaiming the death knell of Chicago's gas monopoly, the *Times* believed that technological competition would enforce a modicum of honesty in city hall's dealings with its franchised utility companies. Rates would fall, the city's utility bills would be slashed, and the overburdened taxpayers would enjoy the savings.[3]

Over the next two weeks, Barrett reinforced a belief in progress by fulfilling his promise to make the electrical system work. "The electric light is no longer an experiment. It is a fact," one daily newspaper proclaimed. "Shortly after twilight . . . the machinery was set in motion," it reported, "and with Fire Marshall Brenner at the lamp, the North Side was illuminated in nearly every quarter." Another paper agreed that the "intensely brilliant and far-reaching . . . power of the light is . . . at present its strongest commendation." Observers as far away as Englewood, nine miles to the south, claimed to have seen it.[4] Although Barrett suspended further demonstrations while awaiting additional equipment, Chicagoans were fed a constant stream of news about electrical experiments across the world. Interest in the new technology remained high, especially after stories began appearing in October about Thomas A. Edison's efforts to perfect the light bulb.[5]

Barrett had shown Chicagoans something truly different from anything they had seen before. Most remarkable was the brilliant illumination of the device, which approached two thousand candlepower, more than 150 times the brightness of the standard twelve- to sixteen-candlepower gas fixture. The arc lamp achieved this unprecedented effect by inducing a direct current to "arc" across a small gap between two carbon rods. This continuous spark of electricity produced a blaze of light. Barrett's tests at the Chicago Water Tower used a device invented by Charles Brush of Cleveland, Ohio. As the steady stream of newspaper stories demonstrated, however, he was only one member of a prestigious international group of electrical inventors that included Edison, who only two years earlier had earned a reputation as the "wizard" for his "talking telegraph" or phonograph player. Over the next three years, Chicagoans would witness the testing of several of these inventors' electric lamps, each having advantages and deficiencies compared with the next.[6]

The intense search for better lighting for public areas and private interiors reflected not only an age-old dream but also a pressing need in the mushrooming industrial cities of the United States and Europe. The *Times*'s trumpeting of the arc lamp as divine inspiration underscored the insufficiency of contemporary sources of illumination: manufactured coal gas and kerosene. (Of course, gaslight was neither available nor affordable in many working-class districts of the city, and few suburban or rural communities were supplied with it at any price.) Moreover, the newspaper's wishful anticipation of electric motors to replace horses in powering transit cars identified another important area of urban public services that had proved woefully inadequate in meeting the demands of Chicago's growth. The *Times* hoped that speeds of twenty-five to thirty miles per hour might be attained by a mechanical streetcar.[7] In fact, this daily journal unknowingly predicted the two main uses of electrical energy for the next forty years.

In contrast, predictions that technological competition would break the gas companies' monopoly grip on city hall and replace it with a new era of honest utility franchises remained doubtful. Since the Civil War, Chicago's two utility companies had adhered to a secret agreement that divided the city into two noncompetitive districts. There is no question that a series of innovations in both electric and gas lighting shattered the virtual monopolies held by the Peoples Gas Light and Coke Company on the West Side and the Chicago Gas Light and Coke Company on the South and North sides.[8] But rather than uplifting standards of public morality, the resulting rush for franchise concessions after 1880 created tremendous political pressure in just the opposite direction. For the next thirty years, franchise "boodle," bribery, and graft would become the order of the day from the neighborhood wards to the governor's mansion. In response, the moral outrage of Chicago's elite would combine with grass-roots demands for better public services to trigger a new style of municipal reform that contemporaries called "progressivism."

To be sure, electricity alone did not "cause" the upheaval in Chicago's municipal politics or patterns of energy consumption. But the new technology played a major role in a basic social and spatial transformation of the city in the fifty years following its introduction at the Water Tower. Our complete dependence today on electrical and electronic technology provides a clue to why the new energy form had such a pervasive impact on the city and its people. The electric revolution not only brought a substitute for and supplement to existing sources of energy but provided entirely new forms of energy consumption as well.

In large measure, the spread of an electrical grid or distribution system throughout the Chicago region merely furthered historical trends toward suburbanization and social segregation. Yet electrification involved much more than a network of power plants, transmission towers, and substations. Beyond the hardware lay the homes, shops, and factories that sustained a rapidly growing and changing urban society. As electricity became cheaper and cheaper, Chicagoans found more and more novel ways to use it to ease the burdens of their work and enhance their daily lives. In the half-century it took to complete the electrification process between 1878 and 1930, Chicago became an energy-intensive society.

The City That Coal Built

To understand the full significance of the far-reaching events that Professor Barrett initiated at the Water Tower, the energy history of the city must be examined, if briefly, in the broadest context. In the case of Chicago, the natural resources of the Midwest provided the foundations for the rise of first a thriving commercial entrepôt and later a great manufacturing center. Whereas the region's grain, hogs, and timber fed the growth of the first city, its abundance of cheap coal fueled the second wave of industrial development. Coal—in the form of illuminating gas—also supplied the best available solution to the ever-increasing need for more light in the rapidly expanding city. Yet chronic shortcomings of coal-gas technology kept the search for better lighting alive among an international group of inventors, including the creators of novel electrical devices. Progress was slow and arduous, but Barrett's experiments proved that their long travail was finally producing solid results. With Thomas A. Edison's successful tests of an incandescent light bulb the following year, a truly new era of energy use in the city had arrived.

Any review of Chicago's past invariably begins with the city's strategic location at the maritime passage or portage between the Great Lakes and the Mississippi River valley. Long the site of an Indian settlement, the mouth of the Chicago River continued to serve as a focus of trade and transportation well into the nineteenth century. Even the move to incorporate the frontier community was directly linked to its maritime functions as a portage and entrepôt. In March 1837, town boosters secured a municipal charter as a prerequisite to winning state and federal aid for waterway improvements. But the opening of the Illinois and Michigan Canal was delayed for eleven long years. By then the railroad had arrived, marking the first major turning point in the city's energy history.[9]

Before 1848, the Midwest's abundant woodlands easily met the energy needs of the commercial port. Since there was initially little manufacturing, Chicago's 20,000 inhabitants required relatively small amounts of fuel for heating and cooking. The hardwood forests of the area also were a rich source of charcoal for blacksmiths and other frontier craftsmen. In contrast, the prairies severely limited the development of waterpower, an energy system that was running an increasing number of factories in the East. Although wood and waterpower would remain important sources of energy for another fifty years, coal and coal-based technology created the foundation for the rise of an industrial society.[10]

In the mid-1830s anthracite or hard coal began to supplant wood in key sectors of the economy. The railroad and its steam engine helped trigger a revolution not only in transportation but also in energy use and technology. Of course, the pivotal role of the railroad in the ascendancy of Chicago over its rivals for the urban crown of the Midwest is one of the best-known aspects of the city's history.[11] The superior ability of Chicago businessmen to exploit this technology is generally recognized as the most important factor in the city's rise over Saint Louis and Milwaukee. Yet even these histories say little about the role cheap coal played in the rapid growth of the region's transportation network.

Although generally ignored by scholars, energy fuels constitute a natural resource that has had a major impact on regional economies, including the growth of their urban centers. With the shift from wood to coal, the Midwest's virtually unlimited supply of that fuel became crucial to maintaining transportation rates on a par with or lower than those in cities farther east. Vast fields of bituminous (soft) coal throughout Illinois would allow Chicago's commerce and manufacturing to develop in step with those of its counterparts in the East. In contrast, regions of relative coal scarcity such as the Southwest would lag behind in manufacturing while high transportation rates added an extra tax on their commerce.[12]

The importance of this natural resource to the growth of the industrial cities in the Midwest cannot be overstressed. A "glut" of cheap coal would act as a magnet attracting a wide array of energy-intensive industries to locate and flourish in Chicago, the transportation hub of the region. The first big consumer of the fuel, the railroads themselves, also furnished the means for transporting it and marketing it in the urban economy.[13] As the number of railroad lines entering the city increased, more and more coal would be thrown on local markets. With further competition from lake steamers and canal barges, the price of energy fuels in Chicago would be kept to a minimum for the rest of the century (see table 1).

Chicago received its first coal shipments by lake steamer from

Table 1   Growth of Population and Coal Consumption in Chicago, 1850–1900

Year	Population	Coal Consumption (tons)	Per Capita Coal Consumption (tons)
1850	30,000	40,000[a]	1.33
1860	122,200	150,000	1.34
1870	299,000	777,000	2.60
1880	503,200	2,084,100	4.14
1890	1,099,800	4,013,000	3.64
1900	1,698,600	7,373,900	4.34

Sources: U.S. Census, Population, 1850–1900; Chicago Board of Trade, *Annual Report*, 1858–1900.

[a]Estimate based on earliest figures of 1852 of 44,800 tons of coal consumed, as reported in A. T. Andreas, *History of Chicago*, 3 vols. (New York: Arno, 1975 [1884–86]), 2:673.

the East, but they languished on the dock because prices were too high compared with wood fuels. After 1848, cheaper supplies began flowing in by railroad and canal boat from the Illinois River valley. Yet a decade later, wood still proved a strong competitor. "The year opened with a large stock [of coal] on hand," the Board of Trade reported, "which with a bountiful supply of wood, caused prices to rule low until the end of the shipping season." [14] Nonetheless, the builders of the Illinois Central Railroad recognized coal's potential as a basic source of energy as well as freight revenues. They hired a geologist, J. W. Foster, to survey the natural resources along the train's planned route through Illinois. In 1856 Foster reported that almost the entire state was underlain by high-quality fuel on a par with bituminous coal from Pennsylvania and Ohio. Future prospects for the railroad looked bright indeed, Foster predicted, because "the great obstacle to the growth of Chicago, as a seat of manufacturing industry, is the high price of fuel." But with huge deposits less than one hundred miles south of the city, the geologist calculated that the marginal fuel cost advantages of rivals like Pittsburgh and Cleveland would be reversed in favor of Chicago. [15]

Foster's predictions were more than fulfilled in the period following the Civil War, when the city's unparalleled transportation network poured coal on local markets. Sharp competition between rail, lake, and canal shippers kept prices to a minimum. At about the time of Barrett's experiments with the Brush arc lamp in 1878, moreover, strip-mining techniques were inaugurated in the Midwest. This innovation doubled the recovery rate of shaft mining techniques and further depressed coal prices in Chicago. As one coal dealer complained, "Severe and, in some respects, unreasonably bitter competition, carried sometimes beyond the boundaries of wisdom, or the simple rules governing supply and demand, ha[s] left quite a large supply of eastern bituminous coal on hand."[16]

An abundance, if not a glut, of coal encouraged certain kinds of manufacturers—those who used great amounts of heat or power, or both—to locate and expand in Chicago. And it was these very energy-intensive industries that represented the vanguard of the industrial revolution. The rise of big business and the jobs it created were in large part responsible for the city's phenomenal growth in the late nineteenth century. Already an important urban center of 250,000 inhabitants at the time of the Great Fire of 1871, Chicago gained over 1.5 million people during the next thirty years, becoming the preeminent "shock city" of the industrial revolution. There is little doubt that Chicago's rise as the second largest metropolis in the nation would not have been possible without the surrounding hinterland's abundant energy resources.

According to Alfred E. Chandler, Jr., economies of speed, not size, were the wellsprings of the modern business corporation and its mass-production techniques. The effort to speed up production first met success by applying more intense heat to chemical processes such as oil refining and liquor distilling. In his benchmark study of the industrial revolution, *The Visible Hand*, Chandler argues convincingly that "coal provided the energy to power the new machines. More important, it generated the high and steady heat needed in the more advanced methods of production in the refining and distilling and in the furnace and foundry industries. The new availability of coal, in turn, permitted the rise of the modern iron industry and with it the modern machine-making and other metal-working industries in the United States."[17]

The Midwest's rich energy resources meant that Chicago became a center for the most rapidly modernizing industries. As Chandler notes, ample supplies of coal were a prerequisite for the establishment of industries that were big consumers of heat energy. Chief among these in the Chicago area was the iron and steel business and its spinoffs: machine tools, railroad cars and equipment, and metal fabricators. This last category included the makers of steam engines and boilers, space heaters, gas pipes and fixtures, and other coal-based technology. Cheap energy in the form of heat also encouraged the growth of a wide range of food processing such as brewing, baking, and canning items like meat and soup. Taken together, this group of heat-dependent enterprises formed the foundation of Chicago's ascendancy as an industrial giant.[18]

A second type of industry was also drawn to this urban center of the Midwest, in part by cheap and reliable sources of coal. In these enterprises fuel accounted for only a minor part of the cost of the finished product, but it was still crucial as a source of power to run rows of the new, faster machines. After the Civil War, improved coal-fired steam boilers and engines soon produced more power than their wood/charcoal-based counterparts. This extra power was needed especially for processes of manufacture that were highly mechanized for their day. These presaged the perfection of mass-production techniques in the twentieth century.

Certain businesses could be adapted to imitate the classic case of the textile factory, with its countless repetition of looms attached by belts, pulleys, and shafts to a source of water or steam power. The key requirement was several machines of the same type lined up in a row. Chicago became a center for these industries, including printing, machine tools, ready-made clothing, boots and shoes, and wood products with standardized parts. Although the small shop continued to operate alongside the big factory in these trades, economies of speed reinforced trends toward larger units of production, more mechanization, and greater use of energy.[19]

A third important group of coal-related enterprises fed the growth of the city. In addition to supplying energy to industries using intensive applications of heat and power, coal was a commodity similar to the region's other natural resources. And businessmen in Chicago grew rich by exploiting each element of the Midwest's cornucopia. As Chandler shows, many railroad companies bought and operated their own coal mines after the Civil War. Their urban-based managers were among the first to pursue this big-business strategy of "vertical integration" because they needed reliable sources of both fuel and freight revenue. Other urban entrepreneurs also profited by acting as middlemen. Coal created a wide range of jobs, from the august heights of the Board of Trade, to the wholesale fuel dealers and processors of furnace coke, down to the lowly ash haulers.[20]

The Gaslight Era

In mid-nineteenth-century cities, coal became the fuel of choice not only for heat and power but for light as well. Introduced in

Baltimore in 1816, illuminating gas manufactured from coal marked a significant improvement over substances such as smelly whale oil and tallow candles made from various animal fats. But despite initial claims that the age-old search for better lighting was over, gas lighting remained limited to the city's public spaces, its shops, and the homes of the well-to-do in areas served by a utility company. The light's open flame, its heat, its flickering glow equaling only twelve to sixteen candles, and its relatively high price presented serious drawbacks that were not easily overcome. These restrictions combined to keep the search for a better method of illumination alive among a large number of Americans and Europeans. The searchers ranged from the managers of the gas companies themselves to electrical inventors like Edison and entrepreneurs like John D. Rockefeller, whose kerosene offered a cheap and superior substitute for other lamp oils and candles.[21]

The story of Chicago's coal-gas services is a typical case of the strengths and weaknesses of this technological approach to supplying the city's demand for more light. Such a review also reveals that the gas companies were largely responsible for setting the economic, institutional, and social framework of urban public utility services. By the time of Barrett's experiments, experience with these peculiar regulated enterprises had accumulated to a point where it would shape the coming battles for political and legal advantage among the various lighting firms. In the process, both gas and electric services would undergo a revolution that would give rise to an energy-intensive society.

The inauguration of gas lighting in Chicago coincides closely with the opening of the Illinois and Michigan Canal in the pivotal year 1848. A few months later, a group of businessmen along the main commercial avenue, Lake Street, formed the Chicago Gas Light and Coke Company (CGLCC). This group of city boosters included hotel proprietors Francis C. Sherman and Hart L. Stewart and real estate speculators Peter L. Updike and Peter Page. Transportation improvements suddenly presented local entrepreneurs with three strong incentives to start an urban utility company. Of first importance, the canal formed an economic link between the city and the coalfields of the Illinois River valley. The fuel became commercially feasible for the first time. Second, the rival cities of Cleveland and Detroit were preparing to begin gaslight service the next year. Today these cities may seem only remote competitors of Chicago, but to contemporaries traveling through the Great Lakes along the main route to the West, they were closely strung together. Each aspiring metropolis was driven by jealousy and pride to keep up with the others. Reflecting this boosterism, the *Chicago Tribune* crowed at the inauguration of service on 4 September 1850, when "for the first time in the history of Chicago, several of the streets were illuminated in a regular city style. Hereafter she will not 'hide her light under a bushel.' "[22]

A third motive for starting a gaslight company was the excellent prospect of making a profit. With over thirty years of business experience in other cities to draw on and cheap coal close at hand, the risks of sinking a large amount of scarce capital in an urban utility venture were reduced to an acceptable level. The founders of the Chicago company wisely invited trained engineers from Philadelphia to build the gas plant, with its complicated chemical processes. Investment risks were further reduced by securing a municipal franchise that effectively guaranteed success to the fledgling concern. In addition to granting it a ten-year monopoly to lay pipes under the streets, the local government agreed to pay a subsidy in the form of a contract for streetlamps and service in the city's only public building. The company also signed up 125 shopkeepers along Lake Street and nearby cross streets before beginning to produce gas, to the universal acclaim of Chicagoans. The *Evening Journal* reported, for example, that "some of the stores on Lake Street . . . made a brilliant appearance. . . . But the City Hall, with its thirty-six burners, is the brightest of all, night being transformed to mimic day."[23]

As the newspapers suggest, Chicagoans perceived gaslight as a dramatic advance over previous forms of artificial illumination, and indeed it was. The most remarkable feature of the new technology was simply a light twelve to fourteen times more powerful than the standard candle or oil lamp. Demands for much brighter illumination were especially acute in public places with large, open interiors, such as theaters, hotel lobbies, meeting halls, and dining rooms. Gaslight gave these establishments a distinctly urban character as nightlife took on a new gaiety and enjoyment. Wall fixtures with clustered burners and hanging chandeliers replaced dim and depressing surroundings with a glowing radiance. Streetlamps also enhanced urban life by giving the city its first regular system of public lighting, at least in the downtown area. Chicagoans could now travel at night with reasonable safety from the everyday hazards of city life, including makeshift plank sidewalks, unpaved streets, and muddy intersections. Public lighting also helped protect people from muggers and other denizens of the dark. Early lists of gas customers reveal that city dwellers also wanted more light in their shops, factories, and government buildings. Chicagoans wanted to install gaslights in their homes too, but high rates and restricted service areas kept this group of consumers limited to the social elite.[24]

In addition to supplying brighter light, the new technology represented a marked advance in safety and convenience over traditional methods. Most important, gas lighting greatly reduced the risk of fire, in spite of its open-flame burners. Because gas pipes were permanently attached to walls and ceilings, the burners could not be accidentally knocked over with disastrous results, as candles and oil lamps could. Lower insurance premiums soon reflected the underwriters' strong preference for coal gas.

Besides being safer, gaslight was the first form of energy that was delivered directly and automatically into building interiors. For the first time in history, city dwellers could have as much or as little illumination as they desired. It was "instantly" available at a turn of the gas cock and the touch of a match. And unlike messy candles and troublesome lamps, the gas burner was relatively clean and free from the drudgery of daily maintenance chores. Public establishments in the city could not afford to be without the new lighting. The bright lights of the city quickly became both a status symbol and a physical manifestation of progress, wealth, and amenities.

The immediate and growing economic success of the utility service was one important measure of contemporary approval. By 1860 the gas company gained two thousand customers as its distribution system of mains and pipes gradually spread from five to fifty-three miles. The utility firm also gained competitors, who in 1855 formed the Peoples Gas Light and Coke Company (PGLCC) in anticipation of the expiration of the CGLCC's monopoly three years later. Although depression and war hamstrung the financing of the new venture, the promise of profits similar to the established firm's fat dividends sustained the promoters until June 1862, when production finally began.[25] The rival firm was organized by city boosters similar to the founders of the first company. In the middle of the nineteenth century, businessmen easily mixed the promotion of private and civil goals without ethical doubts about what would now be called "conflicts of interest."

Common outlooks and shared values help explain why the expected competition never materialized. Instead the two companies secretly agreed to chop the city in half, with the dividing line running along the north and south branches of the Chicago River. The old firm got the built-up areas east of the river, while the new one got the West Side, with its rapidly expanding industrial and residential districts. Together the two gas companies served the city without notable incident, growing in tandem with its economy and wealth until Barrett's tests at the Water Tower shattered the old patterns.

But despite the superiority of coal-gas technology, improved lighting in the home remained a privilege that only Chicago's well-to-do and middle classes could afford. The companies generally did not extend services into working-class neighborhoods, where few residents could pay for the expensive service. It was also unavailable in most of the more affluent residential areas at the city's fringe and beyond in the suburbs. Neither legal obligation nor social ethic imposed a demand on the privately owned companies to serve any but the most profitable districts. In fact the city government itself operated under no sense of duty to deliver an equitable share of modern services to every neighborhood. On the contrary, city hall made gas lighting more difficult to obtain by throwing the cost of light poles and fixtures directly onto the affected property owners in the form of special tax assessments. Residents in lower-income areas could ill afford to add this kind of burden to their strained household budgets. In these and outlying districts, the city provided naphtha oil lamps from a company that also supplied the street posts.[26]

The monopolistic practices that retarded the extension of services in the neighborhoods also help account for the companies' absolute refusal to reduce the price of gas. For over a quarter-century they never allowed the residential rate to fall below the initial price of $3 per thousand cubic feet (MCF). At the same time, production costs were falling because, as we have seen, market forces were depressing the price of coal. The black gold was the main ingredient as well as the fuel of the manufacturing process. In addition, the companies could sell the waste products, including oven coke and coal tars. By the early 1880s, at least 70 percent of the price of gas represented profits for the stockholders.[27] The comfortable duopoly the Chicago utilities enjoyed gave them little incentive to cut rates or to expand the distribution network without a virtual guarantee that every building on the block would install service. Local gas companies had even less reason to pioneer technical innovations or to experiment with radical departures from the established practices of urban utility companies.

Despite mounting business experience and technical expertise,

the gaslight industry remained complacent about serious drawbacks in the utility service until faced with competition from the
electric arc lamp.[28] The search for better lighting was kept alive by
several inherent flaws in the technology of coal gas that detracted
significantly from its usefulness, safety, and convenience. Although a marked advance over other illuminates, coal gas suffered
from the same limits as other hydrocarbon fuels burned with an
open flame. The greatest problem was producing a brighter light,
which depended almost entirely upon the light-giving qualities of
the coal itself. At most, local companies could coax fourteen to
sixteen candlepower from the best-quality fuel. Although this level
was adequate for most purposes in the home, it fell far short of
perceived needs for more light in the city streets, in large interior
spaces, in factories, and in offices. The demand for brighter illumination was becoming especially acute in the downtown office
buildings where white-collar workers pored over rising mountains
of paperwork. Of course gas burners could be clustered, but this
approach quickly ran into the limit of too much heat and the threat
of fire.

   In addition to traditional concerns about light and fire, coal-gas
technology created its own unique set of problems and hazards.
Chief among these were the dangers of asphyxiation and explosion from unattended gas jets. Less dramatic perhaps, but no less
important to contemporaries, was the inescapable soot that ruined
walls, ceilings, and furniture during an era of fashion when upholstery reigned supreme.[29] Constant cleaning and periodic replacement of interior decorations was an especially onerous and
costly problem for proprietors of retail shops, hotels, theaters, and
restaurants, as well as for ordinary housewives. Moreover, gaslights often proved inconvenient because they could not be moved
where they were needed most. The furnishings in rooms used at
night had to be arranged to accommodate the location of the gaslights, or oil lamps had to be added to provide a portable source of
illumination. In sum, contemporaries preferred to use gaslight
when they could afford it while they continued the search for alternatives to hydrocarbon fuels.

The Origins of the Electrical Revolution

If coal-gas technology left much to be desired, the arc lamp only
partially filled the need for more light in the city. It represented a
transitional technology that offered a mixture of old drawbacks

and new advances. The use of electricity to produce a light was novel, but the method of generating it, by burning hydrocarbons in the open air, was traditional. The two-thousand-candlepower blaze of the arc lamp's glowing carbon tips met the most pressing demands of city dwellers for greater illumination in large open spaces. In Chicago, Barrett's successful demonstrations of Charles Brush's arc lamps triggered a rapid proliferation of privately owned systems in the Loop's most exclusive retail shops, hotels, theaters, and men's clubs as well as on the sidewalks in front of these imposing structures. Barrett, moreover, helped convince the municipal government to provide electric light services of its own, beginning more modestly with a few fire stations. Within a decade, a wide variety of private and public circuits were powering about 6,800 arc lamps at an annual cost of over $1 million. The electric suppliers faced a consumer market with no bounds to the demand for more of the lights in the city's shops, public buildings, and streets.[30]

The immediate success of the arc lamp set off separate revolutions in the technology of making gaslight and in the business practices of local utility companies. Despite its faults, the electric arc's shear brilliance posed a potentially fatal challenge to the coal-gas industry that could not be met by direct competition. The plummeting value of gas company stocks shattered old patterns and opened the way for a more ready acceptance of novel ideas and techniques. Chicago's comfortable duopoly suddenly found itself threatened not only by the arc lamp, but also by business rivals armed with new patents that promised to make a superior gas at less than half the going rate. Taken together, the ensuing battle of the technological systems created an entirely novel situation of intense competition to sign up customers and extend service territories, including working-class neighborhoods. The struggle to protect and expand markets for artificial illumination also offered city officials a unique opportunity to advance the public welfare through franchise concessions. Some aldermen, however, were more interested in taking advantage of the franchise seekers for personal gain, ushering in a new era of brazen corruption, blackmail, and "boodle."

Adding to the upheavals in lighting technology and urban politics was the widely heralded announcement of Thomas Edison's perfection of an incandescent bulb. The failure of the arc lamp to eliminate fire hazards or to provide a small unit of light suitable for normal-sized interiors was clearly apparent to Edison and the other electrical inventors. They continued the search for a break-through that could supply a truly superior alternative to the gas jet for the city's innumerable shops, offices, and homes. Within three years of Barrett's first experiments with an arc lamp in 1878, several inventors claimed to have found the solution at long last: a thin filament inside an enclosed glass bulb that was heated to in-candescence by an electrical current.

The almost simultaneous timing of the inventors' announce-ments here and abroad was not coincidental.[31] On the contrary, the multiple inventions of the light bulb reflected the intrinsic na-ture of electrical technology and its historical development. Unlike any previous method of supplying light or power, generating elec-tricity involved a highly interdependent and integrated system. Every one of its several complex components had to be in perfect balance with all the others to maintain an electrical current. The light bulb, then, formed only one part of a symmetrical system that included prime movers, dynamos, regulators, measuring and safety devices, distribution wires, and appliances, all working in instant harmony. To be sure, manufacturing coal gas also involved a complicated series of exacting procedures. But distributing and consuming the lighting fuel was completely separate from manu-facturing, because the gas could be stored in tanks during the day to meet customers' demand at night. Electric supply companies tried to use storage batteries in a similar way to meet periods of peak demand, but they never proved practical or economical on a commercial scale. Instead, utility operators had to keep a perfect balance between demand and the supply of energy flowing through an integrated system of electrical circuitry.

The slow evolution of electric lighting before 1880 underscores the critical importance of system building in the sudden appear-ance of several claims to the invention of the light bulb. Demon-strations proved the feasibility of the arc light as early as 1843, but it took another thirty-five years to bring all the essential parts to the point of practical success reached by the Brush system at the Chicago Water Tower. The problem centered on the difficulty of improving one part without making commensurate gains in all the others. For example, if one inventor created a better arc light de-vice, but without a more powerful generator, he would be unable to test and perfect his innovations. Working independently, a sec-ond inventor would build a better generator that unfortunately was incompatible with the electrical requirements of the first one's lighting device. Lacking a systems approach, the inventors of electrical lighting made only halting progress toward an integrated set of components that worked harmoniously together.[32]

The rise of the telegraph business was especially crucial in training a first generation of technicians who understood the practical workings of an electrical system. It is no surprise that Barrett, like Edison, started with the telegraph and eventually invented various electrical systems for the city. Edison's career is too well known to recap here, but Barrett's deserves attention because of his contributions to public utility services in Chicago.[33] Born in Auburn, New York, John Patrick Barrett arrived in the Gem of the Prairies in 1845, when he was eight years old. The telegraph followed three years later during the first turning point in Chicago's energy history. Barrett received a public education and became a sailor in 1858, but he returned to the city four years later after suffering an injury at sea. He joined the fire department as a watchman and was soon placed in charge of the alarm bells at the top of city hall. When an electric telegraph system of fire boxes was adopted to replace the bells in 1865, Barrett became one of the operators who relayed the alarms from headquarters to the appropriate station houses. Over the next decade he gained experience and technical expertise, rising to the top position in the department—superintendent.

In large part his success stemmed from the practical inventions and improvements he installed to rebuild the system after the Great Fire of 1871 destroyed most of the original equipment. Most important of these, perhaps, was the "Barrett Joker," which automatically registered the exact location of alarms from each firebox directly at the neighborhood stations, eliminating the time-consuming step of relaying the reports from headquarters. By 1880 the superintendent had created a sophisticated network of fire protection for Chicago that included almost 500 alarm boxes, 70 telephones, and 430 miles of overhead and underground wiring. Drawing upon his fire-telegraph experience, the "professor" worked out a similar electrical network of police patrol boxes. An appreciation for system building also helped Barrett devise an integrated set of bridgetender controls to regulate the Loop's increasingly congested traffic on and over the river.[34]

Barrett's reputation as the "professor" was a well-deserved sign of respect from the city's sizable community of telegraph operators and electrical technicians. Handling over 1.5 million messages a year by the mid-eighties, Chicago was the second largest communications center in the nation. To transmit a million messages a year, for example, Western Union had to operate over 2,000 telegraph instruments at its main office alone, in addition to maintaining another 175 branch offices throughout the city and suburbs. The men who serviced this communications system formed a self-conscious community of highly skilled clerical workers. They had their own unions and organizations such as the Chicago Electric Club, which mixed social gatherings with educational demonstrations of new inventions. They also formed an invaluable pool of talent from which an adequate supply of technical experts could be drawn to help install Chicago's first generation of electrical lighting systems.[35]

The importance of Chicago to the nation's electrical industry went beyond the telegraph business into the related realm of equipment manufacturing. Although economically miniscule compared with the production of coal-based technology, these manufacturers provided jobs that added unique elements of technical training and experience to the city's pool of electrical experts. The Western Electric Manufacturing Company was by far the largest firm of this type in the city and one of the biggest in the country. The founder was a famous inventor, Elisha Gray. In 1869 he started the company in Cleveland, but he moved to Chicago the following year to be at the center of the telegraph business in the West. Gray created improved parts, a typewriter keyboard for the telegraph, and other devices such as needle annunciators for hotel rooms and elevators. In 1872 his inventions brought Western Union into partnership with his company for manufacturing most of the equipment used by the communications giant. Two years later, however, he resigned as chief electrician to work full time on the telephone. Although an inventive genius like Gray can never be called "typical," he was representative of the best talent rising out of a community of technical experts.[36]

It was this community of talent that not only ushered in the telephone in Chicago but also helped convince investors from across the nation that the city was the best place to manufacture the new technology. Just two months after Barrett's experiments at the Water Tower, the first telephone exchange in Chicago was opened, using equipment perfected by Gray and Edison. Ten days later, on 28 June 1878, businessmen holding a license to use the patents of Alexander Graham Bell started a second exchange. While the inventors argued over patent claims in the national courts, the two local firms battled for customers. By 1882, corporate consolidations and other financial maneuvers left the Western

Electric Company the exclusive supplier of Bell's telephones and Western Union's telegraph equipment.[37]

The community of electricians also contained men who became as astute in business management as Gray and Barrett became in their areas of technical expertise. In Chicago, General Anson Stager forged important links between the telegraph, manufacturing, and telephone businesses. His career typified the rise of the electrical industry more than any other in the era before the lighting revolution. In fact, his career closely paralleled Edison's, but along business lines rather than inventive ones. Both were farmers' sons who grew up with the telegraph, eventually rising to the tops of their fields.

In the case of Stager, a start in the printer's trade led the young man to take part in stringing the first telegraph lines through his area of western New York. Fascinated with the new technology, the twenty-one-year-old Stager moved to Philadelphia in 1846 to become a telegrapher. Two years later, as a telegraph operator, he was sent to Cincinnati to take charge of the national lines to the West. After helping to consolidate several companies into the Western Union Company, Stager was the natural choice of Lincoln's army to coordinate communications and railroad traffic in the Midwest during the Civil War. Rising to chief of military telegraphy, he gained invaluable managerial experience with large-scale business operations. Stager put this experience to work after the war as the top officer of the company's western operations, which in 1871 were moved to Chicago. When the telegraph company formed a partnership with Gray the following year, Stager was made president of the new venture, the Western Electric Manufacturing Company. Six years later a similar situation developed between Western Union and Edison, leading to the formation of the Chicago Telephone Company with Stager in charge.[38]

Stager, Gray, Barrett, and hundreds of technicians formed an impressive pool of talent that stood ready to usher in the electrical revolution. Barrett's experiments take on added importance because they had special meaning to this community of trained experts. By proving the practical feasibility of the arc light, Barrett signaled Chicago's electricians that a new day of opportunity had arrived. Over the next year and a half, a steady succession of demonstrations and announcements of breakthroughs in electrical technology reinforced the signs of progress for Chicago's technicians and businessmen alike. The opening of the telephone exchanges in June 1878 was followed six months later by a detailed report to the Electrical Society on Edison's work by a close Chicago friend, George Bliss. In February 1880, rapid improvements in the Brush arc system allowed the brilliant lights to be safely moved indoors. At the sixth annual ball of Chicago's telegraphers, two of the devices were suspended from the ceiling and powered by a small generator at the rear of the building. "This was not set in motion until about 10:00," a reporter observed, "and the effect then, as the dancers were whirling about the room in a waltz, was indeed, brilliant. The splendor of the toilets [attire] was very much enhanced, and the rich jewels and diamonds glistened with additional luster."[39] Shortly afterward, the installation of small-scale systems in the city's shops, hotels, and theaters began in earnest.

Of course the electrical men were not the only Chicagoans who were ready to respond quickly to the early signs of a new era of better lighting for the city. The managers of the two gas companies reacted defensively by trying to block the nascent arc light ventures from obtaining municipal franchises to use the public streets. At the same time, some of the major stockholders in the gas companies began to hedge their bets by investing in the alternative technology. Cracks in the solid control of the established duopoly also encouraged other entrepreneurs to enter the lighting market with exclusive patent licenses for new ways of making illuminating gas. As we will see in the next chapter, the result was a fierce battle of the systems that ranged from head-on clashes in the streets to secret and sometimes illegal maneuvers for political and financial advantage in the city council.

This revolution in energy use in Chicago was based on over thirty years of slow, painstaking gains in electrical technology. Spurred by a search for better lighting, the electrical inventors went unchallenged by a coal-gas industry that rested complacently behind traditional monopolistic practices. Yet practical experience and scientific advances were accumulating in coal-gas technology in ways similar to those in the telegraph business. With the announcement of Edison's triumph in 1880, the time was ripe to install some of these innovations in hopes of saving the gas industry from technological extinction. Chicagoans were suddenly offered unprecedented choices. They could now pick among several systems to meet their needs for more light, including arc lamps, incandescent bulbs, kerosene lamps, and different types of illuminating gas. City dwellers also faced novel political challenges. They needed to redefine franchise policies to promote the new forms of technology and, at the same time, protect the public welfare.

2

The Battle of the Lighting Systems, 1880–1893

In the three years following Barrett's tests at the Water Tower, electric lighting systems progressed from experimental demonstrations to commercial applications. The original Brush device, for example, was sold to prominent hotel proprietor John B. Drake, who installed the lamps in his office at the Grand Pacific Hotel. By January 1880, this trendsetter of elite society was ready to try the lights in the "grand exchange" or lobby of the luxury hotel. Claiming they were a "remarkable success," the *Tribune* reported that Drake was immediately ordering $4,000 worth of additional equipment to replace gas fixtures in all the "parlors, dining-rooms, halls and corridors." [1]

Not to be outdone, another pillar of Chicago society and business, Potter Palmer, soon fitted his hotel with arc lamps in a much more lavish fashion. With the Republican national convention arriving in June 1880, the city's number one booster demanded the latest technological amenities for Chicago's influential guests. An electrician from the Western Union company hastily installed twenty light fixtures in the lobby and dining room of the Palmer House. Exactly a year later, P. S. Kingsland initiated the first "central station" service when he put a dynamo in the basement of the YMCA and began renting fifty arc lamps to neighboring stores for $1.50 per night for ten hours of service. [2]

These three installations were typical of the diffusion patterns of lighting services during the early years of electrical systems. Between 1878 and the World's Fair of 1893, small-scale systems proliferated in an endless race to catch up with the demand for more light, especially in the city's central business district. This tightly packed beehive of activity was the focus not only of the city's commerce but also of its high culture and popular amusements. For the very wealthy like Drake, no expense was too great to have the most up-to-date conveniences. Other members of the elite would soon join Drake by installing arc and incandescent lighting in their

own homes and offices. In a similar vein of social exclusivity, it was
no accident that two of the Loop's luxury hotels were among the
first commercial institutions to acquire the new technology. The
most elegant inns of the city had long spearheaded the introduction of semiprivate and household amenities.[3] In the city, the best
hotels came to epitomize the social connection between advanced
technology and class status.

Starting with Drake and Palmer, the electrical revolution would
amplify these cultural trends by reinforcing the links between
modern technology and social prestige. The first use of electricity
by the well-to-do would also promote novel ideas about consumption and leisure that would help create an energy-intensive society
in the twentieth century. In the eighties and nineties, "burning"
the lights was a conspicuous sign of affluence and luxurious
living.[4]

Once established as a highly visible symbol of class status, electric lighting quickly became a necessity in the commercial sector.
In the wake of the hotels there followed other exclusive downtown
establishments, such as the posh men's clubs, the fancy shops and
department stores of the carriage trade, and the theaters of the
legitimate stage. Less prestigious business concerns soon joined
the commercial customers in the Loop who needed modern lighting just to keep abreast of the competition. Although some of
these shopkeepers purchased complete electrical systems, others
chose to pay for light and power services from central station operators like Kingsland. The pioneer venture of this unknown electrician was representative of the entrepreneurial opportunities
spawned by an insatiable demand for better lighting. The same
mix of boosterism, prestige, and profit that explains the proliferation of electrical technology in the Loop helps account for its
spread to the outlying neighborhoods and suburbs of the city.
From its origin, then, the diffusion pattern of electrical systems in
the Chicago region was marked by a rapid polygenesis that it
seemed impossible to reverse.

The result was a period of intense technological and economic
competition that historians have labeled the "battle of the systems." They generally have used the term to refer to the contest
between the direct current (DC) system advocated by Thomas Edison as opposed to the alternating current (AC) system championed
by George Westinghouse. But scholars have defined this concept
of technological rivalry too narrowly. It should also encompass two
closely related contests that were inextricable parts of the energy

revolution. First, electrical utility operators continued to compete against the gas companies, which introduced a new method of making a superior energy fuel at a greatly reduced cost. Second, promoters of electrical systems faced an internal struggle that pitted their sale of complete, self-contained systems, or "isolated" plants, against their efforts to sign up customers for central station service. Some electrical men favored the first sales approach in hopes of duplicating the highly successful diffusion of steam engines to individuals. Others believed this approach was shortsighted compared with the alternative strategy modeled on the steady, long-term profits of gas and water utilities. Both were promising marketing schemes, albeit mutually exclusive. But until one or the other proved to have a decisive advantage in terms of cost and reliability, they tended to undermine each other's economic foundations.

From 1878 to 1893 the battle of the systems raged without resolution because basic technical flaws in all the electrical systems prevented any one of them from achieving a clear economic victory over its rivals. Every shortcoming in a highly integrated technology like electricity became a serious problem. As in the incubation stage, the potential of the whole system was held back by each technical bottleneck. Whether because of underpowered generators, restricted distributors, or faulty appliances, the limitations of technology had the general effect of keeping service areas small and prices high. Only the wealthy elite and the commercial sector could afford the luxury of the new lights. In spite of broad demand for a superior technology, most city dwellers had to continue relying on the cheaper alternatives of gas and kerosene. With the coming of new gas companies in an effort to break the old Chicago duopoly in 1882, the marketplace became crowded with contestants. A growing list of utility companies fought each other over rates and service territories. The electrical firms also had to contend with the internal struggle between self-contained plants and central stations.

Because local utility operators could not resolve the battle of the systems in the marketplace, it quickly spilled over into the realms of political and corporate power. The addition of each new electric or gas company made the economic warfare increasingly unbearable to utility managers and investors. The stakes were exceptionally high because huge amounts of money were involved in plants, equipment, and distributor networks. In contrast to many other types of investment, the utility's capital was permanently fixed in place; it would become worthless if the enterprise fell victim to its rivals. The efforts of utility companies to find political and financial avenues of escape will form the subject of the following chapter, but first it is important to describe the diffusion patterns of the new technology in Chicago. What follows sketches the introduction of arc and incandescent lights in the central business district. This most lucrative market for artificial illumination became the obvious battleground in the commercial contest among the promoters of the early electric lighting systems.

The Fight for the Central Business District

From the onset of the commercial exploitation of electric lighting in 1880, the universal focus of attention was the central business district (CBD). This was the area with the greatest density of buildings and the most wealthy offices and prestigious shops, and the attractions of its nightlife were matchless. In other words, here were clustered the greatest number of potential customers who could afford the improved methods of artificial lighting. The restricted service areas and high cost of the early systems also pointed would-be utility operators to city districts with the highest densities and most potential customers. The unvarying experience of the not-so-distant past, moreover, reinforced the commonsense view that public utilities like gas, water, and streetcars were planted first at the city center. After they were firmly rooted there, they gradually grew outward into the surrounding neighborhoods.

Here, then, was an economic battleground worth fighting for. In Chicago, the gas duopoly fought a rearguard action to hold on to its patrons. Rival electric firms took a more aggressive stand in the sales campaign to sell self-contained systems and to sign up customers for service from primitive central stations. As the Chicago Gas Light and Coke Company lost its grip on the downtown, fierce rate wars broke out in the surrounding residential neighborhoods between the two established firms and upstart competitors. At the same time, separate skirmishes flared in the more affluent suburbs at the city's edge among new gas and electric ventures. The various battles raged for the next twenty years in what amounted to a revolution in the use of energy. Although the ultimate victor long remained in doubt, there was never any question that the decisive conflict would be fought over the heart of the city.

In 1880 the density of Chicago's downtown was extraordinary, even compared with other late nineteenth-century cities. Since Civil War days, the business district had been hemmed in by an iron ring of railroad yards and terminals. Even the Great Fire of 1871 did not alter these land-use patterns. The entrepôt functions of the city as the point of interchange between overland and maritime shipping remained a vital part of its urban economy. Transportation facilities flanked the northern and western boundaries of the river to demarcate two sides of the CBD's expansion. Growth to the east was limited by the lakefront, which in any case was occupied by the yards of the Illinois Central Railroad. Since most railroads entered the city from the south, a jagged line of train stations also defined this southern border of the city's core. With the inauguration of cable car services around an inner circle of the CBD in 1881, the downtown gained not only a new name, the "Loop," but also a new level of congestion. The only way to grow was up, which helps account for Chicago's architectural leadership in the rise of the skyscraper. Perhaps it was no coincidence that the first of these tall commercial structures, the ten-story Montauk Building, was commissioned in the same year the cable car arrived.[5]

Within the tight confines of the Loop, a fantastic mosaic of urban pursuits activities flourished around the clock. Although residential housing had been pushed out of the district, a consistently heavy flow of passenger traffic through the rail hub of the West formed a "permanent," if interchangeable, population. One of the new arrivals was Frank Lloyd Wright. Fresh from the countryside of Wisconsin, the twenty-year-old Wright had been sent to the city by his mother to become an architect. His impressions of Chicago begin with images of its bright lights and its gloomy pall. "Chicago," he wrote in his *Autobiography,*

> Wells Street Station: Six o'clock in late spring, 1887. Drizzling. Puttering white arc-lights in the station and in the streets, dazzling and ugly. I had never seen electric lights before.
> Crowds. Impersonal. Intent on seeing nothing.
> Somehow I didn't like to ask anyone anything. Followed the crowd. Drifted south to the Wells Street Bridge over the Chicago River. The mysterious dark of the river with dim masts, hulks and funnels hung with lights half-smothered in gloom-reflected in black beneath. I stopped to see, holding myself close against the iron rail to avoid the blind, hurrying by.[6]

Like many newcomers, Wright found refuge from the city crowds in the home of relatives. For many others, however, finding temporary housing was the first task upon arriving in Chicago. The need for hotels to be near the train and ship terminals meant the Loop was packed with public accommodations, ranging from the elegance of the Grand Pacific and the Palmer House down to the dubious surroundings afforded by the ubiquitous "men's hotel." Adding significantly to the transient population were the thousands of men employed as railroad workers, teamsters, and sailors, besides the bummers who ended up in Chicago's rail yards. These homeless men hung around the periphery of the Loop among the flophouses of Clark Street on the North Side and the red-light district known as the Levee on the South Side.

Collectively, this sizable and shifting population gave night life in the Loop a special urban vitality. A rough, male-dominated world of restaurants and saloons, theaters and gambling dens thrived on a mix of visitors such as seamen, travelers, immigrants, salesmen, wagon divers, and trainmen. The Loop offered respectable Chicagoans far fewer attractions after working hours. During the 1880s, women could not be seen after dark without an escort. And though mixed groups occasionally enjoyed an evening of dining and dancing at the better hotels or went to see a stage production, most social activities were conducted in more private settings. In the Victorian era, the home still remained the focus of polite society.

In contrast, the Loop came alive with Chicagoans during business hours. Every weekday they poured into this compact area by streetcar, train, foot, and carriage to engage in an incredible range of economic pursuits. The small-scale nature of most manufacturing allowed just about every type of light industry imaginable to operate in close proximity. The variety of these enterprises was matched by the number of wholesalers, jobbers, and retailers who demanded central locations for reasons of prestige or necessity in the highly competitive world of commerce. The Loop also accommodated several open emporiums such as the farmers' hay market on Randolph Street and the city's main fresh produce center on South Water Street. At the other end of the economic spectrum stood the bankers, commodity traders, lawyers, corporate executives, government administrators, and other white-collar workers who occupied the Loop's office buildings.

The sheer density and diversity of economic activity in the Loop created a special need for artificial illumination even during the day. The worst problem was smoke pollution from countless coal-fired steam boilers. Heating plants in nearly every building, steam engines in factories, hot-water systems in public accommodations, and tugboats and steamers on the river all contributed to the black pall. Visibility was frequently cut to three or four blocks. Air pollution became a chronic nuisance during the winter months, when the heaters were working hardest, the days were short, and the skies often became overcast with a solid gray covering of clouds. On these bleak days, extra lighting was needed by 4:00 P.M. in order to conduct business, whether in factory, office, or retail store. The problem was especially acute for street-level shops that depended upon the patronage of commuters on their way home. By 1881 the problem had become so severe that the city council passed a smoke-abatement ordinance, but this was left unenforced, and the situation only became worse after the advent of the skyscraper. During the 1880s these tall buildings turned downtown streets into gloomy canyons filled with smoke from morning to night.[7]

The Introduction of the Arc Lamp

The pent-up demand for more light in the Loop was obvious, but no one knew exactly how to use the new technology to supply this ready-made market. Which type of electric lighting, arc lamps or incandescent bulbs, was best suited for each particular application? Which equipment supplier would first correct the technical flaws that plagued every one of the systems? Should local companies promote the sale of complete self-contained plants or the spread of central station services? If the latter choice proved more profitable, how should these capital-intensive enterprises be financed? And how should central stations be designed to best take advantage of the first generation of small-scale equipment? Closely related to the question of cost was the equally perplexing issue of rates. Since reliable meters were not yet available, how should customers be charged and how much? And what regulations should the municipal government impose on the franchise holders to ensure the public safety and welfare? For example, should the city require that distribution networks be placed underground, or was cheaper overhead wiring acceptable? Chicago was part of an international search for answers to these and many other questions involved in putting the new technology on a solid commercial footing.

It is not surprising, then, that chaos prevailed during the initial burst of commercial applications. The years from 1880 to 1887 represented a period of continuing experimentation, but now on a commercial scale where the stakes were much higher. Even within the tight confines of the Loop, the diffusion of electrical systems was marked by an anarchic process of polygenesis, of many small starts in different directions. In 1883, for example, L. Y. Cowl followed Kingsland's lead by establishing a "central station" with a capacity of forty arc lights in the basement of the Globe Hotel. Nearby, however, the Buckingham Theater also installed an electric plant and began selling its excess capacity to neighboring stores on State Street. A miniature rate war ensued between the two operators, with the nightly rate falling from $1.50 to a ruinous 50¢ per light. Within a year, D. P. Perry started a third small-scale plant on State Street to illuminate the interior and the storefront of his retail shop. The envious merchants next door also wanted street lighting to advertise their stores. They convinced Perry to turn his isolated plant into a "central station" with twenty arc lamps on a single circuit. Other speculators soon bought out Perry and Cowl, forming the Sun Electric Company.[8]

During the early 1880s, similar innovation and imitation occurred block by block throughout the Loop. By 1885 Chicagoans were using over a thousand arc lamps, almost all of them in the central business district. A highly diverse group of operators used nine different patented systems to illuminate large interior spaces and some building fronts as well. An electrician named Elmer A. Sperry operated the most interesting installation during this pioneer era of electric lighting. Sperry clustered all twenty of the lamps his dynamo could run in a tight circle at the top of the Board of Trade Building. The beacon from this symbolic center of Chicago commerce was "one of the most powerful lights ever devised" and reportedly could be seen sixty miles away.[9]

Jerry-built equipment and equally makeshift capital arrangements constituted a large element in most, if not all, of the early "systems." The Kingsland plant was typical. One electrician recalled that it was "liberally capitalized without capital [and] bought first one make of machine and then another, using each until the sheriff took it away, and then buying equipment from some other company."[10] The small scale of the early systems allowed entrepreneurs with little capital to gain entry into the business, but stiff competition soon forced most of these would-be utility operators to either raise more capital and buy out their most immediate threats or be absorbed by better-financed rivals. For example, the need for more capital to expand the Sun Electric Company led to its merger with other small central station companies to form the Merchants Arc Light and Power Company. In a similar way, the Sperry Illuminating Company was consolidated with a second, eighty-arc-light plant to form the Wunder and Abbott Illumination Company. Yet the trend toward larger utility firms did nothing to halt the parade of new shoestring operators into the marketplace, let alone stop the sale of self-contained systems to anyone who could afford one.

In addition to economic competition, utility operators had to face the outspoken opposition of some Chicagoans to the powerful and mysterious force of electricity. Many landlords prohibited the "man killing" wires, which were also considered a dangerous fire hazard. Others simply forbade the installation of the lights on building fronts because they were deemed low class and unseemly. For example, only ten of Cowl's forty lamps for rent on State Street were placed outdoors. Bright lights in front of a theater could be accepted in the big city, but two other nightlife spots with Cowl's arc lamps were politely called "semirespectable establishments." Landlords were not the only ones to express fears about the new technology. The use of electric lights inside a theater also raised sharp protest. The actors went on strike the first night the electricity was turned on at the Academy of Music. Walking out between the first and second acts, they complained that their makeup looked unreal under the bright lights. The strike was soon settled in favor of progress, and by the end of the decade no theater in Chicago would be without the modern convenience.[11]

Although the promoters of electric lighting had to worry a little about public relations, basic technical flaws posed far more serious problems. Each system contained inherent drawbacks—"reverse salients," historian Thomas Hughes calls them—that limited their commercial success by keeping rates at luxury levels and service areas confined to one-quarter of a square mile. In the case of the arc lamp, the most troublesome problem was the design of the lighting appliance itself. Since its carbon rods were gradually consumed by the burning arc, some type of mechanical regulator was needed to maintain the proper gap between the tips of the two rods. The invention of practical regulators by men like Charles Brush in the late 1870s marked the turning point for the arc light from scientific curiosity to commercial opportunity. Yet the regulators remained a weak link in the system, requiring constant and costly maintenance by lamplighters, who also had to turn on each light and replace burned-out rods daily. In addition, the electro-mechanical nature of the regulators made it difficult to put more than one light on a circuit. Although this drawback was gradually overcome, the problem of "subdividing" the brilliant lights remained unresolved.[12]

A closer look at the commercial rise and fall of one system helps show how faulty technology restricted the market potential of electric lighting. In Chicago, the first arc lamps to surge ahead in the battle of the systems were the creation of a local inventor, Charles J. Van Depoele. At the first demonstration of his device in December 1880, the influential John Barrett pronounced it a marked improvement over the Brush lamp. The key improvement was a regulator that moved both carbon rods, keeping them at the correct distance to eliminate the noisy sputtering. At the same time, the regulator kept the tips of the carbons in the same place, which permitted the use of reflectors to focus the light into beams. Six months later, Van Depoele showed how a dynamo of his design could power five arcs on the same circuit without overheating. The reporter for the *Tribune* claimed that "their operation was the embodiment of perfection."[13] Based on these glowing recommendations, the inventor went into business.

But Van Depoele's precipitous fall, like his rise, illustrates the vicissitudes of utility enterprises built on faulty technology. He set up one company to manufacture the equipment and another to operate a central station. In 1884 he began to supply service from two forty-lamp dynamos to businesses around Halsted and Madison on the west bank of the Chicago River. He also sold these units as self-contained systems in the Loop. A year later, almost 400 of the 1,150 arcs in commercial use were Van Depoele's, or more than twice those of the next closest rival.[14] But this success was short-lived. Dr. Chisholm, a physician who used electricity as a therapy, obtained a patent license to use the equipment of a Massachusetts-based firm, the Thomson-Houston Company. The doctor set up a central station just across the river from Van Depoele and ran a cable through the Washington Street Tunnel to his West Side service area. The Thomson-Houston system worked better, and Van Depoele was soon forced to retreat from the utility business. Fortunately the inventor was also making important progress on an electric motor for streetcars. The first leader in this field, Van Depoele secured basic patents that made his small manufacturing company valuable. By the end of the decade, he sold out to the much larger Thomson-Houston Company.[15]

Although the pace of technical innovation accelerated under

market conditions, efforts to "subdivide" the arc light remained problematical. Paradoxically, the very brightness of the device was its greatest limitation. From the first tests in Chicago, the unsuitability of a 1,600 to 2,000-candlepower light for "general household purposes" troubled contemporaries. One newspaper, for example, praised Van Depoele's improvements but bemoaned the fact that the inventor was still "far short of perfecting [a way] to prolong the life of a carbon to such an extent as to render electric lights adaptable to the household economy."[16] These expressions of concern reflected a realization that the search for better lighting was not over. While the arc lamp solved the most immediate problem of illuminating large spaces, a different technology was needed to provide a soft, safe, and self-regulating light for the vast majority of the city's interiors. Thomas Edison's system of incandescent bulbs met all these requirements.

The Coming of the Edison System

The perfection of an incandescent lighting system represented a true technological revolution. During the winter of 1879–80, Edison beat out several rivals by demonstrating a bulb with a carbonized thread of cotton (later bamboo) that lasted two days. The inventor justly deserves a worldwide reputation for finding a filament that would burn long enough to create an acceptable commercial product. This accomplishment was revolutionary because the light produced was electrical in nature and could be safely contained within a glass globe. In contrast, all other illuminates—including the arc lamp—were based on burning hydrocarbons in the open air.

However, a narrow focus on this single accomplishment has overshadowed several other significant aspects of Edison's larger system. Of first importance, he scrupulously followed the model of the urban gasworks, with its central stations, distributors, meters, and light fixtures. He designed the original bulb to give the same amount of illumination as the standard sixteen-candlepower gas jet, and he even called the underground distributors "mains" and "tubes." He believed this marketing approach would promote a ready acceptance of the new technology by the mass of city dwellers. Two years later, in 1882, the historic Pearl Street Station in the financial district of New York City proved the value of this strategy. Although far more expensive than gas, the Edison system of incandescent bulbs was immediately regarded as the perfect solution to better lighting.

Yet Edison's system suffered from two major technical problems that forestalled a clear victory for central station service over the alternative strategy of selling self-contained systems. Like the arc light, Edison's technology was limited by small-scale dynamos. Driven by a 125-horsepower steam engine, the "jumbo" dynamos at Pearl Street could power a maximum of seven hundred bulbs. To design a commercially practical central station with a much larger capacity, Edison had to put several dynamos in a row, each attached by leather belts to its own prime mover. In short, the central station achieved few economies of scale, particularly fuel savings, that could be passed along as lower rates. In contrast, the self-contained plant was an attractive alternative for many businesses that already had steam boilers and engines, such as factories, hotels, and office buildings. The purchase of a dynamo, which could be conveniently attached by a belt to a prime mover, offered a cheap, one-time expense compared with monthly service charges. Until the rates came down much lower, the self-contained system continued to undermine consumer demand for central station service.

The second major flaw of the Edison system also limited the commercial viability of the central station. Edison's system was based on a direct current (DC) circuitry, just like all previous forms of electrical technology, including the telegraph, telephone, and arc lamp. Electricity flowed in one direction around the circuit at low "pressure" or voltage. Direct current worked best with the early lights, batteries, and motors, but it suffered unacceptable energy losses when transmitted at a distance. The only way to offset these losses was by increasing the diameter of the expensive copper conduits—analogous to enlarging the size of the city's water mains. As these pipes get bigger, more and more water can be pumped at a constant pressure to a distant point. In DC systems, however, the need to use copper added dearly to capital costs and the price the utility customer ultimately paid. The price of copper put a practical limit of about a half-mile radius to the service area of a central station. To supply a large city like Chicago would require several stations with overlapping service territories. In the CBD, the dense concentration of energy consumers would justify such an investment. In residential areas, however, the capital outlay for a distribution network seemed prohibitive. Taken together, the limits of small-scale and DC distribution grids encouraged the promoters of the Edison system to emphasize the sale of self-contained plants.[17]

From 1881 to 1887, the introduction of the Edison system in Chicago consisted entirely of these "isolated" systems. Technical shortcomings largely account for a polygenesis similar to the diffusion patterns of the arc light. And the luxury nature of Edison lighting—as with the arc lamp—goes far to pinpoint the first customers: the best hotels, fancy department stores, exclusive men's clubs, and the executive offices of the business elite (see table 2). However, there were important differences between the spread of the two systems that stemmed from the advantages of the incandescent bulb over the arc lamp.

The very first installation of the incandescent bulbs at the U.S. Rolling Stock Company in November 1881 is a good case in point. The railroad car factory, in the Far South Side industrial district of Lake Calumet, purchased 125 eight-candlepower lights and a dynamo that was belted to the main power shaft. The lights were strung in two special rooms, a reporter observed, because "both are packed full of inflammable material and valuable machinery and the problem of how to light them without danger of fire has long occupied the attention of managers." Enchanted with the "new departure," the journalist trumpeted not only the safety of the Edison system, but also its convenience. "The . . . lamps burned steadily and without a flicker," he noted, "[and they] can be hung in any direction and the glass does not become heated." Each of the bamboo filament bulbs cost one dollar and was expected to burn for about six hundred hours. Both the workers and the bosses, the reporter concluded, were "extremely well satisfied."[18]

Besides being safer and more convenient, incandescent lighting fit into virtually any type of interior space. The first residential application illustrates the ready adaptability of the Edison system. In November 1882, John Doane celebrated his silver wedding anniversary by wiring his home with 250 lights for a gala party. The wealthy tea merchant, a director of the local Edison company, had installed a self-contained electrical plant and steam engine in the stable behind his mansion on exclusive Prairie Avenue. Doane shared the electric plant with his two neighbors and fellow company directors, Marshall Field and Joseph Sears. As the cream of Chicago society approached the house on the night of the party, they could see the bright lights. "From the curb to the door of the vestibule there was spread an awning lighted up with electric lamps," a newsman recorded. And as the guests passed inside, the lights "made the house brilliant in the extreme, and brought out the elegant toilets [of the ladies] in all their rich colors." Edison's system supplied a perfect solution to the problems of interior light-

Table 2   Self-Contained Systems Sold by the Western Edison Light Company in Chicago, 1882–87

Year and Purchaser	Number of Lamps	Steam[a]	Number of Dynamos
1882			
Academy of Music	125	x	1
C. W. and E. Pardridge	310	x	1
John Doane residence	558	x	1
American Express	250	x	1
	(1,243)		(4)
1883			
Haverly Theater	637	x	3
Daily News	183		1
Star and Crescent Mills	68		1
First National Bank	250	x	1
Mandel Brothers Department Store	800		2
Marshall Field and Co.	300	x	2
Chicago Packing and Power Co.	174	x	1
Schulder and Holtz	25		1
Anson Stager residence	440	x	1
	(2,877)		(13)
1884			
Pullman Palace Car Co.	2,200	x	2
Commercial National Bank	60	x	1
	(2,260)		(3)
1885			
Charles Counselman	250	x	1
McVickers Theater	1,237	x	3
Board of Trade	859		1
Park District 2	120	x	1
City Hall and tunnels	1,200	x	4
Cheltenham Improvement Co.	300		1
Chicago Opera House	800	x	2
Marshall Field and Co.	190		1
H. M. Kinsley	525	x	2
Pullman Building	0	x	1
Crane Brothers Manufacturing Co.	200		1
Union League Club	800		0
B. D. Eisendrath	353		1
C. T. Reynolds	85		1
U.S. Rolling Stock Co.	671		1
Montauk Building	85		1
J. V. Farwell and Co.	250		1
William Deering and Co.	400		1
	(8,325)		(24)
1886			
Pitkin and Vaughan	250		1
Underwood and Co.	150		1
C. Slack	295		1
National Life Insurance Co.	250		1

(Continued)

Table 2 (Continued)

Year and Purchaser	Number of Lamps	Steam[a]	Number of Dynamos
Woodruff Hotel	250		1
William Deering and Co.	250		1
Demming and Dierkes	75		1
C. D. Wetherall	50		1
W. W. Kimball Co.	250		1
Anglo-American Packing Co.	350		1
Swift and Co.—stockyard	500		1
Montauk Building	250		1
Edgewater Central Station	180		1
Phoenix Insurance Building	1,700		3
Marshall Field—wholesaler	500		2
Grace Hotel	223		0
William Deering and Co.	160		1
William French Co.	360		1
	(5,773)		(20)
1887			
Crane Brothers Manufacturing Co.	600		1
Chicago, Milwaukee and Saint Paul Freight Depot	250		1
Rookery Building	4,000		4
Fair Store	286		1
Nelson Morris and Co.	1,200		2
E. W. Gillett	300		1
F. K. Stevens	60		1
Chicago, Burlington and Quincy Office Building	752		3
Chicago and North Western Railroad Shops	1,060		4
Armour and Co.—stockyard	300		1
Crane Brothers Manufacturing Co.	172		1
Pullman Palace Car Co.	748		1
Crane Brothers Elevator Co.	50		1
Chicago and North Western Railroad Depot	750		1
Selz, Schwab and Co.	600		1
M. M. Bodie Building	175		1
	(11,303)		(25)
Total, 1882–87	30,538		89

Sources: J. M. Clark, "List of Contracts Taken by the Western Edison Light Company," manuscript, n.d., in CEC-LF, F6A-E1; J. H. Goehst, "List of Iolated Plants . . .," manuscript, n.d., in CEC-LF, F6A-E1; "Report to Stockholders, 1888: Plants in the Territory of the Chicago Edison Company," in CEC-HA, boxes 253–54.

[a]An x indicates that a steam plant was provided by the Edison company in addition to electrical equipment.

ing, but only a handful of Chicagoans could afford the $7,000 to $8,000 Doane spent to put a complete electrical system in his backyard.[19]

Although the Edison system was expensive, its local promoters enjoyed a virtual monopoly of incandescent technology. In sharp contrast to the tumultuous rivalry among arc light companies, Edison held key patents that stymied the competition, at least temporarily. In January 1882 one of the inventor's old friends from the telegraph business, George Bliss, and the financial leader Anson Stager were given an exclusive license to sell Edison equipment in the Midwest. Forming the Western Edison Light Company, they drew upon their contacts in the community of electricians to provide customers with expert installation services. In the first five years, Bliss and Stager operated a successful and growing business that sold about 30,000 incandescent lights in the city.[20] They replaced arc lamps in a few cases such as the Palmer House, but far more frequently it was gas that was turned off.

Providing incandescent lighting in the CBD's premier theaters not only established the superiority of electrical technology, it also helped train an emerging generation of architects. Known as the "Chicago school of architecture," the new style of commercial buildings is usually associated with the birth of the skyscraper. Yet meeting the need for more light inside office buildings was a challenge equal to the problem of erecting taller structures. According to the acknowledged father of the modern school, Louis Sullivan, "the immediate problem [in the early 1880s] was increased daylight, the maximum of daylight. This led him [Sullivan] to use slender piers, tending toward a masonry and iron combination, the beginnings of a vertical system."[21] It was the work of Sullivan and his partner, Dankmar Adler, on theaters that provided the testing ground for attaining new levels of interior illumination. Their tremendous success in applying the new technology would earn them many important commissions for other commercial buildings. By the end of the decade, their efforts would culminate in a masterpiece, the Auditorium theater–hotel–office tower complex.

Sullivan's use of the incandescent bulb in a series of theater projects followed a logical progression toward an intensive use of light for both functional and decorative purposes. In 1879, at the dawn of the energy revolution, the young architect of twenty-three entered Adler's office. The slightly older Adler was establishing his reputation for building Chicago theaters with excellent acoustics, such as the Central Music Hall, the Interstate Exposition Auditorium, and the Grand Opera House. To compete with these new centers of culture, J. H. Haverly felt compelled to refurbish his well-known establishment. Sullivan redesigned the entrance areas with 325 lights to give them a "dazzling, jewel-like quality." "Beneath the radiance of the electric lights," the critics agreed, "it is gorgeous in spots."[22] In 1885 J. H. McVicker followed suit by hiring the architects to remodel his concert hall with over 1,200 Edison bulbs. Sullivan recalled that he first made use of electric lights for ornamentation in the McVicker's theater. He replaced the traditional chandelier with lights worked into the ceiling decoration, and critics again applauded the results as "the best" in Chicago, if not the country. The *Real Estate and Building Journal* noted that Adler and Sullivan's motto had become " 'Let there be light.' "[23]

The following year the architects began to design the Auditorium complex, the first true harbinger of the modern skyscraper. Within the theater of this multiuse structure, Adler and Sullivan used the light bulb at a new level of intensity. With over 5,000 houselights decorating the great hall, the effect was overwhelming. The Auditorium Theater was considered then, and remains today, an architectural gem. Contemporaries found the sight "one of the most remarkable of its kind in the world."[24] Although the granite facade was traditional, the erection of the massive building involved several technical innovations, including the first use of electric lighting for construction work at night. Equally important, an aspiring architect named Frank Lloyd Wright joined the firm as a draftsman on the Auditorium project, and Chicago Edison's entire work force gained invaluable on-the-job training as it labored to install a total of 8,600 lights. With the coming of the first office skyscraper in 1889, the sixteen-story Monadnock Building, electric lighting became an intregral part of modern architecture. Designed by another famous partnership, Daniel Burnham and John Root, the Monadnock was the first office building to include electric wiring in its original specifications.[25]

By the late eighties, then, arc and incandescent technology seemed to pose a one-two combination punch that threatened to knock the gaslight business right out of the market. Despite major shortcomings of small scale and restricted distribution, the diffusion of electrical technology had quickly achieved an economic momentum of its own. In 1884 alone, for example, the number of arc lamps had doubled. The Edison system had achieved an analogous record of accelerating growth. Contemporaries had no difficulty resolving the apparent contradiction between such a risky, unreliable technology and its immediate commercial success. "Everywhere more light was needed than could be furnished," one observer explained, "and the industry had secured a foothold which, I believe, can never be weakened." H. Jampolis, a door-to-door salesman of electric service, confirms that accumulated demand for more light overcame the higher costs and the psychological fear of the new technology. Jampolis admitted that arc lighting for a small retail store cost more: $13.50 a month compared with $7.50 for gas service, but he found that shopkeepers were willing to pay this extra amount for brighter illumination. And the example of modern lighting in one store soon secured many other customers nearby, who wanted to keep up with the competition.[26]

The small scale and faulty design of the early electrical systems may have produced chaotic development, but the commercial prospects of the arc and incandescent lights were never in doubt. The spread of new technology in the CBD was steady and cumulative as Chicagoans increasingly linked things electrical with notions of amenity, class, and modernity. It did not take long for the city's gas companies to arrive at the same conclusion.

3

The Crisis of Urban Politics, 1880–1893

In the 1880s, the onset of the energy revolution in Chicago threw the long-established relations between the city's utility companies and local government into turmoil. At base, the commercial application of electric technology unleashed a tremendous demand for more light. As early as 1883, a local architectural journal had already concluded that "electricity for lighting purposes has been declared a success." Although the new technology was still plagued with faulty equipment and high costs, the *Inland Architect* asserted that the business community "have only time to cry 'more light.' The majority have seen enough to satisfy them of the superiority of the electric light." Equally important, the journal reported a radical shift in psychological perceptions of interior lighting levels. What before had seemed adequate now appeared dark, gloomy, and depressing. "The electric light has made a demand for light," the architects concluded. "More gas is used than before electric light [was] introduced."[1] The immediate problem for utility operators was meeting this insatiable demand with flawed technology and limited distribution systems. In the 1880s and 1890s, demand for gas and electric light far outstripped the ability of local utilities to supply the public. As we have seen, the results were a proliferation of self-contained systems, small-scale operators, and a fierce competitive struggle for customers.

The failure of Chicago's utility operators to resolve this "battle of the systems" quickly led them to resort to other methods of restoring stability in the marketplace. They sought both political and corporate avenues of escape from their suddenly risky business ventures. At first the established firms turned to the city council in an effort to forestall any new grants of crucial street right-of-way franchises to upstart rivals, but this approach was doomed to failure in a political culture of democratic anarchy and "honest" graft. The corruption of city politics by the utility companies quickly backfired into franchise blackmail as the aldermen conspired to grant lucrative franchises to their own dummy corporations. The unanticipated addition of new political risks made investors in Chicago utilities even more determined to escape from the competition of an open market.

Beginning in 1887, they turned from the dead end of the public arena to the private realm of corporate finance. A series of interlocking consolidations briefly reestablished monopoly conditions in the areas of gas, arc lamps, and incandescent lighting. Yet the lure of profits and graft soon attracted a new phalanx of speculators and politicians. As Chicagoans prepared for the World's Fair of 1893, the proliferation of small systems seemed to have become an endemic part of the energy revolution in utility services. For the utility operators, the costly struggle to restore secure consumer markets boomeranged, bringing just the opposite result—greater economic and political risk. The corruption involved in granting franchises was stirring up a ground swell of protest that for the first time threatened to blossom into a mass movement for municipal reform.

Although the origins of urban progressivism are complex, the case of Chicago suggests that technological change was an important ingredient in the birth of this era of reform. The electrical revolution in the city's energy, transportation, and communications services upset older patterns of business and politics, creating dynamic instability in the marketplace and in the utility companies' franchise relations with city hall. Competition among the promoters of various types of urban technology presented rich new opportunities for political corruption, corporate takeovers and monopolies, and eventually, municipal reform. In this context historian Robert Wiebe's characterization of progressivism as a "search for order" is a useful concept because it encompasses a broad range of responses to the novel conditions. The drive to restore stability was not confined to readjusting government-business relations. It also included reorganizing business structures and political parties as well as reorienting consumer attitudes toward both these institutions. An examination of the technological roots of change helps reconcile the apparently contradictory evidence that businessmen and politicians were leaders of reform at same time as they were plunging policy formation to unprecedented depths of corruption.[2]

The Corruption of Government-Business Relations

The responses of the gas companies to the coming of the electric light reveals that politics was an integral part of the battle to control Chicago's energy business. Unable to compete directly against a superior technology, the coal-gas makers faced an uncertain future of rearguard actions. They initially attempted to stymie the electrical promoters in obtaining right-of-way franchises, but an anarchic process of polygenesis quickly proved this approach was completely inappropriate for a small-scale technology. At the same time, the established firms faced a more immediate threat from upstart rivals who held patents for new methods of making cheap gas. Both sides felt compelled to use illegal methods to influence the council's votes on franchises because of the multimillion-dollar stakes involved in the large-scale systems of the gas business. To their dismay, however, the businessmen soon learned that the politicians were no one's tools. Playing each side against the other, the aldermen were working chiefly for themselves. This game worked for a while, but the politicians ultimately paid a high price for opening the door to bribery and boodle. Sounding the alarm of moral outrage, reformers found support among city dwellers who now saw a link between the demand for better public services and the need to clean up city hall.

Although technological change furnished one of the root causes of reform in late nineteenth-century American cities, the political culture of each locality helped shape community responses and public policy. In Chicago, a political culture rooted in the neighborhood, the parish, and the ward made it impossible for one individual or faction to control the city council for any length of time. To be sure, political leaders and outside pressure groups regularly built majority blocs to pass local legislation and to logroll appropriations for public works projects. Yet these coalitions seldom lasted long, because party discipline was tenuous at the municipal level. In part this tradition of decentralized power was the product of a city charter that created a weak mayor and a strong city council. If the mayor ruled at all, it was more by the force of his personality than by the power of his office. In a ward-centered system of politics, a charismatic alderman could be more than a match for an unpopular or ineffectual chief executive.[3]

The first victim of the gas companies' political machinations was the original promoter of the electric light in Chicago, Charles Brush. Like Edison, the inventor designed his system of street lighting on the model of the gasworks, with its central station and network of distribution lines. In July 1881 Brush's local agent, Milan C. Bullock, formed a company and applied for a franchise to string wires in the streets.[4] In September, however, his franchise

bill ran into a storm of protest in the city council. Opposition aldermen argued that all electric wires must be buried, in conformity
with similar regulations recently enacted for telegraph and telephone lines. "Professor" Barrett, whose expert advice carried great
weight in the council, was also outspoken in demanding underground conduits for electric cables. The fire department superintendent pointed out that his men's ladders were getting tangled in
the wires. Moreover, the wires themselves would become an increasing hazard if more and more were permitted to crisscross the
city. A month later, a fire caused by an arc lamp wire crossing a
telephone line gave dramatic punctuation to Barrett's warning.[5]

The arguments of the opposition seemed reasonable, but the
franchise debate was ultimately decided behind closed doors. In
December, rumors began to surface about undue pressures on the
council from the gas companies and their allies in the fire insurance business. Despite these efforts by fair and foul means, the
Brush grant passed twenty-four to ten when it came to a final vote
in April 1882. Supporters of the franchise simply stated that the
tremendous demand of the people for better lighting should not
be denied. But the contest was not over yet. Emerging from a private meeting with the mayor after the vote, Albert M. Billings,
founder of the Peoples Gas Light and Coke Company, admitted to
a reporter that he was firmly set against the use of electricity for
lighting. The following day Mayor Carter H. Harrison vetoed the
Brush ordinance on the grounds that overhead wires violated public policy.[6]

The presence of Billings and fellow utility company lobbyists at
city hall effectively stopped Brush, but not the spread of the electric light. The proliferation of jerry-built stations and selfcontained systems continued without interruption because many
operators did not bother to apply for right-of-way grants. By the
close of 1883, however, the dangers of fire and electrocution from
the new technology cried out for minimum installation standards.
Stiff competition among the promoters of the various systems exacerbated the problems. Barrett confirms that "the explanation of
such imperfect and dangerous work may be found in the fact that
sharp and close competition between rival companies required the
most rigid economy in doing the work, and in some cases has led
to the employment of unskilled persons."[7]

In the absence of a powerful or entrenched opposition, the
council acted progressively to bring some order. Barrett's endorsement also helped a general code of regulation for electrical instal

lations sail through the council without dissent. In this area of government-business relations, judicial precedents stretching back half a century had established the parameters of an urban public utility concept. However, many gray areas remained on the borderline between public interests and private rights. For example, the authority of local governments to set consumer rates or service standards stood on the frontier of the "police power" to promote the general welfare. In contrast, the courts had given the cities not only broad rights but also extensive duties to regulate the franchise holders in the name of public health, welfare, and safety. By the mid-eighties, city hall's public works, police, and fire departments had well-established bureaucratic routines to cope with their inspection chores.[8]

Although the electrical promoters had little influence in shaping the regulatory act, the ordinance had a major economic impact on their fledgling enterprises. The most important result was the first effective enforcement of an earlier ordinance requiring that all electrical wires be placed underground in the built-up portions of the city. The need for construction permits and, later, inspection certificates meant that the promoters of electric lighting could no longer ignore public policy. On the one hand, the shoestring operators of central stations could least afford the significant boost in costs from putting distribution lines underground, in addition to upgrading the wiring standards for building interiors. However, the formation of the Chicago Sectional Underground Electric Company by Elisha Gray helped these marginal enterprises hold fixed capital expenditures within bounds by renting conduit space. On the other hand, the municipal regulations affected everyone equally, which helped reduce the competitive pressures that resulted in shoddy work. And the inspection ordinance established a precedent for enforcing safety standards. Within a few years Barrett noted a trend among the electrical men toward policing themselves. Their first priority, he reported, was to "improve the standard of installation. . . . At a conference recently held to consult on this subject, an unqualified condemnation of the cheaper insulations was unanimously expressed."[9]

In the case of the inspection ordinance, the small scale of the electrical technology largely precluded corruption. Chicago's electrical operators were too underfinanced and too competitive to mount a campaign like the gas companies' lobby. In sharp contrast, the presence of Billings and other gas men at city hall was understandable considering that millions of dollars hinged on the council's decision to either encourage or stymie competition by granting franchises to upstart rivals. In April 1882 the aldermen brought up for final consideration not only the Brush grant, but also a sweeping gas franchise bill. The approval of the electrical bill should have warned the businessmen about the aldermen's independence. Nonetheless, both the established firms and the franchise seekers attempted to influence the policy vote.

The politicians again expressed a clear preference for competition over monopoly. This choice seemed natural and self-evident to Chicago's professional politicians. They were not the leaders of polite society; these men came from the rough-and-tumble world of neighborhood streets, institutions, and businesses. They were men on the make, working their way up in a city built on the economic combat of trade and commerce. It should come as no surprise, then, that their sympathies were with the upstarts and newcomers rather than with the guardians of vested rights and special privilege. In the contest over the gas franchise, the aldermen exploited the struggle between the businessmen for their own personal and partisan ends. What no one expected, though, was a public outcry of moral indignation over a utility franchise. In the past, what little attention anyone paid to these matters had been confined to booster statements of praise for more urban services. Now protests against bribery, boodle, and blackmail began to be heard in the daily press and in the elite men's clubs. Although muted at first, the chorus of the reformers would soon rise in a crescendo that would not quiet down for another twenty years.

This movement for municipal reform was rooted in the battle between the gas duopoly and the rival Consumers Gas Fuel and Light Company. From January to April 1882, each side attempted to get its version of a franchise accepted by the council. When test votes showed that the new venture had majority support, the friends of the old companies attempted to add as many onerous financial requirements as possible to the bill. For example, the minimum investment by a new utility was raised from $100,000 to $500,000, but amendments to cut the maximum rate from $1.75 to $1.50 and to give city hall free gas were defeated. The protracted contest over these provisions reflected a realization that the specific terms of the grant could spell the difference between profit and loss.

The high stakes hinging on the franchise created a situation ripe for corruption of businessmen and politicians alike. Gas-making technology had already evolved to a stage of giant central stations, with commensurate increases in capital investment. The promoters of the Consumers company had carefully enlisted key aldermen, including the recent champion of the old companies, Edward F. Cullerton of the Sixth Ward. When questioned about his "queer" change of heart, the alderman protested that he was simply trying to bring cheaper gas to his constituents. But a *Tribune* editorial expressed an uneasy feeling about the motives behind the mysterious switch of the Cullerton bloc. The paper hinted that the politicians were growing rich. The *Daily News* was more direct, calling the turnabout "aldermanic plunder." The chief lobbyist of the new company, Richard S. Tuthill, also complained that the politicians were engaged in a shakedown by playing the two sides against each other for votes.[10] This insider's confession is probably as close to the truth as will ever be known about the matter. In addition to bribery, several people charged that the upstarts intended to blackmail the established firms into buying them out before a single dollar was spent on building a new utility.

Since the advent of electric lighting, the newspapers, the opposition aldermen, and the gas companies' officers kept asking who would invest huge amounts of scarce capital in the gas business. In the case of the Consumers company, the answer supplied two closely connected and convincing reasons why the risks had been reduced to an acceptable level. First, the introduction of electrical technology had the completely unanticipated effect of sharply raising the standard for an acceptable level of illumination. What had before seemed adequate now seemed dark and depressing. After an initial panic in the gas industry, its local managers were among the first to notice these changing perceptions. As one gas man summed it up, electric lamps were "almost compelling people to have more light."[11]

Second, a patent to make gas in a revolutionary new way gave the Consumers company an opportunity to match this increase in demand with a supply of energy that the mass of city dwellers could afford. To be sure, commercial establishments could be expected to switch to electric lighting regardless of any marginal cost savings from gas. But cheap gas could reach a largely untapped mass market of middle- and working-class homes that still depended on kerosene lamps. In the new gas-making process, coal remained the heating fuel of choice. In fact, large-scale equipment was required to create intense heat believed impossible only a generation earlier. Inside high-pressure retorts, steam and petroleum were mixed to produce what was called "water gas." The new process made a brighter light at much less expense because of major savings in the cost of raw materials and labor. The old coal gas now seemed "dirty, expensive, and objectionable in a hundred other ways."[12] Moreover, special mixtures of the new gas were cheap enough to compete with coal as a fuel for heating and cooking. After receiving a franchise in April 1882, the Consumers company began to erect a plant and lay pipes on the South Side, where residential neighborhoods were strung along the lakefront.

The advent of cheap water gas in Chicago caused a radical shift in the battle of the systems from defensive maneuvering to open warfare marked by brutal competition. The introduction of the arc light in 1878 had already forced the two established gas utilities, the Chicago and Peoples companies, to cut their rates from $3 MCF (thousand cubic feet) to $2.25 MCF. When the Consumers company began to offer gas at $1.75 MCF in 1883, it had little trouble finding 7,000 to 8,000 households eager to sign up for service. The Chicago Gas Light and Coke Company (CGLCC) was now caught in an economic vise between electric lighting in the Loop and cheap gas in the neighborhoods of the South Side. The erosion of the market for illuminating gas in the CBD could not be stopped, but the established firm could retaliate with a vengeance against the upstart Consumers company. The CGLCC slashed rates to $1 MCF in the contested residential areas. As one of the key participants in "the war between the rival corporations," Consumers' president Columbus R. Cummings, explained: "The old company . . . did not fancy the idea of having a rival, and made an effort to strangle it by cutting prices to a ridiculously low figure. The old company could afford to do this. It had a large reserve fund that had been accumulating during [a] long period. . . . Not so with a new company which had almost exhausted its resources in erecting its costly plant."[13] By July 1886 this strategy worked, because the Consumers company was driven into bankruptcy. But the gas war was far from over; in fact, it was just beginning.

In May 1886 the established firms had invaded each other's territories, signaling a new phase of general combat among the city's gas suppliers. The CGLCC had started digging a tunnel under the river to the West Side, which previously had been the secure service area of the Peoples Gas Light and Coke Company (PGLCC). The CGLCC also began signing up customers literally in the shadow of the tragic confrontation between labor and capital at Haymarket Square. Adopting the tactics of the upstart Consumers company, the CGLCC offered a price of $1 MCF, 50¢ less than the current rates of the Peoples company. The *Chicago Tribune* gloated over this turn of events: "There is every prospect of a lively gas war on both sides of the river, all to the great amusement of the consumer, and the sorrow of the stockholders."[14]

The inexorable spread of competition throughout the city soon verified the predictions of the daily press. Even the Loop became a battlefield. In the case of the Lunt Building, for example, the CGLCC lost out to the Consumers' cheap rates, but a year later it won back the service contract. When the Consumers company found out and sent a crew to tear out the pipes of the CGLCC, it was met by a similar group of workers. "The employees of the two companies came together under the sidewalk," a journalist reported, "and as they were not particular in the selection of weapons of offense and defense, lumps of coal, chunks of clay, pieces of gas-pipe, broken boards, pots of red-lead, and in fact everything that could be taken into the hands were flying around." A squad of policemen finally broke up the fight and stood guard to protect the CGLCC workers. The clash was comic, but only because no one was badly injured.[15]

By July 1886 the gas war was reaching crisis proportions in the minds of utility investors and the press. By granting a franchise to the Consumers company just four years earlier, the city council had touched off a major upheaval in the supply of gas services in Chicago. The safe and highly profitable days of the utility duopoly were gone, swept away by stiff technological and economic competition. The resulting market pressures forced rates down at the very time when huge amounts of capital were needed to build the new large-scale plants and to expand distribution networks to more consumers. During the summer of 1886, a rapid-fire series of events brought the economic battle to a culmination. Cutthroat competition between the two established firms loomed as each prepared to mount a full-scale invasion of the other's territory. The CGLCC's tunnel to the West Side neared completion, but the Peoples company obtained a preliminary injunction against any further construction. The leader of the West Side utility, Billings, argued that the secret contract of 1862 dividing the city in half was still binding. Founder and president Elias T. Watkins of the South Side CGLCC claimed to the contrary, that Billings had broken the old agreement when he became a silent partner in the formation of yet another new utility enterprise, the Equitable Gas Light and Fuel Company. Certainly there was no disputing that this new venture was tearing up the streets and laying gas mains on the

South Side, where it threatened to sell water gas for as little as 65¢ MCF. The steady inroads being made by electric lighting in the Loop only added to the gas investors' sense of crisis and uncertainty.[16]

The popular press was equally upset by the energy revolution, but for different reasons. With memories of the Consumers franchise still fresh in August 1885, a grant to the Equitable company over the mayor's veto turned the polite protests of the daily newspapers into a moral crusade against the "boodle gang of aldermen." While paying homage to "healthy" competition, Mayor Harrison had denounced the current "warfare," which had thrown the Consumers company into bankruptcy just a month earlier. Rumors about $85,000 in bribe money became the most common explanation of why the aldermen had voted thirty to five to override the mayor's objections to a fourth gas company.[17] A year later, as the Equitable company tore up the streets on the South Side and the stockholders of the Consumers company fought each other over the upcoming receiver's sale, Mayor Harrison threw his support behind the city council's plan to give out even more franchises. The mayor's unexpected turnabout convinced the press that the corruption of city hall was complete.[18]

In July 1886, new perceptions about the value of utility franchises began to crystallize into a full-fledged movement for municipal reform. Daily coverage of urban affairs moved from the back pages to the headlines, while editorials blasted away at both the utilities and the politicians. At immediate issue was a franchise that, in effect, gave free use of the city's La Salle Street tunnel to the cable car company of Charles Yerkes. Despite the *Tribune*'s lecture on "the duties of the mayor," Harrison refused to demand a reasonable compensation for the tunnel grant. More important, however, was the general conclusion drawn from the recent series of grants to gas, electric, telephone, and streetcar companies. The question, "Who got the boodle?" reached page one in the major newspapers and stayed there for the next twenty years. The cause of reform began to change from the narrow concerns of elite groups of taxpayers to a broad-based, popular crusade. After 1886 urban politics in Chicago would be transformed as more and more of the city's organized ethnic and special-interest groups became involved in the movement to root our corruption at city hall.[19]

For utility investors, the newspapers' new posture of moral outrage and political hostility added significantly to the mounting pressures of the marketplace to end the gas war. At the receiver's sale of the Consumers company in July, heavy bidding nearly doubled the price, to $2 million. The winners were led by Cummings, Sidney Kent, and William S. Rayburn. They represented a new generation of stock speculators who no longer took an active part in the daily management of urban utilities. The company was worth $3 million, according to analysts, but the likelihood of a four-way rate war seriously depressed its value. Three months later this prospect seemed inevitable after the court dissolved the injunction against the Chicago company's invasion of the West Side. Judge Shephard of the Superior Court of Cook County ruled that the old contract was void and against public policy. Since gas companies provided an essential urban service, the court refused to enforce any attempt to restrain competition. Left without political or legal remedies in the local arena, the new breed of utility speculators took the initiative in bringing the battle of the systems to a permanent end.[20]

The Consolidation of Corporate Power

In 1887 utility financiers engineered a coordinated series of corporate mergers to establish a monopoly in each of the three fields of gas, arc, and incandescent lighting. In Chicago, a political culture of anarchic democracy seemed to coincide closely with an equally decentralized and competitive marketplace, at least in the energy business. Failing in their search for order in the political arena, utility operators turned to a rising generation of corporate financiers and lawyers for help in controlling the unbridled pace of technological change. This mix of innovation in politics, business, and technology would distinguish urban progressivism as a profound cultural movement, marked by ideals of scientific management, social engineering, and businesslike efficiency in every aspect of city life, including government. Within this broad context, the corporate reorganization of the energy utility business deserves examination as a constituent part of the battle for municipal reform.[21]

In Chicago, a syndicate of Philadelphia speculators acted first by boldly gaining complete control of the gas supply. For the next ten years, creative lawyers kept the resulting "Gas Trust" one step ahead of the law despite setback after setback in the courts and state legislature. However, this arrogant defiance of the law became a raw political nerve that constantly stimulated the forces of reform to strike back at public utility companies. The same group of financiers also consolidated the city's various arc light utilities into a single firm, the Chicago Arc Light and Power Company. This parallel maneuver was aimed at retaining the profits from lucrative municipal contracts for street lighting that had previously gone to the gas duopoly. At the same time, the local backers of the Western Edison company reorganized their venture in order to raise more capital and get a franchise to supply the city with central station service.

The utility speculators were successful in restoring order, particularly in the gas business, but they failed to halt the multiplying of electrical systems. Contrasting rates of technological innovation largely account for these different outcomes. To be sure, the invention of the out-of-state holding company helped the Gas Trust evade reform legislation and court decrees. More important, however, the new water-gas process represented a superior large-scale technology that required huge amounts of fixed capital. In the wider context of the energy revolution, the chances of realizing an adequate return on such huge, long-term investments invariably pointed toward a very limited number of profitable firms. As the first president of the Gas Trust, Norman C. Fay, accurately foretold in his debate with fellow members of the Citizens Association, "No, sir, there will be no permanent dismemberment of the gas combine. That is practically impossible."[22] Furthermore, the Chicago experience of rate wars, distributor duplication, and torn-up streets presented an alternative that was attractive to virtually no one. In contrast, the dynamic pace of technical innovation in the electrical field combined with small scale to promote a continuing increase in low-budget systems. Electrical service remained a luxury, but the demand for more light constantly outpaced the ability of the central station operators to supply this growing market.

In the ten days between 27 April and 7 May 1887, the announcements of the corporate reorganizations struck Chicago like three mighty tidal waves in rapid succession. By far the most unexpected and threatening was the formation of the Chicago Gas Trust Company. Acting as a middleman, Charles Yerkes brought a group of the new breed of utility speculators from his previous hometown of Philadelphia together with the aging owners of the old duopoly. After five years of turmoil, Watkins of the Chicago company was sick and tired, ready to sell out even though his company remained the strongest in the field with over 26,000 customers. His old friend Billings, too, was preparing to retire and pass the leadership of the Peoples company to his son Cornelius. With the sale of these two companies, Yerkes, Kent, and Cummings held all the city's gas franchises, since they had earlier gained control of the Consumers company and the Equitable company. Two suburban companies from the South Side townships of Hyde Park and Lake rounded out the new monopoly. The Gas Trust exchanged $25 million in new stock for about $13 million in old securities, besides offering another $7.65 million in bonds on the open market. A good measure of the profitability of the new venture was the first dividend, which paid $38.25 on each $25 share of stock, or a 153 percent return on a highly inflated security.[23]

Reaction to the news was immediate, widespread, and strident. Emerging from the first meeting of the syndicate on 30 April, Cummings declared that the price of gas was going up from $1 to $1.25 MCF. "This means," the *Tribune* blasted, that "it is proposed to compel Chicago and the towns named to submit to gas extortions." And the syndicate's first president, Norman Fay, recalled that "fierce indeed was the war that burst upon the Gas Trust alike from the press and politicians, reformers and blackmailers."[24] As Fay suggests, the new monopoly galvanized all the previously disparate forces of reform into a united front. Most indicative of future trends was the decision of the prestigious Citizens Association to become involved in the fight against the Gas Trust. Chicago's oldest and most influential civic group, the Citizens Association turned its attention for the first time from taxpayers' concerns about government waste to consumers' interest in the business conduct of privately owned urban utilities. Marking a departure from the past, this group of the city's best men sued the monopoly as an unlawful combination in violation of public policy and the state constitution.[25]

The importance of the Citizens Association's decision cannot be overstressed because it gave birth to a mass movement for municipal reform, called urban progressivism. The civic association had long pioneered the lobby tactics of the special-interest group. Now it spearheaded the reform by enlisting the support of other organized groups to form a powerful political force. A new type of urban politics began to emerge that was characterized by coalition building. Special-interest and ethnic groups united to focus attention on a specific cause such as utility franchises or robber barons like Yerkes. Of course, the Gas Trust also played a key role in this transformation by becoming the perfect epitome of corporate arrogance, beyond reach of the law. The fascinating story of how the monopoly eluded the effects of one defeat after the other in the courts and the state legislature for nearly a decade is outside the scope of this book. Here it is important to stress that the Gas Trust represented a historic turning point in the politics of urban reform.

For more and more Chicagoans, making the companies holding utility franchises responsible to the public became a common cause. Reflecting a new appreciation of essential urban services, the daily newspapers called for the formation of a "gas commission" similar to the state agency regulating the railroads. Such a commission, the *Tribune* asserted, "involves no new principle of constitutional law, nor does it propose a task which is at all impractical. No gas company is strictly a private corporation. On the contrary, its plant can only be made available and profitable by the use of public franchises."[26] Although such an administrative body was nearly twenty years in the making, the Gas Trust's defiance of the law helped to ensure its eventual creation.

Still reeling from the news about the Gas Trust, Chicagoans received a second shock a week later when the same group of speculators revealed that they had bought all the city's arc light companies. For about $300,000, the syndicate purchased nine "central stations," Gray's underground conduit network, and 1,100 arc lamps. In this case the blow of a second "virtual monopoly" was softened by a promise that the nightly rate would not increase from the current level of 50¢ per light. Moreover, the new Chicago Arc Light and Power Company hoped to bring greater efficiency and lower rates in the near future by consolidating all operations at a single location. Reflecting an era of technological incubation, the company planned to use belts made of rope as a unique, albeit primitive, method of transmitting power from the steam engines on one floor to the dynamos on the floor above. Nonetheless, the arc light monopoly began showing both healthy growth and profits after a two-year period of adjustment.[27]

In the wake of the two surprise announcements, news about the reorganization of the Edison company was anticlimactic. After five years of successfully promoting self-contained plants, mounting demands for an incandescent lighting service pointed unmistakably toward a low-risk opportunity to boost the company's profits in a major way. In the Loop, many landlords wanted to replace gas lighting with modern technology, but without the initial expense or ongoing trouble of operating their own electric plants. Many others in the downtown area were simply tenants whose choices were limited to alternative types of utility service. Responding to these needs for more incandescent lighting, the electric company obtained a municipal franchise and a new state charter of incorporation as the Chicago Edison Company, which gave the enterprise a broader financial base. During summer 1887 it began constructing a central station and sixteen miles of underground conduits in the heart of the Loop. The Adams Street station followed the model of Edison's original plant in New York City, with its rows of matching steam engines and dynamos. Each Edison machine could now run 1,250 sixteen-candlepower lights. On 6 August 1888, Chicago Edison inaugurated central station service at a full capacity of about 10,000 bulbs.[28]

Political reaction to the formation of the two electric companies was muted compared with the strident protest against the Gas Trust. For the most part, this disparity reflected an accurate assessment that the city faced no imminent threat of a monopoly in the electric light business. In fact, the battle of the systems remained essentially unaffected by the corporate consolidations, and several new forms of competition appeared over the next two years. Neither of the two large firms could keep up with the demand for central station service. For example, when the Chicago Arc Light company planned to install a 500 horsepower (HP) engine at its new facility, it was heavily criticized on the grounds that this was too big a machine; yet by the time the central station started to deliver services, two of them were needed. Likewise, the Adams Street station was straining at a maximum capacity of 50,000 lights within five years. In the case of the Edison company, the drive to maximize profits encouraged the utility to compete against itself by selling as many self-contained systems as possible. By 1893 these were illuminating 75,000 bulbs, or 50 percent more than central station service. The sale of complete arc lamp systems also continued apace, with the combined output of individual establishments and shoestring operators on scattered blocks providing as much light as the big downtown company.[29]

The small scale of electrical technology also helps account for the failure of Chicago's two large firms to halt the process of polygenesis. In contrast to the gas business, entry into the electrical field required relatively little investment capital. Even a city government chronically strapped for funds could experiment with a public system for lighting the bridges. Upon Barrett's urging, the city council approved such a scheme, and in 1887 the system began to supply power to 104 lamps and steadily expanded to provide about 1,000 arc lights and an equal number of incandescent

Table 3   Chicago Central Station Companies, 1887–93

Name of Company	Year Established	Year Absorbed
Underground Wire District[a]		
Chicago Edison Co.	1887	1907
Chicago Arc Light and Power Co.	1887	1893
King Electric Light Co.	1887	?
National Electric Construction Co.	1887	1894
Archer Avenue Electric Light Co.	1888?	1898
Central Electric Light Co.	1888	1892
Cooperative Electric Light and Power Co.	1888	1894
Consumers Electric Light Co.	1888	?
Merchants Arc Light and Power Co.	1888	1898
Northwestern Electric Light and Power Co.	1888?	1894
Milwaukee Avenue Electric Co.	1889	?
Central Illuminating Co.	1889	?
Thirty-first Street Merchants Electric Light Co.	1890	1903
Auburn Park Electric Light Co.	1891	?
Overhead Wire District		
Calumet Electric Lighting Co.	1887	1898
Economic Electric Light and Gas Co.	1887	1898
Lake View Electric Light Co.	1887	1888
Edgewater Light Co.	1888	1898
South Side Electric Light Co.	1888	1897
Englewood Electric Light Co.	1888	1898
Peoples Electric Light and Motor Power Co.	1888	1898
Western Light and Power Co.	1888	1898
Hyde Park Thomson-Houston Co.	1889	1892
Calumet Gas and Electric Co.	1890?	1898
Chicago Illuminating Co.	1890	1898
Stiger and Newhall Electric Light Co.	1890?	1893
Hyde Park Electric Light Works	1891	1892
Mutual Electric Light and Power Co.	1892	1897

Source: "Chicago Central Station Development Chart," in CEC-LF, F6C-E4.
[a]Area bounded by 3900 south, 1600 west, 1600 north, and the lakefront on the east.

lights within the next five years.[30] More threatening to the two major firms was the rapid proliferation of rivals on the fringes of the Loop and in the outlying neighborhoods and the suburbs. Between 1887 and 1893, no fewer than twenty-four central station companies were established within the city limits (see table 3). Most were legitimate enterprises that were started to meet commercial demand for modern lighting, especially by small merchants who banded together to provide street lighting along their neighborhood business strips. But a few of the new utilities posed a threat that was more political than economic.[31]

In the late 1880s, mounting expectations concerning a sweeping annexation of Chicago's suburbs led several groups of speculators to obtain franchises from these residential communities. Legal precedent strongly suggested that these special licenses would soon be valid throughout the metropolitan jurisdiction. This liberal interpretation of private vested rights probably accounts for the inclusion of two suburban companies in the Gas Trust. It certainly explains the timing and the origins of analogous electric ventures in Hyde Park, Lake, Lake View, and other outlying communities on the eve of the great 1889 annexation. The most important concession in these grants was the right to string overhead distribution wires, whereas the city required the much more expensive underground conduits. Contemporary journals predicted that the new franchise holders would "soon reap the reward of their shrewdness."[32] Although the use of overhead wires in the Loop remained in doubt, there was no question that the two major companies there were now surrounded by rival firms.

Besides competition from the municipal government and the overhead wire companies, new uses of electrical energy created additional forms of economic rivalry. After the mid-eighties, electric motors came into general use to power a broad range of machines from 1/8 HP fans to 15 HP elevators. In 1890 the city electrical inspector reported that "the variety of small industries to which this species of power is applicable is constantly and rapidly widening. A few years ago, the only use for which a motor seemed fit was its employment as a fan, and only the smaller models were in demand. Today the electric motor is crowding out the steam plant and taking its place for driving small machinery, printing presses, elevators, restaurant fans, embossing and stamping machinery, polishing, etc."[33] Since the motors used direct current (DC), the Edison and the Chicago Arc companies found themselves competing against each other. With underground distributors in the Loop running along many of the same streets, a battle for customers became unavoidable (see map 1).

Once the contest was joined, each utility strove to broaden its range of services and undercut the advantages of the other. In 1889 the arc light company began to provide incandescent bulbs to its customers. The utility introduced the new service "for the purpose of covering the territory threatened by a new incandescent company [the Edison company] which we feared might furnish arc lights." A year later, Chicago Edison retaliated in kind by offering arc lamp service (see tables 4 and 5).[34]

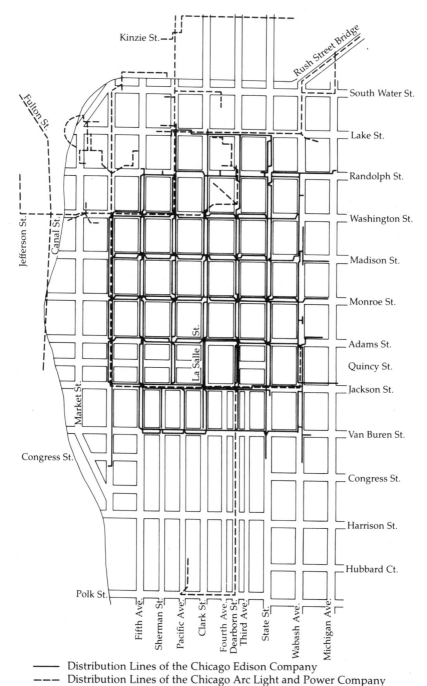

Kinzie St.

Fulton St.

Rush Street Bridge

South Water St.

Lake St.

Randolph St.

Washington St.

Jefferson St.

Canal St.

Madison St.

Monroe St.

La Salle St.

Adams St.

Quincy St.

Jackson St.

Market St.

Van Buren St.

Congress St.

Congress St.

Harrison St.

Hubbard Ct.

Polk St.

Fifth Ave.

Sherman St.

Pacific Ave.

Clark St.

Fourth Ave.

Dearborn St.

Third Ave.

State St.

Wabash Ave.

Michigan Ave.

——— Distribution Lines of the Chicago Edison Company
– – – Distribution Lines of the Chicago Arc Light and Power Company

MAP 1. Distribution systems of the Chicago Edison Company and the Chicago Arc Light and Power Company, 1983. *Sources:* "Important Paper Files: Chicago Edison Company and Chicago Arc Light and Power Company," in CEC-HA, boxes 1–2.

Table 4   Growth of Central Station Services by the Chicago Edison
Company, 1888–92

Year	Generator Capacity (kw)	Lights Capacity	Maximum Load (kw)	Output (kwh)
1888	640	9,600	313	164,000
1889	960	14,400	618	827,000
1890	1,540	21,600	1,280	2,221,000
1891	2,300	24,000	2,140	2,846,000
1892	3,200	43,200	3,200	5,827,000

Sources: Chicago Edison Company, *Annual Report,* 1887–93; H. A. Seymour, "History of Commonwealth Edison," manuscript, 1932, 192, in CEC-HA, box 9001.

Table 5   Growth of Central Station Service by the Chicago Arc Light and Power
Company, 1889–92

Year	Maximum Arc Lights	Maximum Incandescent Lights	Maximum Motor (HP)	Net Profit
1889	1,808	769	0	$ 91,900
1890	2,301	1,279	0	$120,300
1891	2,480	1,856	397	$127,900
1892	2,310	2,595	792	$ 85,700

Source: Chicago Arc Light and Power Company letter to Samuel Insull, 31 January 1893, in Chicago Arc Light and Power Company file, CEC-HA, box 1.

After 1890, then, the battle of the systems became even more intense than before the corporate consolidations. While economic and technological competition had accelerated the pace of innovation, it had also churned a stable market for energy into a state of flux. Political and financial schemes to stop the multiplying of electrical systems had failed miserably. On the contrary, it had seemed to gain momentum with each increase in the technology's reliability and range of applications. At base, the demand for modern technology had constantly outpaced the ability of the utility companies, including the two well-financed Loop firms, to meet this need for more light. Hamstrung by short-range distributors and small-scale generating equipment, even the most efficient suppliers of central station services could do little to put either the self-contained system or the shoestring operators out of business. Electric lighting remained a luxury in spite of its successful transition from an experiment to a commercial proposition. As Chicagoans built a "dream city" for the approaching World's Fair of 1893, the energy revolution raged on without resolution in the real world.

4

The Rise of Samuel Insull, 1893–1898

The Chicago World's Fair of 1893 was unique; there has never been anything quite like it before or since. In part, the spectacular architectural unity of the White City accounts for its singular impression on the American mind. Lavish displays of electrical technology also help explain why over 22 million people—one-fifth of the nation—were attracted to the City of Light. As one visitor concluded, "The architecture . . . was its chief beauty; the lighting . . . was its crowning glory."[1] Yet the timing of the World's Columbian Exposition is the key to its special importance. The fair took place at a major turning point in the history of the United States, appropriately called the "crisis of the nineties." During this decade, America made a wrenching transition from a rural, agricultural society to an urban, industrial one. As the perfect embodiment of this critical period, the World's Fair represented both the culmination of the Victorian era and the promise of the machine age.[2]

The symbolism of the fair as a juncture between the past and the future was not lost on contemporaries. Daniel Burnham and the other planners of the exposition insisted on classical styles of architecture to create a soothing and familiar environment. The fair left an indelible impression on visitors because it looked exactly as they expected a "dream city" to look: monumental, orderly, beautiful, and clean. In a similar vein, the electricians hoped to overwhelm fairgoers by opening a window to the coming age of high technology. "Everything that is done in the shape of power will be by electrical transmission," a representative reported to the applause of the 1891 meeting of the National Electric Light Association (NELA).[3] Dazzling displays of light and power, including a futuristic electric kitchen, had the intended effect. Perhaps Henry Adams, one of the fair's keenest observers, best summed up its meaning: "Chicago asked in 1893 for the first time the question whether the American people knew where they were driving."[4]

Arriving in the city at the time of the fair was an immigrant named Samuel Insull, who believed he knew the answer to Adams's question. The thirty-four-year-old Englishman came armed not only with a vision of the future but with a plan of how to get there. After serving an apprenticeship under Thomas Edison, Insull became the chief apostle of the central station concept. In 1892 he accepted an offer to become president of the Chicago Edison Company so as to prove that central station service could supply electricity to every type of urban consumer at low rates. Over the next six years, he emerged as the industry's leading spokesman of a new gospel of mass consumption. In the process, the young executive also resolved the economic dilemmas of polygenesis and the political crises of Chicago's energy utilities. In 1898 both the gas business and the electric business would begin to operate on new commercial principles based on the idea that "low rates may mean good business." The spectacular success of Insull's plan helped fulfill the American dream by promoting the growth of an energy-intensive society.

The rise of Samuel Insull personified the triumph of a culture of technology that was propelling America toward unprecedented material affluence, especially in the cities. Guided by the experiences of the recent past, a new generation of urban public utility operators came to maturity during the crisis of the nineties. They took the cultural significance of the World's Fair as a starting point in plotting new directions for their business enterprises. Insull rapidly climbed to success because he seemed to harness technology in the cause of urban progressivism. Since the 1880s, municipal reformers had been turning to environmental improvements and the efficient delivery of public services. Enhancing the quality of urban life, they believed, was a practical step toward restoring social harmony and political democracy. After taking over Chicago Edison, Insull immediately began to implement a bold plan for supplying central station service throughout the city. Pursuing this vision with relentless determination, he was the first to master the peculiar economic characteristics of the electric utility business. This insight led him to restructure rates in a way that strongly encouraged consumption by sharply reducing customer's bills. In 1898 Insull announced his gospel of consumption to the NELA, opening a new era of low-cost energy for the city.

The Culture of Technology

The potential of electrical energy to improve everyday life in the city was demonstrated for the first time at the World's Fair. Here the engines of social progress were vividly defined in terms of a technology of power and a technique of environmental planning. The incredible number of things electrical at the exposition was not chance but a by-product of a vibrant cultural tradition. Since the onset of the industrial revolution a half-century earlier, Americans had been inclined to view technology as a benevolent agent of social progress and republican values. Steam power—the engine of the factory, steamboat, and railroad—became one of the most important cultural symbols of the industrial age. During the 1880s, Americans easily translated the gospel of steam into electrical terms, and the new technology evoked even more fantastic images of mysterious, invisible forces. The myths surrounding Thomas Edison reinforced a constellation of beliefs that equated technological innovation with the steady march of material improvement, moral uplift, and social democracy. Even before Chicago's fair in 1893, then, Americans were disposed to look to technology to solve their economic and political problems.[5]

The first commercial applications of electricity added important new dimensions to the cultural symbolism of technology. As we have seen, electric lights quickly became associated with urban modernity, luxury, and class status. Whereas steam power was confined to work spaces, electrical appliances were designed for everyday life. Frustrated by the high cost of the early electrical systems, Americans gave this technology a special meaning. During the 1880s the elite men's clubs, fancy department stores, downtown theaters, and Prairie Avenue mansions turned the novel technology into a symbol of convenience, comfort, and conspicuous consumption. Contemporary images of the city lights created desires for things electrical that went far beyond utility. American cultural and social orientations predisposed the planners of the 1893 fair to envision the city of the future as a technological utopia.[6]

In this context, a lavish use of electricity at the World's Fair became crucial to a "dream city." The initial planning session of the NELA in 1890 centered on the lessons of previous international expositions. One of the electrical men argued that the Philadelphia Centennial of 1876 had accelerated the commercial success of the telephone by ten years. A consensus emerged that the exhibits on electricity had to exceed those of every other industry in size and grandeur. More important, however, the Paris exposition of 1889 pointed the way toward an extensive integration of electrical technology into the environment of the fair itself. The planners considered several components of the infrastructure essential, including artificial illumination for all outside and interior spaces, an electric street railroad, and lights built into water fountains to produce dramatic shows of dancing colors. In addition, power had to be provided for several electric communications systems, including telephones, telegraphs, and fire alarms. In nominating John Barrett to be chief electrician of the fair, the NELA entrusted him to fulfill its great expectations.[7]

By the time the exposition opened in May 1893, Barrett had outdone even the grandiose schemes of the electrical fraternity. In a place "where are gathered the glories and mysteries of human achievement, rises the Electrical building, stored with the most marvelous of the marvels of the age. The potentialities and splendors of electricity were never before so exhibited as under this picturesque roof," a writer for *Cosmopolitan* magazine exclaimed. Inside, the most impressive display was the General Electric Company's seventy-foot Tower of Light, which contained 10,000 Edison bulbs enclosed in colored glass fixtures. Almost equaling the total number of lights at the Paris exposition, the tower gave electrical men deep satisfaction. "Probably nothing more brilliant," their trade journal boasted, "nothing more gorgeous has ever been attempted before. It may be likened to an electric fountain, immobile crystal—petrified, as it were, in the midst of its play."[8]

Indeed, most visitors agreed that the "wonderous enchantment of the night illumination" was the White City's greatest spectacle. Providing over ten times as much artificial illumination as the Paris show four years earlier, the 1893 fair used over 90,000 incandescent bulbs and 5,000 arc lights. The Ferris wheel alone had 1,340 lights on the rotating section and another 1,100 bulbs on the supports. Again, contemporary description best conveys the impact of this technological tour de force: "The Fair, considered as an electrical exposition only, would be well worthy of the attention of the world. Look from a distance at night, upon the broad spaces it fills, and the majestic sweep of the searching lights, and it is as if the earth and sky were transformed by the immeasurable wands of colossal magicians. . . . It is electricity! When the whole casket is illuminated, the cornices of the palaces of the White City are defined with celestial fire."[9]

Electricity also played an important role in making a visit to the White City a pleasant experience that suggested a future world of efficiency, convenience, and leisure. Circling the grounds was an elevated railway. Its speed and grace amazed visitors because it seemed to be driven by invisible forces. The railway used an innovative third-rail system instead of the standard overhead trolley. A fleet of fifty battery-powered gondolas also transported fairgoers around the park's Venicelike canals. In addition, a seven-mile string of electrical buoys guided larger lake steamers that shuttled visitors between the Loop and the South Side site. Several communications systems rounded out the contributions of this new technology to an urban environment of safety, cleanliness, and comfort.[10] Americans spending a day and an evening at the White City could hardly avoid being dazzled by high technology, and they carried away a lasting impression, if not faith, that there was an electrical device for every purpose.

Henry Adams was one of those visitors. This perceptive observer of contemporary society summed up the meaning of the 1893 exposition and its culture of technology in an influential essay titled "The Dynamo and the Virgin." For Americans, the new source of energy had supplanted nature as a "symbol of infinity." In what may have been the first attempt to define high technology, Adams compared the electric dynamo to a religious object or "occult mechanism." Standing before such a powerful but barely audible machine, "one began to pray to it; inherited instinct taught the natural expression of man before silent and infinite force. Among the thousand symbols of ultimate energy, the dynamo was not so human as some, but it was the most expressive." Although the social critic argued in favor of retaining traditional values, he recognized that his antimodernism represented a dissenting point of view.[11] In fact, the crisis of the nineties added tremendous impetus to a technological vision of progress.

Many who attended the fair undoubtedly felt more comfortable and relaxed inside the White City than in the chaotic urban environment surrounding it. Ironically, as historian Jackson Lears points out, "the city became the emblem of modern unreality" during the Gilded Age. The Victorians were slow to adjust to the industrial revolution until a rapid-fire series of violent shocks jarred them into facing the social and political implications of the great transformation. The mushroomlike growth of Chicago represents a prime example of this modernization, which helped double the population in a single decade, to over a million inhabitants at the time of the 1893 exposition. Seven years earlier, the tragic clash at Haymarket Square had begun an alarming rise in violent confrontations between workers and capital, culminating in the Pullman railroad strike of 1894. In the midst of a national depression, labor unrest in the cities was more than matched by a farmers' revolt in

the countryside. In 1892 the People's party scored impressive electoral gains, giving the rural-based Populists substantial political
momentum toward victory in the next presidential contest. The
frightening feeling that America was coming apart at the seams
gave the 1890s a sharp sense of social tension and cultural ma-
laise.[12]

In the midst of this general crisis, Americans were ready, even
eager, to embrace high technology as the best hope for the future.
The World's Fair could not have been better timed to build faith in
the belief that electricity was the new force of progress. And as
with the earlier gospel of steam, Americans readily attached a
democratic promise to this materialistic definition of the advance
of modern civilization. The new technology, contemporaries believed, would spread its benefits among all the people, raising
standards of living and freeing the human spirit from the grim
necessities of survival.

The birth of the home economics movement at the fair illustrates the American penchant for linking electrical appliances to
liberation and modernity. Reformers of domestic life who were
meeting in Chicago called for the adoption not only of the industrial techniques of scientific management but also of the hardware
of the factory world. As Gwendolyn Wright notes in her study of
these feminists, "New domestic technology was central to the aesthetic and the cultural redefinition of the model home." An experimental electric kitchen presented women with an alluring glimpse
into the future. The exhibit was composed of a rather odd collection of primitive appliances, but compared with the equipment
then available, the model kitchen created an enticing impression
of efficiency, convenience, and fashionable status. Besides lighting
fixtures, it displayed a thermostatically controlled oven, water
heater, coffeepot, and chafing dish, as well as a hot plate that
boiled water without smoke or smell in half the usual time.[13]

If the perfection of a servantless, all-electric kitchen lay somewhere in the future, the big-city department store presented a
ready-made environment where women could act out their eman-
cipation from Victorianism into a new culture of technology. Like
the World's Fair, palaces of consumption such as Marshall Field's
offered their patrons everything under one roof. And as with the
World's Fair, electricity played an integral part in creating these
"festival environments" that enveloped shoppers in a dream world
of luxury, leisure, and comfort. "Artificial and natural lighting,"
cultural historian William Leach contends, "transfigured the [de-

partment] stores into 'refined Coney Islands.' . . . The window [displays] were often more important than the goods within: they communicated festivity, vitality, beauty, and fantasy, revealing the . . . inner possibilities of store life." [14]

The World's Fair of 1893 and modern institutions like the department store strongly reinforced America's faith in technology at a time of cultural crisis. Anxious to restore social harmony and political order, visitors flocked to the "dream city" because it promised an urban-industrial future worth striving for. Coming during a period of violent unrest and collective self-doubt, the exposition made the path to a technological utopia seem accessible and familiar. Moreover, the fair helped make electricity the ultimate symbol of progress. Already associated with urban modernity and class status, the new high technology became firmly fixed in the American imagination as a benevolent agent of moral uplift and democratic values. After 1893, the desire for electrical service in Chicago, especially for artificial lighting, approached universal demand.

The Perils of Progress

The very success of the new technology at the White City underscored the basic problem confronting utility companies in the real city. In spite of a decade of rapid commercial growth and copious financial investment, central station operators had made virtually no headway in overcoming supply-side constraints. Small generators and restricted distribution grids seemed insurmountable hurdles to utility companies. Faulty technology meant that the insatiable demand for more light at a reasonable price went largely unfulfilled.

Faulty technology created equally frustrating economic problems for electric utilities. In spite of Edison's faith in the central station concept, electrical technology seemed to contain inherent diseconomies of scale. Practical experience continued to accumulate in favor of the self-contained, "isolated" plant for the large consumer of electric lighting. Unit costs (and hence rates) were highest for small consumers, since they made the least efficient use of the station's generating equipment. As more lights were burned for longer periods, the cost of a unit of electricity—expressed as a kilowatt-hour (kwh)—declined. When the equipment was used more fully, the utility's huge capital costs were spread over more units of electricity, making each unit cheaper to generate. The resulting equation between rates and costs appeared to point irreversibly toward the eventual triumph of the self-contained system. As a customer's use increased, it would reach a point where significant savings could be gained from disconnecting the utility lines and purchasing a self-contained system of the appropriate size. In response, utility companies could offer discounts to large consumers, but at the expense of shifting a heavier burden of capital costs onto the small consumers. The result would be rates beyond the reach of most city dwellers who badly wanted to install the new technology.

More than any other individual, Samuel Insull solved the problems inherent in electrical technology and put the central station on a sound economic footing. Great ambition, international background, apprenticeship with Edison, and faith in technological progress combined to give the English immigrant a unique perspective on the problems facing local utility operators. The education of Samuel Insull pointed toward an exceptionally comprehensive view of the electrical industry in the years leading up to his arrival in Chicago in 1892.

Born near London on 11 November 1859, Insull was raised in a religious middle-class family. Spurning his father's urging to join the clergy, the young man sought a career in business. He learned shorthand to gain a competitive edge in the London job market, and in 1879 he secured a position as secretary to an American, George E. Gouraud. A friend and representative of Thomas Edison, the banker had come to England to set up telephone companies licensed to use the inventor's patented equipment. It did not take long for the ambitious twenty-year-old secretary to impress Gouraud, who recommended him for a similar position with Edison. Insull sailed for New York City, where he first met the wizard on 28 February 1881, a date he fondly remembered almost fifty years later. He recalled that he was "thoroughly imbued with the idea that I had met one of the master minds of the world. I was young and enthusiastic, and it is true that Edison [had] a peculiar gift of magnetism. But I have never changed my mind from that day to this."[15]

The timing of the immigrant's apprenticeship with the "old chief" could not have been more opportune. Arriving during the construction of the first Edison central station in the financial district at Pearl Street in New York City, Insull was involved in every phase of the electrical industry, from conducting the inventor's business affairs during the day to laying underground cables at night. Edison had no doubts about the technical superiority of the incandescent light bulb, and he taught his aid to concentrate on the economic challenge presented by gas, a much less expensive source of artificial illumination. Writing home to a friend in 1881, Insull explained that "Edison [central station service] will work just as the gas companies do." Following the much-heralded opening of the Pearl Street Station on 5 September 1882, the immediate problem was to perfect hardware that would allow local Edison companies to compete commercially against the gas companies.[16]

For the next eleven years, Insull devoted himself to making electricity from central stations cheap enough to pose a viable economic threat to gas. During these years, the rising young executive became Edison's chief apostle of the central station concept. He traveled across the country to sell local entrepreneurs on the idea, and in 1886 he moved to Schenectady, New York, to oversee the building of a manufacturing plant for Edison equipment. Reflecting the instant success of the incandescent light, the Schenectady plant grew from 200 to 6,000 workers during the six years Insull supervised the operation.[17]

In the winter of 1891–92, the executive faced "the first great crisis of [his] career in America." The consolidation of the Edison Electric and the Thomson-Houston interests elevated Insull to an envious third place in the corporate hierarchy of the new General Electric Company (GE). In reporting the merger, *Electric World* noted that Insull was "so well known to those who have had dealings with the Edison Company that his selection as Mr. Edison's representative among the financial managers of the new organization has not been a surprise."[18] But the second vice president was dissatisfied; perhaps he believed his promotion was actually a step down from the top position at the Schenectady works to a middle-level slot in a company dominated by men from the Thomson-Houston side of the merger. Or perhaps, Insull's crisis arose because another promising opportunity came to him during this transitional period of high anxiety: Insull knew that the Chicago Edison Company was seeking a new president, because he had been asked to recommend a candidate.

In the spring of 1892 he resolved the crisis by putting his own name forward for the position. Insull later justified his choice by complaining that the new executives of GE strongly favored a sales strategy based on the self-contained plant rather than on central stations. On the contrary, he reasoned, "it was in that branch of the Electric business that my education and experience under Mr.

Edison had been gained." Coming to Chicago in March 1892 to negotiate an employment contract, Insull brought to the job a comprehensive understanding of the industry second to none. In little over a decade he had learned every aspect of the business firsthand, from the art of invention, the science of engineering, and the techniques of manufacturing to the finances of Wall Street and the economics of utility operators. Moreover, Insull maintained contact with his homeland and kept abreast of the progress of electrical technology in Europe. Although he took a hefty pay cut in picking Chicago Edison over GE, he gained a significant stake in the future of the fledgling utility. Marshall Field, a corporate director, lent the new president $250,000 to purchase Chicago Edison stock.[19] On 1 July 1892 Insull assumed his place as chief executive officer, a position he would hold for the next forty years.

The state of the electrical industry in Chicago at the time of Insull's arrival can be described simply as one of chaotic growth. He would have to use all his considerable skills and talents to prove the superiority of the central station approach to meeting the city's electrical energy needs. A brief survey of the electric supply in the metropolitan area will illustrate the challenges facing the utility executive in his quest to beat the competition of both the self-contained system and alternative energy supplies such as gas and kerosene. In the years following the inauguration of central station service by the Edison company in 1887, the proliferation of franchise holders, shoestring utility companies, and "isolated" plant operators had continued without relief, at both the center and the periphery of the city. Within municipal borders alone, the city inspector's report of 1892 listed 18 central stations and 498 self-contained systems that were powering a total of 273,600 incandescent lights and 16,415 arc lamps.[20]

Direct competition in the central business district (CBD) from the Chicago Arc Light and Power Company (CALPC) posed the most immediate threat to the Chicago Edison Company. Since the great utility reorganization of 1887, both central station enterprises had experienced rapid growth and rising profits (see table 6). However, the companies had come into more and more direct competition with each other as their distribution grids overlapped and their businesses expanded to offer the same full array of incandescent light, arc lamp, and power services (see map 1). At the same time, Edison and other companies continued to sell self-contained plants. In the Loop, these systems were used both as individual units and as new central stations. Would-be utility op-

Table 6   Comparison of the Growth of the Chicago Edison Company (CEC) and
the Chicago Arc Light and Power Company (CALPC), 1889–92

Year and Company	Incandescent Lights	Arc Lights	Motors (Hp)	Net Profit
1889				
CEC	17,812	234		74,000
CALPC	337	1,559		92,000
1890				
CEC	29,139		925	116,000
CALPC	915	1,845		106,000
1891				
CEC	53,135	486	1,258	154,000
CALPC	1,519	2,284	224	131,000
1892				
CEC	88,259	3,417	1,826	—ᵃ
CALPC	2,112	2,191	639	149,000

Sources: "Net Increase of Contracted Business from 1889 to 1925 Inclusive," in CEC-
LF, F4-E5; Important Papers Files, in CEC-HA, boxes 1–2.
ᵃA fire in November 1891 resulted in a temporary reduction of business.

erators encountered little difficulty in obtaining franchises from
the city council. In 1892, for example, the promoter of the Northwestern Electric Light and Power Company boldly promised prospective investors that his partner "has assurances from prominent
members of the city council that he will be able to secure other
franchises that he may desire, so that easy access will be had to
the choicest portions of the business center."[21]

The challenge of the rapid proliferation of electric light services
in the middle-class residential sections of the city was less immediate but equally fatal to Insull's long-term plans. Within a few
years Chicago Edison could find itself confined to a relatively small
service territory, where city regulations and company franchises
required underground wiring.[22] The potential for rival operators
to surround and isolate the Loop with much cheaper overhead distribution grids was real enough. By the late 1880s, demand among
middle-class households for better lighting was virtually unlimited. Although at first real estate developers offered to build homes
with electric wiring as a special incentive, it quickly became a necessity.

Chicago's well-to-do quickly added electric lighting to the list of
essential improvements that no modern house could be without.
In 1881, for example, J. Lewis Cochran began building an upscale
subdivision called Edgewater on 380 acres of farmland about four

miles north of the settled area of the North Side. Far from any gas
lines, Cochran purchased an Edison self-contained system to furnish street and interior lighting, as well as providing other essential amenities such as sewer connections, sidewalks, curbing,
paved streets, and landscaping. Cochran also hired an architect,
Joseph L. Silsbee, to ensure that the houses in the project would
meet upper-class standards of exterior appearance and interior
convenience. To help complete the design work, Silsbee gave a
young man, Frank Lloyd Wright, his first job as an architectural
draftsman.[23]

Edgewater quickly became the "latest society 'fad.'" With ten
houses and two neighborhood stores completed by 1894, a newspaper critic agreed that it was "one of the most beautiful and exclusive suburbs of Chicago." A pioneer resident of Edgewater
leaves little doubt that electric lighting was one of the most important attractions of the subdivision. "Our lighting facilities," he recalled, "were the pride of the day then. It was pretty grand to have
electric light up here when Argyle [Street, to the south] was still
using kerosene lamps. . . . I used to have forty-one lights in my
house. The others had plenty of lights too, so that at night it was
a well lighted district around here."[24]

A similar demand for electric lighting greeted home builders
and utility operators in the more established, affluent communities. Hyde Park on the South Side represents a typical case of the
multiplying of electric companies and the rapid spread of services
into the home. As noted in the previous chapter, several speculators had obtained utility franchises in 1888–89 from the town's local government in anticipation of the area's annexation to Chicago.
The rush of franchise seekers, of course, had stemmed from an
appreciation of the large capital costs of burying distributor wires
as required within the city of Chicago. The key concession the
speculators inserted into the grants was a right to string wires
overhead. On 18 October 1890 one of the new enterprises, the
Hyde Park Thomson-Houston Light Company (HPTHLC), began
generating electricity and "hustling" residential service. The company had set aside $2,500 to bankroll a free wiring offer but soon
discovered it was unnecessary. Two home builders who agreed to
install wires and fixtures found buyers eager to purchase these
modern houses equipped with convenient electric services. Consumer demand was immediate and universal, a company officer
noted, so that "now every builder in our section is compelled to
wire houses in order to be on as good a footing as his competitor."

In less than a year, the utility company was connected to 196 new houses that contained about 7,500 incandescent lights.[25]

In the early 1890s a similar process of rapid growth from small beginnings was occurring throughout the metropolitan area. The sheer number of these shoestring operations reinforces the contention that there was a deep-seated popular fascination with high technology as a sign of social status and a symbol of progress. In the Hyde Park district alone there were at least five utility companies and perhaps as many as thirty small, self-contained "block" plants competing against the HPTHLC. A small-scale technology kept initial investment risks down to a reasonable level, while pent-up demand for better lighting promised a rich reward. Within the city's expanded borders, utility companies sprang up in several more middle-class neighborhoods, such as Englewood, Woodlawn, and Lake View. During this period, the suburbs of the well-to-do also began to recieve electric streetlights and residential service. The first overhead lines were usually strung along the Main Street business strip and the surrounding residential streets. In Oak Park on the West Side and Evanston and Highland Park on the North Shore, for example, electricity provided a highly visible community emblem of modernity and urban amenities. Farther out from the city center, the origins of electric utilities followed similar patterns in several of Chicago's satellite cities such as Joliet and Waukegan.[26]

The resulting crazy quilt of small distribution grids growing out from jerry-built central stations raised serious doubts about the ability of any single utility company to attain the economies of scale necessary to beat the competition of the self-contained system and cheaper sources of light and power. Perhaps the greatest shortcoming of the Edison system was its DC distributors. They used low voltage or "pressure," which made transmission at a distance extremely costly in terms of copper wiring and energy losses. The DC grid works much like a system of water mains. Larger and larger mains are needed to pump water to more distant points at a constant pressure. In a similar way, the diameter of the copper cables had to be enlarged in proportion to increases in both the size of the electrical load and the distance of transmission. For example, the Chicago Edison underground system was originally nineteen miles long and cost $197,100, or over $10,000 per mile. In comparison, the company spent $113,900 to build the Adams Street Station and only $49,500 for eight 1,200-light dynamos. In addition, the DC grid was inefficient, suffering an average energy loss of about 25 percent (in kwh) between the generator station and the customers' meters. These technological limits made it economically impractical to extend the DC system beyond a mile and a half radius from the central station.[27]

Before 1892, the constraints imposed by an underground system of DC distributors had already become painfully evident to Edison company managers. A steady stream of petitions requesting residential service in the Prairie Avenue district forced the company to confront the question of expanding services outside the CBD, but the company could not simply extend its lines from its first central station in the heart of the Loop at 138 (now 120) West Adams Street. The city's most prestigious residential street, Prairie Avenue, was just beyond the limits of the Edison station's distributor system. In 1891 the company instead had to spend a large amount of capital to build a second generating station about three miles south of the Loop on Wabash Avenue at Twenty-seventh Street. A similar powerhouse had to be erected at Clark and Oak streets to serve the well-to-do on the Near North Side.[28] This approach solved the immediate problem of responding to the demands of the city's business elite and their wives, but it offered no hope of cutting consumer rates to compete with gas in the long run.

In fact, the growth of central station services pointed in just the opposite direction of rising unit costs as generator plants and distribution grids were located in less and less lucrative sections of the city. The millionaires of Prairie Avenue agreed to help defer the heavy expense of burying the service lines in Edison tubes. But it was highly unlikely that any other neighborhood could afford to subsidize the underground conduits. In a similar way, the expansion of the company's business in the Loop suggested a future of diminishing profits. As Edison's distributor grid increasingly overlapped the lines of the CALPC, competition could be expected to cut into the dividends of both utilities. In effect, the technical limits of the Edison and other DC systems created an unusual case of diseconomies of scale. Although these supply-side constraints had appeared early in Chicago, the local managers of the utility company had not found a way to resolve the strange paradox of economic growth that promised to yield a shrinking rate of return.

Operators of Edison systems faced a challenge even more threatening to their futures after the appearance in 1886 of an entirely new electrical technology using alternating current (AC). As the name implies, a generator creates a flow of electricity that rapidly alternates back and forth in a circuit. In contrast, in a DC system electricity flows in a single direction around a circuit. The critical, practical difference between the two was the unequivocal superiority of AC for transmitting electricity at a distance. Alternating current could be sent efficiently at high voltage or "pressure," which meant that comparatively inexpensive thin copper cables could be used without suffering unacceptable energy losses. The transmission of energy became increasingly efficient as the voltage was raised. To achieve high-voltage transmission, a passive device called a transformer was used, first to boost the voltage at the generator station for long-distance transmission and then to reduce it for local distribution at the service area.[29]

Championed by George Westinghouse, the AC system threatened a fatal economic blow to the backers of Edison's DC technology. In the late 1880s, local companies using Westinghouse equipment began operating in middle-class communities such as Englewood on the South Side and Evanston on the northern border of the city.[30] The applications of the early AC systems were restricted to incandescent lights and did not include practical motors or streetlamps, but the obvious advantages of the AC distribution grids made the technology ideal for the metropoitan area's vast expanse of low-density residential housing. To be sure, Edison distributors could also be strung overhead, as was done in Edgewater. Yet the cost of the copper wires and the energy losses of these short-range grids was much higher than for their AC counterparts. The technological limits of the Edison central station seemed insurmountable in the face of the economic competition offered by cheaper sources of artificial illumination, self-contained plants, and alternative electric light systems.

Besides the appearance of AC technology, the energy revolution spawned other challenges to central station utilities like the Chicago Edison Company. Innovation in the electrical industry continued at a rapid pace, encompassing a widening range of urban pursuits and public services. For example, technological and manufacturing improvements in the DC motor were quickly translated into the production of a broad spectrum of appliances, from small personal fans to office building elevators and powerful railroad motors. In 1892 Chicago's streetcar companies finally began switching from horse and cable cars to electric trolleys. This belated conversion was restricted initially to outlying neighborhoods on the South Side, but the entry of electric traction into the Loop was inevitable within a short time. Concluding another long search for a better urban technology, the trolley car quickly became the largest consumer of electrical energy in the United States. Chicago's traction companies could soon pose yet another major source of competition against the central station type of utility.[31] The future direction of the Chicago Edison Company was indeed doubtful at the time Insull arrived in Chicago.

The Economics of "Natural" Monopoly

In hiring Samuel Insull, Chicago Edison retained a man who shared Edison's faith that the economic equation of the electrical supply business could be reversed in favor of large-scale central station service. Even before assuming his new position in July 1892, Insull had begun to implement a plan to monopolize the electric light and power business in the CBD. He had extracted two promises from the utility's corporate directors. First, they promised to finance the construction of a new central station on an unprecedented scale. Second, they agreed to buy out the CALPC to end the ruinous duplication of underground DC distribution lines in the CBD. Insull's strategy represented a practical application of a novel economic and constitutional theory of "natural" monopoly. This fashionable notion proposed that market mechanisms did not work in public utility enterprises with heavy capital investments, including urban services, railroads, and communications. Instead, such services could be provided efficiently only by a single supplier, because they were "natural" monopolies.[32]

Although Insull's achievement of a monopoly in Chicago did not significantly reduce consumer rates, it did provide him with the insight to gain this result. The new president of the Chicago Edison Company soon learned that the technology of production was inextricably tied to the economics of consumption. A narrow focus on removing the technical limits of small-scale generators and distribution areas did little to stop the proliferation of either self-contained systems or utility companies. Over the next six years, he worked out new equations of utility company costs and customer rates that finally overcame the economic constraints of central station service. In 1898 Insull inaugurated a novel rate structure that promised to cut electric bills down to an attractive level for both the small and the large consumer. He also propounded a radical gospel of consumption to the NELA, which predicted a coming age of intensive energy use.

Of course the local officers of the Edison enterprise had already learned some important lessons about the peculiar technological constraints of electricity. During the five years of commerical operations before Insull's arrival, a first generation of managers emerged to help the new chief executive officer meet the demand for electric light and power at a competitive price. The limits of DC distribution grids and small generators began to retard economic growth as soon as the company opened its first generating station on Adams Street. In 1887 the choice of where to locate it was obvious: in the very center of the city's most lucrative market, the CBD. But operating a coal-burning steam plant in the heart of the financial district immediately proved a mistake. Despite the efforts of the chief engineer, Frederick Sargent, to design an especially strong foundation, the vibrations produced by the pounding 200-HP steam engines soon brought a chorus of complaints from the tenants of neighboring office buildings.

Moreover, the Adams Street station was overloaded from the start, increasing from a 10,000-light capacity to a limit of 50,000 lights by the time Insull arrived. With engines crammed in the basement, belts running through the ceiling to the generators on the floor above, and coal bins and offices on the top level, there was little room to expand. A separate steam engine was required to run two of the largest available Edison dynamos, each with a capacity of 1,200 lights, producing excessive heat that created a constant risk of disaster for workers and equipment. As one company employee recalled it,

> The appearance of the Adams Street Station suggested a glimpse of Dante's Inferno. The engines were constantly pushed to their utmost capacity under the heavy overload. In the roaring dynamo room, the smell of shellac and varnish from the hot armatures told the same story. Switches and bus-bars were frequently too hot for the bare hand, while in the boiler room the half-naked firemen, shoveling coal with feverish energy, made one feel as if an explosion might furnish a climax at any moment.[33]

As early as December 1890, the secretary-treasurer of the Chicago company, Frank Gorton, began discussing with Insull—then in New York—the need to move the central station to a different location. Gorton, son-in-law of General Anson Stager, had made an easy transition from the telephone to the electric business, where he acted as chief executive officer. Insull was asked for help in persuading his close friend, financier Henry Villard, to sell a piece of unused railroad land on the west bank of the Chicago River at Harrison Street. Although the extra copper transmission lines to connect the new station to the distribution grid might cost $500,000, the Harrison Street land offered three promising advantages. First, the industrial site was large enough for expansion. "The history of all electrical enterprises that serve the public is this," Gorton lectured Insull, "that they must be increased all the time, and new construction constantly going on." Second, a riverfront location meant the company could take advantage of the fuel-efficient condenser type of steam engine. These prime movers had originally been developed for use by steamships, which had limited storage space for coal and unlimited access to cold water. The condenser worked like a high-pressure boiler in reverse. By rapidly cooling the steam after it had been used in the engine, the device created a vacuum that improved the efficiency of the engine and used less fuel. As a third advantage, the site promised major savings on fuel costs. Since it was between the rail yards and river docks, hauling twenty to fifty tons a day through the crowded streets of the CBD would be eliminated.[34]

In July 1892, Insull's first priority was to complete the construction of the Harrison Street Station as fast as possible. The new chief executive entrusted the job to his fellow Englishman and former New York colleague, Fred Sargent. Two years earlier, Sargent had resigned from the General Electric Company to set up his own electrical engineering firm in Chicago with Ayres D. Lundy. Unlike Sargent, Lundy was college educated, and his early training had focused on the electrification of the street railway. Between September 1892 and August 1894, the two men designed and constructed the most modern powerhouse of its time. Abandoning the effort to disguise a generator station as an office building, the engineers adopted architect Louis Sullivan's dictum to let "form follow function."

Inside the spacious plant, Insull installed the largest available steam engines and dynamos to give a maximum output of 2,400 kilowatts (KW). The equipment was about three times as large as the machinery at Adams Street, which was permanently shut down. The new station included 1,250-HP engines and 400-KW generators. In comparison, the Adams Street Station had 450-HP engines and 160-KW dynamos. The efficiency of the high-pressure condenser engines at the Harrison Street Station was impressive, dramatically cutting the amount of fuel needed to generate a kilowatt-hour from about 10–14 to 3–5 pounds of coal. Massive copper cables with a combined diameter of thirty-six inches linked the two stations.[35]

The need for the new station stemmed from Insull's plan to replace not only the overloaded Adams Street plant but the generators of the rival CALPC as well. Even before leaving Schenectady, New York, Insull had begun negotiating with the new president of GE, Charles A. Coffin, to secure a monopoly of the electric business in the CBD. A month after taking over the Chicago company, Insull outlined a shrewd scheme to gain control of all the major patent licenses for electrical equipment. In a letter to Coffin, he asked GE to agree not to sell any machinery, especially self-contained "isolated" plants, to anyone else within Cook County. The purpose of this plan, he revealed in a separate memo, was to gain "exclusive Central Station rights . . . [and] if possible, exclusive rights to all isolated [plant] business in Cook County." Since the manufacturing giant now held almost all the important electrical patents except those owned by Westinghouse such an agreement would effectively preclude any additional local competition. New utility franchises from the Chicago city council would be worthless without access to generating equipment and light bulbs.[36]

To implement the plan, Insull needed to win Coffin's consent and to purchase several key local companies that already held patent licenses. Although obtaining Coffin's agreement was easy enough, acquiring a local monopoly involved more intricate political and financial maneuvers. The most crucial step was the purchase of the CALPC, because of its competitive threat and its rights to sell Thomson-Houston equipment. Writing directly to the manufacturing firm in early August, Insull asked for exclusive rights and received a reply promising "liberal" terms in his favor. He then approached the rival utility with the help of the company's sharp young legal adviser, William K. Beale of Isham, Lincoln, and Beale. Two months later rumors of a merger began surfacing, but they were initially dismissed as "improbable."[37] In January 1893, however, the stories were confirmed as the securities of both companies began to rise wildly on the Chicago Stock Exchange. Eventually Chicago Edison paid $2.2 million to buy out the CALPC, though only $1.2 million of the purchase price represented tangible assets. The deal was a bargain nonetheless, Beale declared, since Chicago Edison would now have a "practical monopoly" of electrical services in Chicago.[38]

Table 7   Companies Purchased by the Chicago Edison Company, 1892–98

Year and Company	Purchase Price ($)	Tangible Property ($)	Goodwill ($)
1892			
Chicago Arc Light and Power Co.	2,195,000	1,195,000	1,000,000
1894			
National Electric Construction Co.	96,000	10,000	86,000
Northwestern Light and Power Co.	76,000	14,000	62,000
Co-operative Light and Power Co.	131,000	63,000	68,000
City Illuminating Co.	6,000	2,000	4,000
Madison Light and Power Co.	2,500	500	2,000
1895			
W. J. Fagan plant	2,000	1,000	1,000
Dearborn Light Co.	5,500	1,000	4,500
1896			
American Gas Engine Co.	2,000	2,000	—
1897			
Chicago Illuminating Co.	96,000	44,000	52,000
Allen plant	12,000	2,000	10,000
1898			
Merchants Arc Light and Power Co.	25,000	6,000	19,000
Archer Avenue plant	10,000	5,000	5,000
Bachelle Power Co.	17,000	10,000	7,000
Miller plant	30,000	12,000	18,000
Total	2,706,000	1,367,500	1,338,500

Sources: H. A. Seymour, "History of Commonwealth Edison Company," manuscript, 1934, 270, CEC-HA, box 9001; "Chicago Central Station Development Chart," in CEC-LF, F6C-E4.

Over the next five years of depression and crisis between 1893 and 1898, Insull completed his plan to acquire all the central stations and their franchises in the Loop. Hard times undoubtedly helped persuade the small companies to sell out, expecially since Insull followed a policy of offering generous terms (see table 7). In most cases the equipment of these shoestring operators was simply junked, and the customers were connected to the Edison company lines. Added together with new light and power business in the Loop, the growth of the utility was extremely impressive during a period of general economic distress (see table 8). For example, the Harrison Street Station had to be expanded by 167 percent from a maximum output of 2,400 to 6,400 KW to keep up with the peak demand. The use of electricity rose from 6.6 million kwh in 1893 to almost 11 million kwh in 1897, an average annual increase in consumption of about 18 percent.[39]

Table 8   Growth of Electrical Output of the Chicago Edison Company, 1889–98
(All Electrical Figures in Thousands)

Year	Customers	Connected Load (kw)	Incandescent Lights (kwh)	Arc Lights (kwh)	Power (kwh)	Total (kwh)	Lost[a] (%)
1889			615.1		150.4	765.6	20
1890			1,096.2		638.0	1,734.2	21
1891			1,702.7		1,102.8	2,805.5	25
1892			2,856.8	241.7	1,481.5	4,580.0	21
1893	4,452	12.1	3,851.7	838.9	1,861.1	6,551.8	16
1894	4,645	13.4	3,857.6	929.2	1,866.0	6,652.7	15
1895	5,289	15.8	4,238.7	1,050.9	2,059.7	7,349.2	18
1896	6,173	17.4	5,244.1	1,670.4	2,105.0	9,019.5	19
			(171.2)[b]	(18.8)	(559.5)	(737.4)	
			5,415.3	1,689.2	2,664.5	9,756.9	
1897	7,333	19.9	5,857.8	1,948.6	2,467.2	10,273.5	19
			(167.8)	(18.8)	(522.2)	(708.8)	
			6,025.6	1,967.4	2,989.4	10,982.3	
1898	10,535	30.0	7,934.6	2,264.8	3,193.4	13,392.6	17
			(215.3)	(40.5)	(623.7)	(879.5)	
			8,149.9	2,305.3	3,817.1	14,272.1	

Sources: H. A. Seymour, "History of Commonwealth Edison," manuscript, 1934,
287, 321, in CEC-HA, box 9001; "Kilowatt Hour Output Data, 1888–1902," in
CEC-HA, box 205.
[a]Percentage of electrical power losses in amps. The power losses are not included
in the figures, which record actual consumption.
[b]AC circuits in the overhead distribution districts.

But to Insull's dismay, generating all the CBD's electric business
by a modern central station did not save enough to undercut the
competion of self-contained systems, let alone gas lighting in the
home. To be sure, the Harrison Street Station, with its more economical engines, generators, and fuel-handling equipment, could
be expected to reduce the company's operating costs. And the
growth of demand in the Loop would probably repay the investment capital sunk in the copper transmission line to link the riverfront plant to the Adams Street building. Yet, as Insull admitted,
"no one in the central station business at that time really understood its fundamental economics."[40]

During the depression, rates remained at luxury levels. Complaints from businessmen about high prices replaced the initial
enthusiasm for the novel technology. Even in the homes of the
affluent middle classes, the common practice was to use electric
illumination only in the parlor when guests were present. After
they left the light bulbs would be turned off, leaving the gas jets lit
in the living quarters. The popularity of dual gas/electric chande-

liers was a testament to the high cost of better lighting in the home.[41]

Since the inauguration of electric light services in Chicago, utility operators had imitated the practice of the gas companies and charged a flat rate for each unit of energy (a thousand cubic feet of gas or a kilowatt-hour of electricity) as measured by a meter. Beginning in 1887, for example, Chicago Edison charged 1¢ per hour for each sixteen-candlepower light, or an equivalent of 20¢ per kwh. Charges for power were also at a flat rate, with three separate categories for continuous, intermittent, and elevator use. For both light and power, the company offered a schedule of discounts, based on the amount of the customer's bill. And in some cases salesmen had to give secret low-rate contracts to lure potential customers away from purchasing a self-contained system. In 1891, for instance, a young MIT-trained electrical engineer named Louis Ferguson secured a contract with the Great Northern Hotel for 4,100 lights and a 7.5-HP motor at rates that "seemed ridiculously low at that time," according to Insull.[42] Although this type of expedient might prove effective in a few cases, it could not be adopted universally without turning company profits into losses. In 1892, central station service in Chicago remained hamstrung by the same two constraints of small-scale generating equipment and limited distribution grids.

The World's Fair of 1893, however, provided important clues on how to remove these supply-side constraints. For the electricians and the scientists, Chicago's exposition had technical as well as cultural lessons. Here was a unique opportunity to build a model of an energy-intensive city—a self-contained city within a city. A separate Electricity Building housed the industry's commercial products and special educational exhibits. Moreover, almost all the equipment used to generate the White City's electrical energy was proudly displayed over a seven-acre area in Machinery Hall. At a time when Americans believed that bigger was better, the electric plant featured a gigantic 2,000-HP steam engine that was belted to "two of the mammoth new type Westinghouse alternating incandescent dynamos of 10,000 lights capacity each."[43]

But what impressed the experts most was how all this hardware was put together to supply such a widely diverse array of energy needs and appliances. The White City demonstrated the principles of a universal system of distribution. Westinghouse employed a recent invention, the rotary convertor, to convert AC to DC for use by the intramural elevated railway. When coupled with the transformer, the rotary convertor provided a technological means of re-moving the constraints of restricted distribution areas. Electricity generated at a central station could now be boosted to high voltage by a transformer for long-distance transmission and then changed at local substations by transformers and convertors to meet the particular needs of the distribution area. The model city of 1893 taught Insull and other electrical men in Chicago that there was no real "battle of the systems" between AC and DC. The most impor-tant lesson of the World's Fair was that the two systems could be harnessed together to deliver a complete range of electrical ser-vices throughout the city.[44]

Insull soon applied the lessons of the model city to the real one, making Chicago Edison first or second to put the rotary convertor into commercial use. In August 1897 electricity from the Harrison Street Station was transmitted at 2,300 volts to the Twenty-seventh Street Station, converted back to DC, and fed to the homes and businesses of the Near South Side. Two years later the first true substation, containing no generating equipment, was jerry-built on the North Side to help handle periods of peak demand from 4:00 P.M. to midnight. Within a few months, the cost efficiencies convinced Insull to shut down the nearby Clark Steet Station and convert it into a full-time substation.[45]

More important, these experiments proved the feasibility of a metropolitan power network. Insull was quick to appreciate the economic implications of the new technology for supplying elec-trical energy at a price lower than previously possible. The com-bined AC-DC system suggested a hierarchy composed of a few efficient power plants and a citywide network of substations where electricity would be transformed to meet the needs of each local district. This novel concept of central station service led Insull almost immediately to expand his monopoly plans from the CBD to encompass the entire city. In just two years between 1897 and 1898, he was able to achieve this major goal owing to his control of key patent licenses and a measure of incredible good luck.

By the mid-nineties, reform agitation in Chicago was moving rapidly toward a climax over utility franchises. The organized forces of reform had shifted popular attention from the Gas Trust to the notorious owner of the street railways, Charles Yerkes. The story of the thwarted bid by the "robber baron" for a fifty-year franchise in 1897–98 is too well known to repeat here,[46] but it is important to place Insull's back-room political maneuvering in the context of the uproar over Yerkes. While masses of people took to the streets and newspaper headlines blasted the arrogance of the traction boss, the city council quietly approved a fifty-year franchise for one of its paper creations, the Commonwealth Electric Company. The politicians hoped to blackmail Insull's Chicago Edison Company into paying them a high price for the company's franchise. Repeating the "Ogden gas scheme," the "gray wolves" of the council threatened to sponsor a new company, which would have the advantage of being able to use overhead distribution wires in all sections of the city except the CBD. However, Insull's patent licenses put him one step ahead of the politicians, and they sold the company and its valuable franchise to him for a nominal amount.[47] In the midst of massive protests against Yerkes, the Commonwealth Electric grant and its transfer to Chicago Edison went virtually unnoticed. Armed with this franchise, Insull proceeded to buy out all the utility companies that were operating overhead systems in the residential sections of the city (see table 9).

By the end of 1898, the utility manager could finally prove that central station service was the most economical method of supplying energy to the city. Inventions like the rotary convertor had eliminated the technological limits of restricted distribution areas, and similar innovations were making rapid progress in removing the constraints of small-scale generating equipment. The resulting concept of a central station hierarchy, together with the monopoly he had obtained, put Insull on the threshold of creating a metropolitan network of power. Yet all this technical and business success would mean little unless Insull also found an answer to the riddle of rates for the large consumer as well as the small residential customer. Electric rates had to be cut sufficiently to compete with the self-contained system and cheaper sources of artificial illumination. Only then would the demand for electrical energy reach a point where true economies of scale could be achieved.

The Gospel of Consumption

During the pivotal period 1893–98, Insull grew to appreciate that the economics of consumption were more important than the technology of production. A single-minded determination to prove the superiority of central station service drove him to shift attention from the problems of the supply side to the equally vexing questions of customer demand and rate making. What he learned ws that supply and demand were inseparably linked in electrical sys-

Table 9 Companies Purchased by the Commonwealth Electric Company, 1897

Company	Purchase Price ($)	Tangible Property ($)	Goodwill ($)
Hyde Park Thomson-Houston Co.			
Hyde Park Electric Light Co.			
Englewood Electric Light Co.	—	279,000	—
Mutual Electric Light Co.			
Peoples Light and Power Co.	—	627,000	—
Total	2,125,000	906,000	1,219,000
Western Light and Power Co.	250,000	194,000	56,000
West Chicago Light and Power Co.	64,000	41,000	23,000
Edgewater Light Co.	15,000	9,000	6,000
Total	329,000	244,000	85,000
Commonwealth Electric Co.	170,000	—	170,000
Grand Total	2,624,000	1,150,000	1,474,000

Source: "Summary of Original and Purchased Companies of Commonwealth Electric Company," in CEC-LF, F6C-E4.

tems. Breaking with the teachings of his "old chief" Edison, the former secretary discarded the model of the gas business and forged new principles of ratemaking that coresponded more closely to the unique characteristics of electrical utilities. During two highly charged meetings of the NELA in 1897 and 1898, the Chicago executive and his lieutenants announced the far-reaching findings of their search for a way to deliver energy to every type of consumer at a reasonable price. Marking a climax of the energy revolution, they proclaimed a new era of intensive use of electricity in everyday life.

As with the rotary convertor, an invention helped Insull solve the equation of utility costs and consumer rates. But in contrast to the hardware displayed at the World's Fair, the Wright demand meter did not directly answer the question of how to undersell both the self-contained plant and the alternative sources of artificial illumination. Instead, this ingenious device provided Insull with an "aha" experience that suddenly made the pieces of the puzzle fit together. All the economic requirements to beat the competition fell into place once he grasped the peculiar, instantaneous relationship between supply and demand that is inherent in electrical systems. As Insull testified, the demand meter "first taught us how to sell electricity."[48]

Insull first heard about the innovative measuring instrument in 1894 while visiting his homeland. Recently invented by Arthur Wright of Brighton, England, the meter recorded not only a customer's amount of consumption but also the timing and the maximum level of his demand. Intrigued by the device, Insull returned to Chicago but sent his chief electrician, Louis Ferguson, to make a thorough study of its use in Brighton. The engineer returned with an enthusiasm for the metering system that soon infected Insull. In September 1897 the Chicago Edison Company started making a practical test of the measuring device. Within a few years, Wright's invention replaced most of the utility's other meters.[49]

The demand meter showed Insull that electric companies did not work "just as the gas companies do." On the contrary, the two were fundamentally different. The storage tanks of the gas companies allowed them to even out their production schedules and make maximum use of the equipment on a twenty-four hour cycle. In this way gas companies could keep their capital investments in central station machinery to a minimum, because there was no need for expensive but little-used equipment to meet brief periods of peak demand on the system. In contrast, electric companies had to keep an instant balance between demand and supply or suffer service blackouts and equipment damage. To be sure, storage batteries could help meet periods of peak demand, but they had serious drawbacks of their own: they were expense, cumbersome, dangerous, and very inefficient. Whether to purchase batteries or extra generating equipment to meet periods of maximum demand remained a problem.[50]

Helping Insull break free of his mentor's teachings, the demand meter gave him the insight to recalculate the equation between the electric utility's costs and the customers' bills. Wright's invention suggested a new method of ratemaking with a two-part bill to replace the traditional flat charge. A customer's energy consumption (the number of kilowatt-hours of electricity recorded by the meter) corresponded to the company's operating expenses. The measure of peak demand on the meter represented the customer's share of the capital invested in generating equipment that had to stand ready to serve him. The utility would use each customer's maximum demand to apportion equally the cost of financing the utility's plants and equipment. This primary charge serviced the interest payments on the company's bonded indebtedness, which constituted about 70 percent of its total expenses. The demand meter would determine the number of kilowatt-hours for which to charge each customer at the higher primary rate. A much lower secondary charge would be levied for any consumption beyond that monthly minimum number of units of electricity. The primary rates were fixed for all customers, but the secondary rates were discounted on a sliding scale to encourage greater consumption.

In this way a two-tier rate structure promised to cut the bills of both the small household and the large commercial enterprise. In effect, residential customers would pay a smaller proportion of the utility's financial costs because they had relatively small peak demands. The net result of the two-tier system would be a lower effective rate for each kilowatt-hour of energy. For the most part, the heavy burden of the utility's interest payments would be absorbed by the large consumers of light and power. At the same time, they would benefit from the discounts for heavy use of the equipment, which would progressively reduce their net rate of charge for each kilowatt-hour of electricity.[51] In this way a two-tier system of ratemaking was structured to encourage every type of consumer to use more energy.

At the NELA meeting in June 1897, Wright first outlined the far-reaching implications of a two-tier rate structure for the creation of a mass market for electrical energy. Invited by Insull, the inventor and manager of Brighton's municipal facilities shocked the American utility operators by claiming that ratemaking, not technical progress, was the key to unleashing a "practically boundless" expansion in the use of energy in the city. "It is possible," he continued, "that more profit may be derived from the supply of electricity to small, long hour consumers at low rates than from the supply of large consumers, such as extensive stores, at a much higher rate." Wright had confidence in his radical proposals because the demand meter had shown him that the best way to reduce the unit cost of electricity was to spread out the central station's load over the entire day. "The consumers have a community of interest," he explained, "in flattening out the station load curve, and they are bound to do it on a tariff [rate schedule] that encourages them to wire their premises throughout." This meant that station operators needed to restructure their rates to encourage diverse groups of consumers to use energy at times other than the usual peak demand for light between 4:00 and 8:00 P.M. Rate schedules that improved the diversity and load factors, or average use of the generating equipment, would result in a steady decline in unit costs and a commensurate rise in profits. Passing some of this surplus back to consumers in the form of rate cuts would encourage them to use even more energy. Wright concluded by calling for rates low enough to undersell gas not only in the homes of the middle classes but in the dwellings of "the masses" as well.[52]

The discussion that followed was revealing in a number of ways. On the one hand, the objections of several members of the audience demonstrated just how radical Wright's position seemed to the utility fraternity. Moreover, there was a general sense that Wright's paper was a theoretical proposal, not a workable solution to their problems. Wright had to spend considerable time explaining the practical application of the two-tier rate structure in Brighton and other English cities.

On the other hand, the defense of Wright's arguments by Insull and Ferguson displayed a sophisticated understanding of the primary importance of building the size and diversity of demand on the generating equipment. Only then could the central station achieve true economies of scale and undersell the competition of gaslight and self-contained systems. For instance, Ferguson suggested that "Mr. Wright could go a step farther in his work, and make a difference [in rates] depending on the time of day at which the maximum load comes." Utility companies, he proposed, could offer special low rates to late-night customers, because they would be using generating equipment that would otherwise stand idle. Wright himself did not immediately follow the logic of the engineer's arguments.[53] Certainly many members of the NELA were confused by the discussion of ratemaking, and it carried over to become the all-absorbing topic of debate at the next annual meeting.

The 1898 convention of the NELA marked a major crossroads in the energy revolution. The members elected Insull president, and he led them a long way toward accepting the new economics of central station service. Most important, Insull and his close associates turned the very terms of the debate completely around from the supply side to the demand side of the equation. As Insull proclaimed, the subject of ratemaking had become "the most important one that we can consider; I devote more of my time to it than to any other one subject connected with our business. The way you sell the current has more bearing on . . . cost and profit than whether you have the alternating or direct-current system, or a more economical or less economical steam plant." Ferguson, the engineer, concurred with the business executive's new priorities, declaring that "there is more money to be made in the intelligent selling of your product than in attempting to introduce further economies in operation."[54] After 1898, selling energy became the main business of urban utilities. Electricity was no longer a mysterious technology but a mere commodity—energy—to be sold like other products in an emerging urban culture of middle-class affluence and leisure.

The new economics of the mass consumption of energy also had profound political repercussions. In his presidential address, Insull made a bold statement in favor of the monopoly of urban utility services under public regulation. Representing a logical corollary of "natural" monopoly theories, the call by a business executive for government controls reflected a keen appreciation of the novel accounting methods of calculating utility company costs and consumer rates. "In order to protect the public," Insull argued, "exclusive franchises should be coupled with the conditions of public control, requiring all charges for services fixed by public bodies to be based on cost plus a reasonable profit." The association leader explained, for example, that demands by city councils for special taxes on franchised companies could now be easily accommodated without the usual political friction. Municipal taxes simply would be added to the utilities' expense ledgers and passed on to consumers.[55]

Insull predicted that his economic and political approaches would result in a self-perpetuating cycle of rising consumption and falling rates. Charts tracing the experience of English cities were shown at the meeting to demonstrate this very effect (see chart 1). The charts showed a direct correlation between cutting rates and increasing output. To Insull this equation meant that every consumer of energy, large and small, was a desirable customer of the utility company. When one of the NELA members asked him how small a consumer he could afford to serve, Insull replied that he would sign up a customer who used a single, twenty-five-watt light bulb if they made one that small. Ferguson agreed that "we take any and every customer on the Wright demand meter basis." The company could not get the meters fast enough to keep up with homeowners' requests for new service connections. Furthermore, the engineer claimed that the demand meter pointed out the way to attract large power consumers and to put the self-contained system on the technological scrap heap. Even a 50 percent discount of 10¢ per kwh would not be enough. A "wholesale discount" beyond that, equivalent to a net rate of 4–5¢ per kwh, would be necessary.[56]

By the middle of 1898, Insull's faith in central station service had

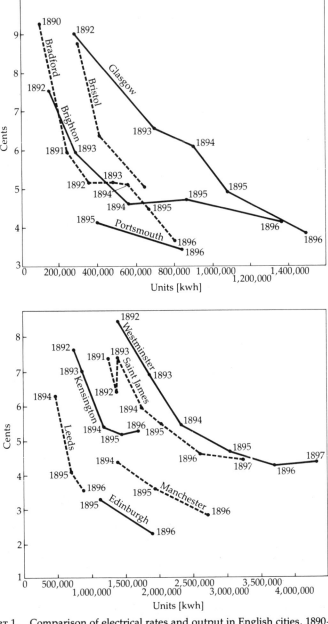

CHART 1. Comparison of electrical rates and output in English cities, 1890–96. *Source:* National Electric Light Association, *Proceedings of the Twenty-first Annual Meeting,* Chicago, 1898, 89–90.

evolved into a full-blown gospel of consumption. He used a lecture at Purdue University in May to present for the first time a detailed vision of the massive use of energy in everyday life. Perhaps he felt more comfortable making predictions about the future to engineering students than to his fellow utility executives. In any case, Insull drew upon the tremendous progress of the recent past to foster belief in a coming age when change would be even faster and more fundamental. "Is it too much to predict," he asked, "that in a far less time than the succeeding twenty years electricity for all purposes will be within the reach of the smallest householder and the poorest citizen?" He answered that electric lighting at low rates would soon capture that 80 percent of the urban market for artificial illumination that was then supplied by gas. In addition, he pointed out, "if you bear in mind the many other purposes to which electricity can be adapted throughout a city and supplied to customers in small quantities, you may get some faint conception of the possible consumption of electrical energy in the not-far-distant future."[57]

During the second half of 1898, Insull not only talked about the future, he began taking practical steps to get there as soon as possible. Since the impact of these new directions will form the subject of the following chapters, here it is enough to list the most important first steps. The Chicago utility announced a two-tier system of rates for light and power. For the average residential customer, the new method of billing translated into an immediate 32 percent savings, or a net reduction from 19.5¢ per kwh to 13.33¢. At the same time, the utility extended a special incentive to new customers by offering to install six lighting outlets free of charge. In addition, the company underwent a significant reorganization that relected Insull's shift in focus toward marketing and consumption. For example, he promoted a station records keeper, E. J. Fowler, to head a statistical department. Using data from the demand meters, Fowler soon began supplying extremely valuable reports to company managers on the energy-use patterns of the city's different consumer groups.

Insull's sense that the energy business stood on "the threshold of a great development" was shared by the leaders of Chicago's gas companies.[58] The year 1898 also marked a major revision in the self-image of these utilities, from manufactures of coal gas to suppliers of energy fuel. Since the formation of the Gas Trust in 1887, a series of technological, economic, and political events had caused revolutionary changes equivalent to those in the electrical industry. The introduction of an incandescent gas mantle had helped slow the retreat of the industry from the field of artificial lighting. More important, however, the incandescent mantle had triggered a fundamental transformation in the desirable qualities of the fuels used to make illuminating gas. With the traditional gas jet, the light-giving value of the coal determined the brightness of the illuminate. In contrast, the mantle worked best at high temperature, making the fuel's traditional role as an illuminate compatible with its new uses for heating and cooking.[59] In the face of competition against a superior lighting technology, the gas utilities were forced to redefine themselves as suppliers of energy fuel for a variety of power, heating, and cooking applications. From its inception, for example, the Gas Trust had fostered the use of cookstoves in the home, with sales accumulating to over twenty thousands ovens and burners in the first eleven years of the promotional campaign. "By 1898," a gas company historian concluded, "gas stoves had become so salable and the effect of their use upon company income had become so obvious that pushing the sale of them soon took rank as a major gas company activity. That was a turning point in the gas business in Chicago."[60]

In that same year, the reform battle against the gas monopoly also reached a climax. As its first president, Norman Fay, had foretold in his address to the Citizens Association in 1889, the reformers could destroy the Gas Trust but not the monopoly. For almost a decade, cracks in the federal system had allowed clever lawyers to escape the effects of a steady stream of state court decrees, antitrust laws, and executive vetoes of obnoxious bills sponsored by the monopoly. During the uproar over Yerkes and the streetcar franchise in 1897, the Gas Trust lobbied yet another tailor-made bill through the legislature to allow gas companies in Illinois to consolidate. But this time the bill went for final approval to a sympathetic governor, John R. Tanner, instead of his antimonopolist predecessor John P. Altgeld. Two months after Tanner signed the legislation in June 1897, the monopoly underwent a final rebirth as the Peoples Gas Light and Coke Company.[61]

In 1898, then, both the gas and electric monopolies in Chicago were well prepared to meet expectations for more intensive energy use in everyday life. In the electrical industry, innovative technology and economic concepts had largely removed the constraints on the supply of light and power at a reasonable price. Since arriving in Chicago six years earlier, Insull had helped perfect larger-scale generating equipment, a universal system of distribution, and novel principles of ratemaking. More than any other individual, he had advanced the industry toward meeting the city's insatiable demand for better lighting.

In the gas industry, too, profound changes in the processes of generating and marketing the fuel had triggered a revolution in urban energy supplies. The importance of the incandescent mantle in transforming the energy fuel from light giving to heat producing cannot be overemphasized. With the coming of the Spindletop, Texas, gushers in 1901, a virtual glut of cheap natural gas and oil would begin flowing from the Southwest into the nation's urban centers. Rich in heat value, natural gas would gradually replace coal gas, coal, and wood for heating the city's homes, shops, and factories. Although the decline in the importance of gas for lighting was irreversible, the gas business made a successful transition to become a partner of the electrical utilities in meeting the city's growing demand for energy.

At the turn of the century, perceptive observers in Chicago had little trouble finding signs of where urban Americans were heading. They were trying to create an affluent society of consumption and leisure. In the Loop, for example, the bright lights shining from the palaces of consumption along State Street would have been difficult to miss. "The department stores," one reporter noted in 1900, "[are] quick to seize upon any new scheme for advertising. . . . [They] have developed the electrically lighted shop window almost to a state of perfection, and [they] vie with each other in lavish displays of electric illumination." Reflecting a moment, the journalist realized he had witnessed a profound change in the city's lighting, a revolution "little short of marvelous. The field where but yesterday the flickering gas flame held full sway now blazes nightly in the glow of myriads of electric lamps, aggregating in intensity the illuminating power of 15,000,000 of candles."[62]

For many Chicagoans, the bright lights of the city had come to symbolize modern times and the steady march of progress into the new century. Starting with Edison and the invention of the light bulb, the mysterious force of electricity held a special place in the American imagination. During the crisis of the nineties, rapid improvements in central station service offered comforting reassurance to a nation torn with doubt about the power of technology to promote social justice and democracy. The costly nature of the modern utility services added an important element of snob appeal that helped make electricity an emblem not only of high technology but of conspicuous consumption. As city dwellers entered the twentieth century, historian John R. Stilgoe contends, the Harrison Street Station and others like it were themselves becoming important cultural icons. "In the great industrial zones ringing the nation's cities," he concludes, "no structure loomed more majestically or thrust its smokestacks higher. As a symbol of efficient power, the station knew no equal. . . . As the very generator of the nation's emerging electric vision, the power station attracted the scrutiny of technical and popular writers, artists, and the general public it ceaselessly and mysteriously served. In the power station, Americans glimpsed the electric future."[63]

Part II

Metropolitan Webs of Power, 1898–1914

5

The "Massing of Production," 1898–1908

In 1898 Samuel Insull began offering Chicagoans attractive incentives to install electric service in their homes. Insull's utility companies would wire six lighting outlets free of charge in new or older houses. Together with the announcement of substantial rate cuts, the free wiring offer stimulated strong psychological impulses for consumers to fulfill their desire for the technologically superior and socially prestigious source of lighting. For many well-to-do homeowners, Insull's enticements were sufficient to overcome a natural reluctance to pay the initial cost of installation and fixtures. Over the next decade, about 5,000 homeowners took advantage of this marketing ploy, other giveaways, and further rate reductions of nearly 25 percent. They joined an urban elite that had already turned objects of high technology into cultural icons, fashionable and conspicuous symbols of success.[1]

By 1909 these homeowners constituted part of an expanding community of 50,000 residential and 34,000 commercial and industrial energy consumers who had increased the use of electricity in Chicago by over 2,000 percent in a mere decade.[2] At base, this phenomenal growth stemmed from accumulated demand for better lighting and more efficient applications of power in transportation and industry. At the same time, inventors and manufacturers continued to produce more efficient generating equipment, improved lighting devices, and novel appliances at a revolutionary pace. Utility operators like Samuel Insull also helped solve the problem of supplying energy at a reasonable price and stimulating a psychological desire for things electrical.

After 1898, Insull was impelled to prove his theory of electrical utility economics. Based on the technological imperative of an instantaneous balance between supply and demand in electrical systems, Insull's economics provided a clear policy direction for central station operators. His gospel of consumption placed first priority on selling energy to every type of consumer in order to build up the utility's load and spread it out over the entire day. Creating a widely diverse community of consumers was the key to achieving economies of scale in the generation of electricity. Yet he did not ignore the engineer's quest for a truly efficient generator. In the early 1900s, he would take a leading part in eliminating this last technological constraint on supply by sponsoring the construction of the first large-scale steam turbine. Taken together, the savings resulting from the "massing of production"—as he put it in 1910—permitted rate reductions that in turn encouraged additional use. In this way, Insull predicted, utility operators could set in motion a self-perpetuating cycle of rising consumption and falling rates, while enhancing corporate profits at the same time.

It took Insull nearly ten years to prove the validity of his economic theory. Reflecting on the historic debates at the 1898 meeting of the National Electric Light Association (NELA), the Chicago executive admitted that "we used to talk very glibly at that time about the diversity factor, but we, of course, knew nothing about it, as we had no experience in selling large volumes of energy to a few customers."[3] During the intervening decade, however, he used the demand meter to conduct an elaborate study of the energy consumption patterns of the city's different groups of householders, shopkeepers, and manufacturers. The statistics quickly revealed that his marketing campaigns aimed at the residential and commercial sectors were highly successful in encouraging Chicagoans to become customers of central station service. But in the process of raising consumption levels, he was also exacerbating the utility's costly problems of meeting the peak demand for energy during the early evening. In 1902 Insull turned to the street railway companies, because they had two periods of peak demand—during the morning and evening rush hours. The utility manager began signing up the trolley and elevated lines at incredibly low rates in order to spread and enlarge the load in wholesale fashion. By 1909 a handful of these transit companies would consume almost twice as much energy as all the utility's other light and power customers combined.[4]

Yet the traction contracts pulled Insull back into the storm center of Chicago's franchise politics and municipal reform. After 1898, crusades to stop the council "boodlers" and the utility "barons" like Charles Yerkes evolved into efforts to restructure the basic relationship between government and business. Whereas conservative progressives called for municipal regulation under the guidance of expert engineers, more radical Chicagoans advocated public ownership of the transit lines. In a similar vein, the structural reformers also argued over how far rate and service regulations should be imposed on the other public utilities, such as the gas, electric, and telephone companies.[5]

The battle for municipal reform moved toward a climax in 1905 after the radical, albeit erratic, Edward F. Dunne was elected mayor. As he entered city hall, several pivotal issues were pressing for resolution after years of heated controversy and deadlocked litigation. These included transit policy and finances, utility rates and service standards, and an elite-sponsored "home rule" charter for the city. Elected on a municipal ownership platform, Dunne demanded a public takeover of the transit lines as well as strict regulation of the other franchise holders.

The energy contracts of traction companies and the rates charged lighting customers for gas and electricity drew Insull into this political maelstrom. With his bold declaration in favor of private monopoly under government supervision, the utility executive found broad support for his position among conservative progressives such as elite businessmen, academic experts, practicing engineers, influential journalists, and some popular politicians. Insull adroitly cultivated these powerful allies while overwhelming the opposition with statistical evidence. He emerged stronger than ever from the policy debate over the future of the city's public services. In effect, Chicago adopted a mainstream compromise of regulated capitalism. In the case of the transit companies, the city council created a Board of Supervising Engineers that soon gave official approval to Insull's power contracts with all the remaining streetcar, elevated, and interurban companies entering the city. In the case of his electric utilities, the city government not only ratified his differential rate structures but also accepted his calculations in setting maximum rates for lighting services.

By 1909 Insull's spectacular business and political success made his theories seem like immutable laws of economics. It appeared that his merchandising campaigns had triggered a self-perpetuating cycle of falling rates and rising output. A sophisticated array of statistical studies on group consumption patterns reinforced his conclusion that "low rates may be good business."[6] And the enviable growth of the company's profits supplied proof beyond a reasonable doubt that utility operators should place first importance on building up and diversifying the demand for energy. After a decade of experience, Insull began to institutionalize his marketing techniques into a systematic campaign to reach more and more consumer groups over an expanding area of the Chicago region.

With all the big transit companies signed up, Insull again concentrated on selling electricity in the homes and shops of Chicago's vast number of small consumers of energy. In fact, he had been conducting experiments since 1902 in the residential areas of the affluent North Shore. These tests showed him that weaving the city and suburbs into a single metropolitan web of power would result in both significant savings for consumers and profits for investors. Insull began building an empire of electric companies as well as gas, water, and interurban railway companies that in 1912 embraced a 4,500-square-mile territory and 239,000 customers. This number grew rapidly to over 500,000 customers by 1917, when the United States plunged into World War I and its first national energy crisis.[7]

Building a Sales Campaign

Chicago entered the twentieth century with a boomtown spirit that was sustained by thirty years of unprecedented urban growth. Since the mid-1880s, annexation movements had expanded the city from 80 to almost 200 square miles and had added 300,000 people. These suburbanites, together with an equal number of newcomers to the city during the decade, gave Chicago a population in 1890 of 1.1 million inhabitants. In spite of the depression, labor violence, and political turmoil of the nineties, the metropolis of the midwest continued to act as a magnet for an additional 600,000 foreign immigrants and native farmers by the turn of the century. In large measure, the promise of jobs proved Chicago's chief attraction. The manufacturing sector, for example, increased its work force by 231 percent during the last twenty years of the nineteenth century. Other employment grew apace in established fields such as commerce and transportation, while many new jobs were created, including white-collar positions for professionals, office workers, and telephone operators. By 1910, steady growth boosted the population of the city to 2,185,000 people.[8]

With a booming urban economy, Chicago offered ambitious businessmen like Samuel Insull many promising opportunities to achieve success. In 1898 Insull's insights into the economics of electrical utilities helped him establish a new set of priorities, define operating guidelines, and clarify long-term goals for central station operators. To be sure, he continued to draw on the industry's first twenty years of commercial experience, but these lessons were now reformulated with a single-minded sense of purpose. Insull initially applied the gospel of consumption among the central station's traditional customers: the affluent homeowners, the shop-keepers, and the businessmen. Given their unsatisfied demand for better lighting, these consumer groups provided relatively safe test markets for trying out novel sales gambits and incentive schemes. Within two years, the success of this sales campaign began to strain the capacity of the utility's generators.

Insull's initial efforts to build up the use of electric lighting in the home reflect the emergence of a basic approach to marketing energy. In 1898 he inaugurated a sales program that matched a rate cut with a special incentive. The switch to a two-tier system of rates meant that the average residential customer paid one-third less than the previous flat charge for each unit of energy (kilowatt-hour, or kwh). More important, perhaps, the new rate structure contained a built-in psychology of consumption. It encouraged customers to take advantage of a comparatively low secondary rate by burning more lights for longer periods.[9] To this substantial rate cut, Insull added a free wiring offer that had already been employed successfully for signing up business offices. This kind of giveaway proved a powerful incentive in overcoming the potential customer's natural reluctance to make the necessary immediate outlays for installation and fixtures, which would also entail future energy bills higher than those for gas lighting. Once this psychological resistance was broken down by an extra inducement, however, the homeowner often had the entire dwelling wired.

Insull and his top salesmen soon began elaborating several novel nuances to augment the basic one-two punch of rate cuts and special incentives for residential customers. The head of the sales department, John Gilchrist, compiled a list of the city's most expensive homes. He then organized a corps of solicitors to approach each of these affluent households. The door-to-door salesmen were authorized not only to install six outlets free of charge, but to retrofit the entire house at cost. The solicitors could also make these initial outlays relatively painless by setting up a deferred-payment plan. In exchange for a two-year service contract, the utility would spread out the installation charges in monthly payments added to the regular electric bill. To encourage the installation of electricity in new home construction, Gilchrist compiled separate lists of architects, lawyers, and developers, who were supplied with a constant stream of literature on model plans, safety standards, and contract specifications. For instance, home lighting plans subtly promoted the installation of electric fixtures in closets, pantries, and other confined places where gaslights were impractical. According to the company literature, these extra amenities would give the developer a competitive edge in selling new homes.[10]

Signing up apartment dwellers presented a unique challenge, because first the landlord had to be convinced to permit the installation of the new service. In these income-producing properties, the lower cost of lighting the hallways with gas made overcoming the initial resistance to installation much more difficult than in private homes. Yet this kind of steady, late-night business was exactly what the utility was seeking in order to boost the load on its generating equipment during the off-peak hours.

Gilchrist conceived an ingenious ploy to offer the building owners a special discount for hall lighting without violating legal prohibitions against discriminatory rates. The landlord signed a contract that turned him into a special agent of the utility company. For each tenant he recruited for service, he would receive a 5 percent "commission," or discount, on his hall lighting bill. The company set the maximum reduction at about 60 percent—amounting to a rate of 1/2¢ per kwh—in order to undercut the competition from gas. "The result of the introduction of this scheme," Gilchrist crowed, "was marvelous. From having hall-lighting in a comparatively small percentage of the buildings, we obtained it in nearly all of them [which have been solicited]. The attitude of the owners changed from one of cold indifference to one of interested friendliness."[11]

The apartment-building deal was representative of Insull's marketing strategy. Experience taught his lieutenants how to elicit positive responses from potential customers who were engaged in commercial enterprises. The utility needed to focus its sales campaign on demonstrating how using energy could enhance profits. To a large extent this discovery confirmed what Chicago's shopkeepers had already concluded: electric lighting was a necessity of competition in the modern age. During the early nineties, the fancy shops and department stores in the CBD established new standards of illumination for interior areas, window displays, and street lighting. By 1895 the merchants along Clark Street between the Chicago River and North Avenue banded together to petition Insull to provide modern lighting in front of their stores. City hall promised eventually to supply electricity from its municipal plant for this purpose, but the members of the North Clark Street Business Men's Association were willing to pay extra for an immediate response to their pleas for more light.[12] Insull and his chief engineer, Louis Ferguson, proposed to cut the usual rate in half if the merchants could guarantee a minimum annual contract of 100 arc lights. Within a few weeks, 250 lights were turned on in a well-publicized event that drew crowds to the area. The success of this neighborhood "Great White Way" in attracting shoppers quickly spurred other merchants to organize similar associations along their commercial strips, including North Avenue, Wells Street, and Division Street on the North Side and Cottage Grove Avenue and Thirty-first Street on the South Side.[13]

After 1898, the demand for modern lighting among shopkeepers became so universal that Insull concluded no further incentives were needed to sign up new business in the central portions of the city.[14] Instead, he instructed Gilchrist to concentrate on building up the off-peak load during the evening hours among merchants in the outlying districts. The manager of the sales department cleverly concocted several schemes to persuade store owners that spending more on lighting was good for business. For example, his salesmen would go to one of these commercial strips, rent a store, and convert it into a model of modern lighting with brilliant window displays, appealing advertising signs, and ornamental streetlamps. They would then offer neighboring merchants special deals on similar fixtures. Giveaways, deferred-payment plans, and wholesale discounts were some of the bargaining ploys used to get potential customers to sign one-to-two year contracts. Gilchrist even supplied special services without charge, such as sign painting and cleaning, and hired window-trimming consultants to give shopkeepers a sense that the utility company really cared about making their storefronts as attractive and profitable as possible.[15]

In the central business district, Insull concentrated on meeting the needs of large commercial structures for a full range of light, heat, and power services. Here he had to make his best sales pitch, because he came up against a competitor tougher than cheap gas lighting—the self-contained electrical system. This alternative technology continued to make good economic sense in large buildings and factories that had to have steam boilers for heat during the winter. Building owners and consulting engineers often considered the "isolated" plants simply an extra attachment to the required infrastructure. They preferred to have their own electric plants because they helped make modern buildings completely self-contained. During the first decade of the twentieth century, the number of self-contained systems increased steadily in the skyscrapers, hotels, and department stores of the CBD, as well as in

Table 10 Growth in Number of Self-Contained Electrical Plants, 1898–1910

Year	New Plants	Year	New Plants
1898	73	1905	76
1899	120	1906	36
1900	121	1907	32
1901	93	1908	32
1902	100	1909	22
1903	107	1910	17
1904	68		

Source: Chicago Department of Electricity, *Annual Report*, 1898–1911.

Note: After 1900 no records were kept of the number of self-contained plants that had been removed or replaced with central station service. This omission makes it impossible to calculate the total number of self-contained plants and their connected load of lights and motors.

well as in the manufacturing plants of the surrounding areas (see table 10).[16]

In 1898 Insull launched a more aggressive campaign to defeat the self-contained plant. An effective, if indirect, way to achieve this goal was to undercut the self-contained plant's economic foundations. After studying the problem for a couple of years, he set up the Illinois Maintenance Company to supply steam heat to buildings in the densely packed CBD through underground pipes from a central station. Insull also provided an extra level of free services to relieve building owners' anxieties about undertaking such a departure from normal business practices. For example, the subsidiary company was prepared to purchase old boiler equipment for resale and to keep its customers' remaining machinery in good working order. The early converts to this arrangement included the Schlesinger and Mayer department store (the predecessor of Carson Pirie Scott), the Merchants Loan and Trust Building, the Pullman Building, Orchestra Hall, and the University Club. Although steam heating service was restricted to a small district, it helped Insull tip the economic scales in favor of becoming a utility customer among the owners of some of the grandest and most prestigious buildings in the city.[17]

Of course, the most direct way to beat the competition of the self-contained system was with low rates. In 1898 the institution of the two-tier structure of rates brought a significant reduction in the charge for power, but exactly how deep the cut went is difficult to determine. The crazy-quilt pattern of individual contracts and rate schedules precludes calculating an average unit price before the adoption of the demand meter method of billing. From 1893 to 1898, a comparison of rates for large consumers shows that the

Edison company charged about half as much for power as for light. In 1898 this ratio was maintained for the primary charge, but a sharply sliding scale of secondary charges for power brought the average unit price down to less than half the cost of a kilowatt-hour of lighting.[18]

From the origins of central station service in Chicago, utility companies had differentiated between power rates and light rates, for two reasons. First, electric lighting was a superior and unique source of illumination that did not need to be discounted. In contrast, power delivered from steam, gasoline, and electric sources was essentially the same for most industrialized purposes. In this type of head-on competition, the utility company had to set its rates low enough to win a share of the market. Second, and perhaps more important, the power business represented the first potential consumer of large amounts of electricity during daylight, when most of the central station's equipment was lying idle. Until Insull's arrival in 1893, however, electric motors were restricted to DC and worked well only in small, fanlike sizes of less than 3 HP. In that year, for example, the Chicago Edison Company had about 400 power customers with a connected load of approximately 2,500 HP. In comparison, electric light customers numbered over 4,050, and their connected load was almost 3.5 times as large. But Edison's power consumers used over 28 percent of the utility's output while representing fewer than 10 percent of its customers.[19] The potential of this business to build up and diversify the load of the central station seemed unlimited in an industrial city like Chicago.

During the mid-nineties, revolutionary improvements in electric motors brought this promise much closer to realization. The pace of innovation was remarkable. The motors' efficiency, quality, size, and adaptability to a full range of applications advanced to the point of unmistakable commercial success. The genius of Nicoli Tesla in perfecting an AC motor stands out as the most important contribution to what one contemporary expert would soon call a golden age of invention.[20] In Chicago and other big cities, these new sources of power were immediately pressed into service to help meet a crying need for more rapid transportation through the streets and vertical movement in the soaring office towers. The role rapid transit played in achieving Insull's goal of a "massing of production" will be considered in the following sections. The elevator deserves attention here because it supplied Insull with a powerful new weapon to knock self-contained systems out of the commercial buildings of the CBD.

Like the trolley car, the electric elevator represents a classic example of a practical solution to a demand awaiting a supply. The key technical problem posed by this essential piece of machinery in tall buildings was the intermittent consumption of a large amount of energy. Heavy-duty equipment had to be purchased at great expense and operated full time to meet very brief periods of demand. In other words, the elevator had an extremely poor load factor. It seemed to pose an economic loss regardless whether the power was supplied from a central station, a self-contained plant, or other sources such as steam, gasoline, and hydraulic systems. Most central station operators shied away from accepting elevator contracts even after the rapid strides of the mid-nineties made the electric motor the obvious choice of engineers and architects in new construction.[21] But the insight Insull gained into the economics of electrical utilities led him and his lieutenants to reach just the opposite conclusion.

In 1898 Ferguson asserted that "the electric elevator is . . . the key to the solution [in large commercial buildings] of the much-discussed question, 'central station supply versus the isolated plant.' " A new emphasis on spreading the load—the diversity factor—allowed him to look beyond the diseconomies of individual elevators to the promise of profits from their collective consumption of large amounts of energy during the off-peak daylight hours. In any case, Insull's economics left absolutely no doubt that the central station could supply power to many elevators at a cheaper unit cost than a self-contained system in a single building. Moreover, according to Ferguson, the technological superiority of the electric elevator had gone far to convince architects to use motors to meet an array of other infrastructural requirements of tall buildings. Among the utility service needs of the skyscraper that were supplied by electricity, the engineer listed blowers and fans for ventilation, pumps for water supply, and ejectors for sewage.[22]

Two years later, in 1901, Chicago Edison's chief salesman Gilchrist confirmed that the elevator had done more to promote central station service in large commercial buildings than any other device. At a meeting of utility operators, he reiterated Ferguson's observations on the cumulative effect that adopting one electrical technology had upon others. Gilchrist proclaimed that "now it is comparatively easy" to prevent the installation of self-contained plants in new skyscrapers. A decision for electric elevators often had the added benefit of persuading building owners to cancel plans for a self-contained lighting system in favor of central station service. The salesman reported the results of a statistical study that supported the rapid and widespread shift of architects and engineers in favor of the electric motor. For example, the investigation showed that the cost of operating an electric passenger elevator was $53.04 per month compared with $276.26 for a hydraulic counterpart. Gilchrist also displayed graphic proof of the new importance of the electric motor in order to reverse "the cool indifference or picayunish apprehension of short-sighted station managers." Reflecting the sales effort in Chicago, he displayed a map showing over four hundred elevators served by the company in the CBD (see map 2).[23]

The elevator was just one example of accumulated demand for power awaiting a practical supply. The coming of age of the electric motor in the mid-nineties triggered rapid adoption and conversion in several industries, especially those that were already highly mechanized. The steam engine had ushered in the industrial revolution, but its use in the factory had always been difficult and unsatisfactory. Electric motors were a ready-made substitute for cumbersome steam systems where manufacturers used many small machines or several machines of the same type. In both instances the steam engine required a costly investment in a power plant and a complicated transmission system of shafts, belts, and pulleys that reduced its efficiency to a mere 30 percent or less. The use of steam technology in the factory also precluded orderly work flows, because a mechanical transmission system dictated that the biggest machines be placed closest to the steam engine. The ceilings of factory buildings, moreover, had to be reinforced at great expense to support the heavy load of the power system's shafts, belts, and pulleys. A final drawback of the industrial use of steam power was the fire hazard created by transmission shafts in multi-story buildings. By the turn of the century, insurance underwriters were beginning to levy higher premiums on these factories than on modern one-story structures.[24]

In Chicago, the first industries to convert from steam to electric power were the most mechanized. The first studies of the utility's power customers give prominence to printing, woodworking, clothing, and machine shops. Other businesses quickly adopted the electric motor because steam technology had never proved practical in handling heavy materials outside the factory. More mobile sources of power were needed in such places as steel mills and stone quarries. The refrigeration machine was another technology ready-made for harnessing to the electric motor. It was soon pressed into service in Chicago's candy factories, bakeries, and

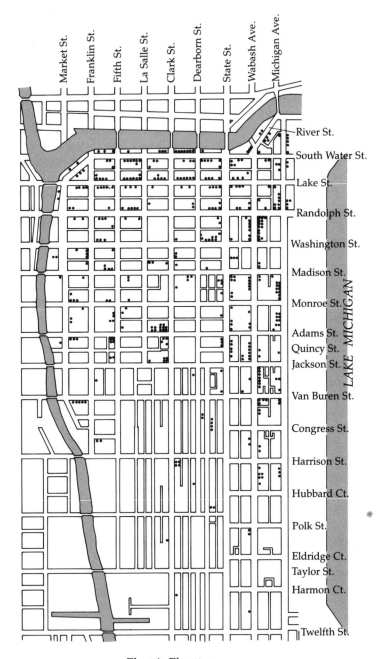

Market St.
Franklin St.
Fifth St.
La Salle St.
Clark St.
Dearborn St.
State St.
Wabash Ave.
Michigan Ave.

River St.
South Water St.
Lake St.
Randolph St.
Washington St.
Madison St.
Monroe St.
Adams St.
Quincy St.
Jackson St.
Van Buren St.
Congress St.
Harrison St.
Hubbard Ct.
Polk St.
Eldridge Ct.
Taylor St.
Harmon Ct.
Twelfth St.

LAKE MICHIGAN

• Electric Elevators

MAP 2.   Electric elevators in the central business district, 1900. *Source:* John F. Gilchrist, "Electric Elevators and Conveyors," *Proceedings of the Association of Edison Illuminating Companies,* 1900, 7.

breweries. The coming-of-age of the electric motor during the mid-1890s was reflected in the mushrooming growth of the power load connected to the utility's generators. In 1895 the company supplied power to approximately 3,000 HP worth of motors, a number that rose rapidly over the next five years to 19,000 HP and in 1905, to 69,000 HP.[25]

During the mid-1890s, electricity began to supply a similar demand for power in a second type of industrial application. In these cases electrical technology provided original solutions to long-standing problems and opened new frontiers of commercial enterprise. Insull's relatively low power rates helped several nascent industries become well-established ventures. In Chicago these made use of such processes as electrochemicals, electroplating, and arc welding, which helped set the stage for the mass production of the bicycle and the automobile.[26] Taken together, novel and ready-made uses of electric power promised to improve the central station's daytime load while becoming increasingly important consumers of energy in their own right.

During the early years of the sales campaign, however, the outcome of the economic battle between self-contained systems and central station service remained doubtful. An industry's adoption of electric power technology was no guarantee that it would purchase energy from a utility company. For this class of customer, Insull and Gilchrist realized they would have to groom an elite corps of salesmen who had engineering training as well as marketing skills. Technical experts like Ferguson had already developed basic methods of investigating the economics of energy use in the factory. They also had learned how to respond to factory managers' needs by offering special incentives such as custom designing motors and preparing plans for reorganizing work in the shop. Although these sales techniques won some notable converts, many other industrial concerns continued to prefer self-contained electric plants for light and power (see table 10).[27]

The ability of Insull's salesmen to win over all of Chicago's building owners and factory managers to central station service may have remained in doubt, but there was no question about the steady and rapid increase in the demand for the utility's electric power. Between 1896 and 1902, in fact, the consumption of electricity for power increased about 50 percent faster than the growth of the Edison company's lighting business. The easing of the depression, the improvement of the electric motor, and the rate cuts of 1898 helped turn a relatively stagnant branch of the utility business into its most dynamic sector. In the five years preceding 1896, the demand for power grew at an anemic rate of 8.4 percent a year in comparison with a 41.4 percent annual increase for incandescent lighting. During these years, power consumption declined from about a third to a quarter of the company's total output. In contrast, the years following 1896 witnessed a vigorous growth rate in the power business, averaging 65.7 percent a year. By 1902 these increases had boosted the consumption of power back up to a third of the utility's total output.[28] The dramatic reversal in the trends offered Insull convincing proof that concerted efforts to sell energy were effective in building up and diversifying the demand for central station service.

The early figures for the combined growth of light and power strongly reinforced belief in Insull's gospel of consumption. Comparing the net increase in the utility's connected load in the five years before and after 1898 shows a whopping 400 percent annual gain. Although Insull's companies added the equivalent of 49,000 incandescent lights per year in the first five years of his leadership, the annual increase in the next period averaged the equivalent of over 200,000 lights.[29] The striking contrast between the two periods helped convince Insull, Gilchrist, and other top managers that the one-two punch of rate cuts and special incentives worked.

Most of the new demand for energy remained confined to the utility's traditional commercial customers in the CBD, but the refinement of the electric motor put the utility at the threshold of a great leap forward in the use of power in the factory. The ability to match each machine with an appropriate motor opened endless possibilities for factories to mechanize and reorganize work according to the popular principles of scientific management. Moreover, the success of the sales gambits among homeowners and landlords near the CBD suggested that greater efforts would tap a huge reservoir of demand among the masses for better lighting in the home.

The spectacular success of the early sales strategies encouraged the utility executive to forge these various efforts into a systematic and persistent marketing campaign. With the establishment of an advertising department in 1901, Insull initiated a process of institutional growth and specialization to reach out to every group of potential energy consumers in the Chicago area. The new department was put in charge of mounting a general assault, informing and enticing the largest consumer groups: the city's householders, office workers, and shopkeepers. To be sure, it is virtually impossible to determine the effectiveness of Insull's mass-marketing campaign, especially during this early period of modern advertising, when the emphasis was still on providing product information, not on tapping psychological impulses. The newspaper copy probably worked in combination with a complicated mix of other sources of consumer demand. Insull also continued to expand the established departments involved in selling energy to special target groups. In both cases the marketing campaign rested on a solid foundation of social statistics gathered from the demand meter and the door-to-door solicitations of a swelling army of salesmen.

Since the impact of this sales campaign on the daily lives of Chicagoans is the subject of the next chapter, here a few examples of the advertising department's activities will illustrate how it sought to reach a mass audience.[30] Only two years after its inception, the department started publishing a glossy magazine called *Electric City* to supplement increasing advertising copy in the daily newspapers. A special deal was offered to drugstores throughout the city to circulate the free monthly in exchange for free wiring and discounts on lighting bills. The department also took the model shop idea a step further by creating a mobile "Electric Cottage" packed with the latest lighting fixtures and home appliances. This showroom on wheels was moved from one neighborhood to the next in the wake of advance publicity to ensure a large turnout of curious residents. Teams of solicitors would soon follow to make personal calls on each household and present several special offers. In 1908 these efforts culminated in the ballyhooed giveaway of ten thousand irons for a free trial to new customers.[31]

The very success of the early sales campaign forced Insull to consider once again the problem of supply. In January 1901, for example, he reported to the corporate board of directors that no new business could be accepted on the North Side of the city. The capacity of the generators and the distribution system had been exceeded. The demand for energy on the South Side, he reported, had also taxed the utility's lines to the limit.[32] During the following year, statistical reports confirmed that both the company's output of energy and its connected load were growing faster than its generating capacity. Even more disturbing, the increase in energy consumption was pushing the point of maximum demand up at the fastest rate of all (see table 11 and chart 2). This meant the utility had to face the prospect of adding more and more generating equipment that would sit idle much of the time. The sales efforts seemed to be exacerbating the utility's heavy burden of capital investment, which would forestall further rate cuts and higher profits. To prove his theory of a self-perpetuating cycle of falling rates

Table 11 Growth of Central Station Service in Chicago, 1898–1902

Year	Kilowatt Generating Capacity		Total Connected Load		Consumption		Maximum Load	
	Amount (kw)	% Increase	Amount (kw)	% Increase	Amount (000s kwh)	% Increase	Amount (kw)	% Increase
1898	14,800	35.8	30,000	50.8	26,084	53.3	11,460	—
1899	15,600	5.4	37,000	23.3	32,755	25.6	14,260	24.4
1900	16,200	3.8	46,300	25.1	39,084	19.3	16,420	15.1
1901	20,600	27.2	58,900	27.2	46,726	19.6	21,060	28.2
1902	25,700	24.8	79,500	35.0	65,656	40.5	29,080	38.1
Total increase, 1898–1902	10,900	73.6	49,500	165.0	39,572	151.7	17,620	153.8
Average annual increase	2,725	18.4	12,375	41.2	9,893	37.9	4,405	38.4

Sources: H. A. Seymour, "History of Commonwealth Edison," manuscript, 1934, 321, in CEC-HA, box 9001; William E. Keily, ed., Central-Station Electrical Service: Its Commercial Develpment and Economic Significance as Set Forth in the Public Addresses (1897–1914) of Samuel Insull (Chicago: privately printed, 1915), 325.

Portraits of Samuel Insull.

1881 (University Archives, Loyola University of Chicago).

1888 (Commonwealth Edison Company).

1920 (University Archives, Loyola University of Chicago).

1935 (University Archives, Loyola University of Chicago).

Central manufacturing block, 28 North Market Street (now Wacker Drive at Madison Street). Early home of central station services, including the Chicago Arc Light and Power Company. Picture about 1910, after Commonwealth Edison Company converted the building into a repair facility (Commonwealth Edison Company).

Electric lineman on a horse-drawn truck in a residential neigborhood, 1885 (Commonwealth Edison Company).

First central station of the Chicago Edison Company (built in 1887), disguised as an office building in the heart of the financial district at 120 West Adams Street (Commonwealth Edison Company).

Chicago Edison Company crew laying the first network of underground "tubes" at Jackson and La Salle streets, 1887 (Commonwealth Edison Company).

(*Opposite, top*) Dynamo room of the Chicago Edison Company, 1887. Rows of small-scale machines were belted to engines on the floor below (Commonwealth Edison Company).

(*Opposite, bottom*) Engine room of the Chicago Edison Company, 1887 (Commonwealth Edison Company).

Model electric kitchen at the World's Fair of 1893 (John P. Barrett, *Electricity at the Columbian Exposition* [Chicago: Donnelly, 1894], 403).

Night illumination at the World's Fair of 1893, including brilliant spotlights, dancing fountains, and decorative lights on the buildings of the Court of Honor (John P. Barrett, *Electricity at the Columbian Exposition* [Chicago: Donnelly, 1894], 323).

Main engine room of the Harrison Street Station, 1902. Behemoth 5,000-HP engine at rear (Commonwealth Edison Company).

Harrison Street Station (built in 1893). The central station has taken on a more functional, industrial appearance (Commonwealth Edison Company).

Generator room of Fisk Street Station, about 1910. First two original turbogenerators (1903–4) in foreground (Commonwealth Edison Company).

Fisk Street Station (built in 1903). The central station now has a distinctive design embodying unlimited power and modern efficiency (Commonwealth Edison Company).

Chicago: The Electric City cover, August 1904 (Commonwealth Edison Company).

Hyde's gristmill and electric station, Joliet, Illinois, 1900. Waterpower site on left; auxiliary steam power plant on right (Commonwealth Edison Company).

Bluff City Electric Street Railway Company, about 1894–96. This was the earliest interurban service on the North Shore and the first large consumer of electricity in the suburbs (Commonwealth Edison Company).

New central station of the North Shore Electric Company, Highland Park, Illinois, 1905. Older, abandoned facility at rear (Commonwealth Edison Company).

Interior of the Highland Park station, 1905 (Commonwealth Edison Company).

Construction of the Bloomington, Pontiac, and Joliet Railroad Company interurban line in a rural area, 1905 (Commonwealth Edison Company).

Celebration of the inauguration of service by the Bloomington, Pontiac, and Joliet Railroad Company, 1905 (Commonwealth Edison Company).

White City Amusement Park, on the South Side of Chicago, 1906 (Commonwealth Edison Company).

Commonwealth Edison Company crew laying underground cables in the neighborhoods, 1910 (Commonwealth Edison Company).

Mobile "Electric Cottage" of the Commonwealth Edison Company in a residential neighborhood, 1910 (Commonwealth Edison Company).

Marketing Over Thirty Tons of G-E Electric Flatirons

The Commonwealth Edison Company of Chicago is now placing 10,000 G-E Flatirons in as many homes. These irons are sold on the free-trial small-monthly-payment basis.

This result-getting local campaign takes full advantage of the interest and demand aroused in Chicago ;by the General Electric Company' *continuous*, national advertising in [millions of copies of the best known magazines.

Companies in smaller cities can obtain equally gratifying results now by conducting similar campaigns suited to local conditions.

General Electric Company
Principal Office: Schenectady, N. Y.

The Guarantee of Excellence~   on Goods Electrical.

General Electric Advertisement, 1912. One of Sam Insull's earliest and most memorable mass marketing schemes (*Electric Merchandise* 11 [April 1912]: 341).

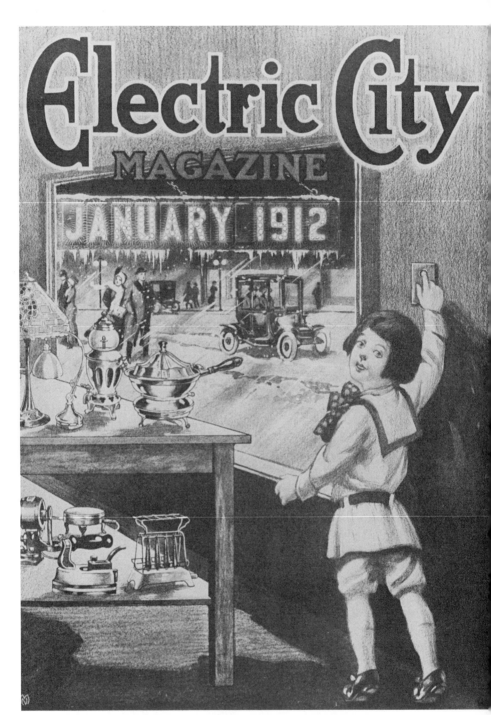

Electric City cover, January 1912 (Commonwealth Edison Company).

Electric City cover, October 1912 (Commonwealth Edison Company).

Electric City cover, January 1913. The city lights created a new type of modern urban environment (Commonwealth Edison Company).

Electric City cover, May 1917. As America enters World War I, the housewife confronts modern household appliances (Commonwealth Edison Company).

Statistical department of the Commonwealth Edison Company, 1910s (Commonwealth Edison Company).

Load dispatchers' office, Commonwealth Edison Company, 1919 (Commonwealth Edison Company).

Exhibition of electric vehicles, 1915. No amount of promotion could convince the public to buy electric vehicles rather than gasoline-powered ones (Commonwealth Edison Company).

A crazy promotional scheme: the first air delivery of an electric appliance on 11 March 1919. The plane flew from Grant Park in the downtown area to the home of Rufus C. Dawes in Evanston, Illinois (Commonwealth Edison Company).

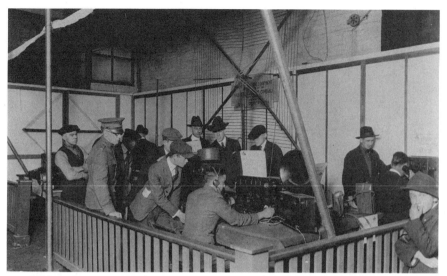

Soldiers demonstrate radio equipment at the Chicago Electrical Exhibition, 1919 (Commonwealth Edison Company).

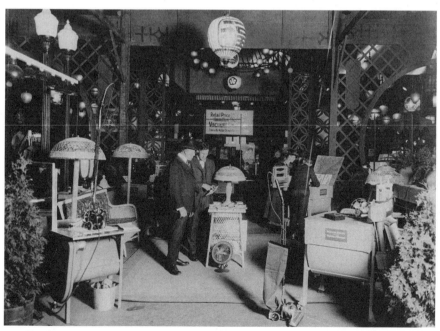

Display of household appliances at the Chicago Electrical Exhibition, 1919 (Commonwealth Edison Company).

Parade truck used in neighborhood festivals to attract commercial customers, 1915 (Commonwealth Edison Company).

After World War I, parade trucks in working-class districts appealed to home consumers, 1923 (Commonwealth Edison Company).

ELECTRICITY
—the modern Emancipator

PUBLIC SERVICE COMPANY
OF NORTHERN ILLINOIS

(*Opposite and above*) Billboard posters of the Public Service Company of Northern
Illinois, 1924 (Commonwealth Edison Company).

Mrs. L. Maier, 5021 Quincy Street, Chicago, Illinois, gets a new washing machine, 1922 (Commonwealth Edison Company).

Model electric home, Waukegan, Illinois, 1924 (Commonwealth Edison Company).

Phonograph department of the Electric Shop, Edison Building, Chicago, Illinois, 1925 [Commonwealth Edison Company].

Neighborhood Electric Shop on Southwest Side, 4834 South Ashland Avenue, Chicago, Illinois, about 1928 (Commonwealth Edison Company).

Cooking demonstration at the downtown Edison Building, 1925 (Commonwealth Edison Company).

Door-to-door solicitors of the Commonwealth Edison Company, 1925 (Commonwealth Edison Company).

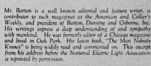

"How long should a wife live?"

By BRUCE BARTON

Mr. Barton is a well known editorial and feature writer, a contributor to such magazines as the American and Collier's Weekly, and president of Barton, Durstine and Osborne, Inc. His writings express a deep understanding of and sympathy with mankind. He was formerly editor of a Chicago magazine and lived in Oak Park. His latest book, "The Man Nobody Knows" is being widely read and commented on. This excerpt from his address before the National Electric Light Association is reprinted by permission.

"Some years ago there was a celebration in Boston in honor of the landing of the Pilgrim Fathers. After several laudatory speeches had been made by men, a bright and vivacious woman was called on. Said she:

"'I am tired of hearing so many praises of the Pilgrim Fathers. I want to say a word about the Pilgrim Mothers. They had to endure all that the Pilgrim Fathers endured, and they had to endure the Pilgrim Fathers besides.'"

"Do you know what happened to the Pilgrim Mothers, my friends? I will tell you. They died. They died young. It took two or three of them to bring up one family. The fathers were tough and lived long, but work and hardship made short work of the wives. Listen a minute:

"Between 1701 and 1745 there were 418 graduates of Yale who got married. What happened to their wives?

33 died before they were 25 years old.
55 died before they were 35 years old.
59 died before they were 45 years old.

"Those 418 husbands lost 147 wives before full middle age.

"Harvard was no better. Take the class of 1671, which was typical. It had eleven graduates, of whom one died a bachelor at the age of twenty-four. Of the remaining ten,

4 were married twice,
2 were married three times.

"For ten husbands, therefore, there were eighteen wives.

"It has been truly said that you can measure the height of any civilization by the plane upon which its women live. Measured by that standard, we have made great progress in the United States, but we have not made enough. An electric motor which runs a washing machine or a vacuum cleaner, works for 3 to 5 cents an hour. There are still millions of women doing this work which motors can do,

working away at coolie wages of 3 cents an hour and having to neglect the highest work entrusted to human beings, the work of motherhood.

Women who work for 3 cents an hour

"Some day you expect to have every home in the United States electrified. My friends, why should we wait until some day? Why don't we do it immediately, next year, within the next twelve months? Does that seem impossible? I tell you that I believe it would be possible, . . . to arouse such sentiment in the mindset of the women of this country that every woman would realize that it is beneath the dignity of human life for her to work for 3 cents an hour.

"The time in the life of a child when a mother can exert her influence is terribly brief. 'Give me a child until he is seven years old,' a great philosopher said, 'and I care not who has him afterward.' Seven years in which to mold character; seven short, fleeting years. What a tragedy that a single moment of those years should be wasted in work which an electric machine can do!

Squandering the National Wealth

"I make no apology for growing emphatic. It is a subject worth being emphatic about. It opens a whole new world of opportunity to us; it gives us a new interest, a new enthusiasm. Every day that we lose in this business of electrifying homes costs the nation its richest wealth—the training of children, the lives and happiness of mothers.

"'How Long Should a Wife Live?' The answer, in the old days was, not very long.' The homes of those days had two or three mothers and no motion. The home of the future will lay all of its tiresome, routine burdens on the shoulders of electrical machines, freeing mothers for their real work, which is motherhood. The mothers of the future will live to a good old age and keep their youth and beauty to the end."

Commonwealth Edison Company
72 West Adams Street

Newspaper advertisement for the Commonwealth Edison Company by the famous adman Bruce Barton, 1925.

Model electric farm of the Public Service Company of Northern Illinois, Libertyville, Illinois, 1928 (Commonwealth Edison Company).

Installation of the one millionth meter of the Commonwealth Edison Company during the opening ceremonies at the Daily News Building, Chicago, 1929. *Left to right:* James L. Hougheling, Louis A. Ferguson, John Root, Homer E. Niesg, O. J. Bushnell, Irving Stone, W. L. Abbott, P. J. Smith (Commonwealth Edison Company).

Refrigeration section of the Electric Shop, Edison Building, 1929. The electric refrigerator heralded a new era of "load-building" consumer products (Commonwealth Edison Company).

Illuminated buildings at the Century of Progress Exposition, Chicago, 1933 (Commonwealth Edison Company).

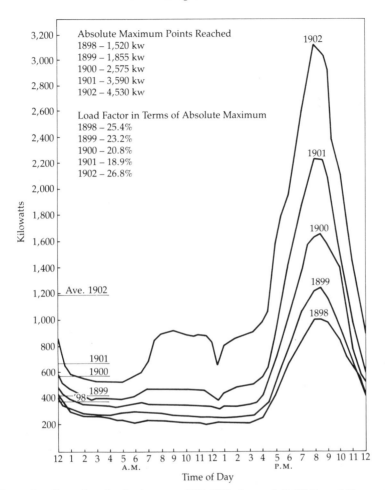

CHART 2.   Central station load curves, 1898–1902. *Source:* J. B. White and Company, *Report on the Commonwealth Electric Company* (London: White, 1902), in CEC-HA, box 1.

and rising consumption, Insull would have to diversify the load by finding major consumers of energy during daylight. He would also have to find a way to overcome the technological constraints that were preventing true economies of scale in the generation of electrical energy.

Building the Load

Between 1902 and 1908, Insull achieved a "massing of production" that appeared to finally verify his gospel of consumption. To meet the immediate need for more energy, he built an ultramodern central station containing the first large-scale steam turbine generators. The new technology was crucial to completing the electrical revolution and opening a new phase of steady improvement in the size and efficiency of central stations. The steam turbine quickly proved a practical solution to the economical supply of electrical energy. Early results with the experimental equipment showed substantial savings in capital investment, labor, and fuel. The turbogenerator also represented the capstone of a unified system of power plants and distribution substations for the metropolitan area.

The virtually unlimited promise of the turbogenerator renewed Insull's faith in the gospel of consumption. Guided by his economic theories, he was determined to win over all the city's large energy consumers to central station service. The recent coming-of-age of the electric motor pointed the utility operator to the major demand potential for power during the daytime. The elevator, for example, helped diversify the utility's load by using energy during weekday mornings and early afternoons, when most of the utility's equipment stood idle. Insull had no difficulty drawing an analogy to the traction companies, which were already the city's largest consumers of electricity.[33] Insull used his sales formula to offer them special incentives and rates so low they could not afford to reject his contracts. By 1908 the transit business had built up and diversified the load sufficiently to trigger a self-perpetuating cycle of falling rates and rising demand.

The origins of the first modern generator station were rooted in the problems Insull faced at the turn of the century. The rate cuts of 1898 had tapped more demand than the utility company could supply with the technology then available. The sales campaign had succeeded in building up the load but not in diversifying it to any significant extent. As chart 2 shows, a trend toward a declining load factor reflected less and less efficient use of the generating equipment. Satisfying the demand for better lighting remained the main business of the central station. In 1898 the installation of the world's largest storage battery in the old Adams Street Station had helped meet the maximum demand on the system between 5:00 and 8:00 on dark December evenings. Yet this type of solution was expensive and could serve only as a stopgap. The steady growth in demand would soon exceed the limits of the generators at the company's main power plant.[34]

In 1902 the practical limit of the reciprocating steam engine was reached at the Harrison Street Station. Since the plant opened a decade earlier, one piece of equipment after the next had been added until its original capacity of 800 kw reached 10,900 kw. The power plant's eleven steam engines used up all the space set aside in the plans for future expansion. To meet the pressing need for additional capacity, a huge addition had to be built to house a gargantuan machine of 5,000 HP and its companion, a 3,500-kw generator. The gigantic engine was four times as large as the original equipment and created pounding vibrations that put the structural strength of building materials and metal parts to the ultimate test. Many visitors came to see the giant engine, including a rising architect named Frank Lloyd Wright. He seemed as awestruck as Henry Adams had been by the technology of electrical energy. Describing his impressions of the power plant in terms of an organic metaphor of the city's "nerve ganglia" in 1901, Wright wrote that

> the peerless Corliss [engine] tandems [were] whirling their hundred ton fly-wheels . . . while beyond, the incessant clicking, dropping, waiting-lifting, waiting, shifting of the governor gear controlling these modern Goliaths seems a visible brain in intelligent action, registered infallibly in the enormous magnets, purring in the giant embrace of great induction coils, generating the vital current meeting with instant response in the rolling cars on elevated tracks ten miles away, where the glare of the Bessemer steel converter makes a conflagration of the clouds.[35]

Yet the cost of the steam engine and a foundation strong enough to hold it pushed this technology to the breaking point—one of practical economics. In most respects, engineers and inventors had anticipated the problems of designing larger and more efficient prime movers. But until the rapid growth of demand for electricity in Chicago forced Insull to find a solution, technological innovation had proceeded at the slow pace of cautious experimentation. In the case of the turbine engine, the exploitation of waterpower sites had given impetus to its technical improvement and kept interest focused on its commercial promise. But the steam turbine was different, and its development was restricted to small demonstration models. In Europe turbogenerators of 250–500 kw were being tested under commercial conditions. American efforts lagged behind in the laboratories of General Electric, Westinghouse, and other manufacturers.[36]

In December 1901 Insull decided to purchase land for a new

powerhouse that would mark a historic departure from the current standards of the industry. The utility executive had already received technical reports on the European experience with steam turbines from his chief engineer, Louis Ferguson, and his trusted consultant, Fred Sargent. Insull also conferred with the president of GE, Charles A. Coffin, who suggested that the largest available model of 1,000 kw be given a commercial test in Chicago. Insull, however, was persuaded by Ferguson and Sargent to insist on something larger still. He demanded a machine that would exceed the limits of existing technology, a machine of at least 5,000 kw. Coffin agreed to assume the risks of attempting to build such a generator while Insull took responsibility for making it commercially operational. The experimental device was installed in the new Fisk Street Station, on the south branch of the Chicago River about two miles beyond the congestion of the CBD. Started on 2 October 1903, the turbogenerator was so successful that it was eventually returned to the GE works in Schenectady, New York, where it stands today as a monument to engineering genius. It would take another decade for the steam turbine to evolve into its present design, but its practical value was immediately evident.[37]

The introduction of large steam turbines into commercial use ushered in a halcyon era of rapid improvements in the size and the efficiency of central stations. The replacement of the first generation of turbogenerators within four years was emblematic of this dynamic process of innovation. After only two years of commercial experience with the first four engines, the engineers had learned enough about the peculiar characteristics of the steam turbine to modify it and enlarge it to 9,000 kw. The first two of these units were found to be so efficient that they were further modified a year later, in 1907, to generate 12,000 kw. Though producing a 140 percent boost in output, the second-generation turbine was the same size and cost as the first. The improved model cut investment costs even further because it required no additional boiler equipment. Savings in capital were matched by similar economies in operation. Most important, it used about one-third less coal per unit of electricity. When added to several laborsaving innovations introduced at the Fisk Street Station, the result was a marked decline in the cost of energy.[38]

At the same time, the development of a distribution network of high-voltage transmission lines and substations paralleled the growth of the utility's generating facilities. Only five years after the establishment of the first substation in 1897, twenty-two of these distribution points dotted the city (see map 3), creating an integrated service grid that covered approximately two-thirds of the built-up sections of Chicago's 200-square-mile territory. By 1908 the number of these distribution points would climb to thirty-three. During this period, the elaboration of specialized substations would begin in response to the city's increasingly diverse needs for more light, heat, and power. A critical new position, the "load dispatcher," would also emerge to keep a complex and far-flung system in balance between demand and supply.[39]

The system Insull built represented a special kind of technological breakthrough. The turbogenerator became more and more cost effective as it became larger and larger. The size limit of this technology has only recently been reached in machines exceeding 1 million kw. In effect, then, the introduction of the turbogenerator set in motion a process that led inexorably to the goal of cheap electricity. Perhaps the best way to illustrate the economic savings resulting from the turbogenerator is to reproduce some of the diagrams Insull used in preaching the gospel of consumption. The simplest one (see chart 3) superimposed the falling unit cost of electricity on the rising volume of output. This diagram seemed to present irrefutable proof of Insull's theory about the "massing of production." A second illustration (see chart 4) provided a clear-cut example of how higher output helped attain economies of scale. In contrast to the reciprocating engines at the Harrison Street Station, the turbogenerators at the Fisk Street Station cut the amount of coal burned per kilowatt-hour in a steady, step-by-step progression. The widening gap on the chart between energy output and coal consumption was reflected in increasing benefits from the conservation of fuel and the abatement of smoke pollution in the city.[40]

Insull's conclusion that "low rates may be good business" was given graphic expression in a third illustration (see chart 5). To the businessman, of course, its most important feature was the utility company's rising line of "net from operations," or profits. This enviable record of corporate success was achieved in spite of rate cuts that reduced the income from each customer. The interaction of two factors largely explained the basis of Insull's theories. Falling rates allowed more and more consumers of energy to enjoy central station service. More customers, in turn, reduced each one's share of the utility's heavy burden of fixed costs for investment capital. The growth in demand stimulated by lower rates also helped achieve significant economies of scale and additional savings from

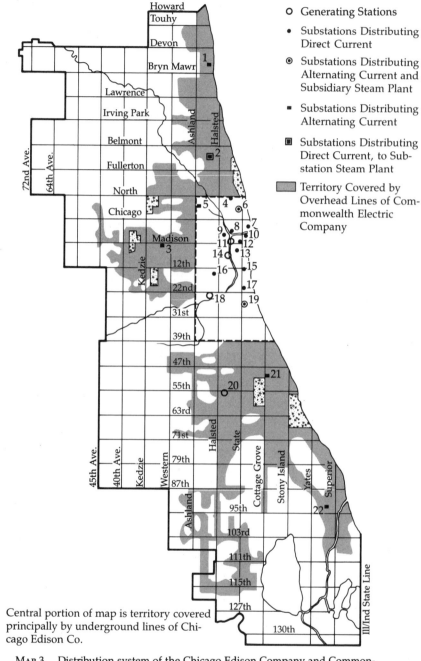

Central portion of map is territory covered
principally by underground lines of Chicago Edison Co.

MAP 3.   Distribution system of the Chicago Edison Company and Commonwealth Electric Company, 1903. *Source:* Ernest F. Smith, "The Development, Equipment, and Operation of the Sub-stations . . . ," *Journal of the Western Society of Engineers* 8(March–April 1903): 214.

KEY TO MAP 3

	Stations	Location
1	Edgewater	Ardmore & C.M. ST.P. TRACTS
2	Lake View	A1 S. of Diversey-E. of Lincoln
3	W. Madison	1096 W. Madison St.
4	Sedgwick	444 Sedgwick St.
5	W. Division	162 W. Division St.
6	Station 4	340 N. Clark St.
7	Illinois	15 Illinois St.
8	Kinzie St.	S.W. cor. Kinzie-Kingsbury
9	Lydia St.	144 W. Lake St.
10	Randolph	Basem't-Masonic Temple
11	Station 5	S.W. cor. Market & Washington
12	Dearborn	112–114 Dearborn St.
13	Adams St.	139 Adams St.
14	Station No. 1	W. Harrison St. & Chicago Riv.
15	State St.	1235 State St.
16	W. 14th St.	311 W. 14th St.
17	21st St.	145 E. 21st St.
18	Fisk St.	Fisk St. S. of 22nd St.
19	Station No. 2	2640 Wabash Ave.
20	Station 'A'	56th St. & Wallace
21	Hyde Park	A1 S. of 50th & S. of Cot. Grove
22	S. Chicago	9235 S. Chicago Ave.

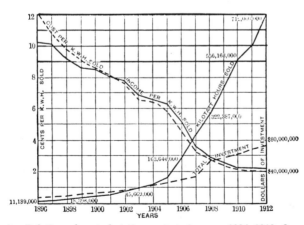

CHART 3. Relation of central station costs to income, 1896–1912. *Source:* William E. Keily, ed., *Central-Station Electrical Service: Its Commercial Development and Economic Significance as Set Forth in the Public Addresses (1897–1914) of Samuel Insull* (Chicago: privately printed, 1915), 437.

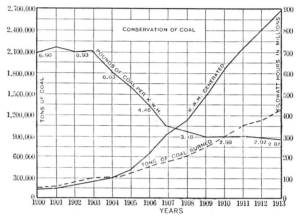

Chart 4. Central station conservation of coal, 1900–1913. *Source:* Keily, *Central-Station Service*, 469.

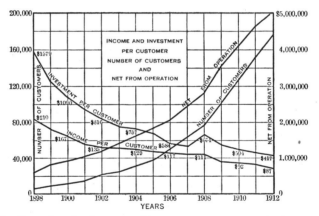

Chart 5. Central station income and investment per customer, 1898–1912. *Source:* Keily, *Central-Station Service*, 436.

spreading the load over the entire day. Thus the income collected from an increasing number of customers more than made up for any decline in the profits from each one.[41]

Insull did not have to distort the statistics to make his gospel of consumption appear to be an iron law of economics. During the first fifty years of central station service, a peculiar configuration of conditions produced a cycle of falling rates and increasing output that no longer prevails. Unique to Insull's time was a historic buildup of demand in the city for better sources of light and power that sought a supply at a reasonable price. This reservoir of demand was large enough to absorb all the available supply as the technological revolution in the electrical industry permitted a steady reduction in price. A glut of low-priced coal was another major factor helping to make Insull's equation work. But eventually the pool of demand would exhaust itself, and the revolutionary pace of improvement in basic components such as generators, motors, and light bulbs would slow down. The price of fuel could also be expected to rise as demand increased, natural resources were depleted, labor unions won wage concessions, and other market conditions changed. In fact, a national coal strike in 1902 first raised the specter of a fuel shortage. Insull reacted by forming a partnership with the Peabody Coal Company to acquire huge reserves in central Illinois and Indiana.[42] In this way he was able to take advantage of a highly favorable set of circumstances as the head of a vast utility empire until the end of his tenure during the opening years of the Great Depression.

Although it took some time to accumulate statistical data on the turbogenerator after the opening of the Fisk Street Station in October 1903, immediate indications of a technological triumph strongly bolstered Insull's belief in theories of mass-consumption economics. The new generators spun ten times faster while weighing only one-tenth as much as the old reciprocating engines, for instance.[43] The loosening of the final constraints on the supply side allowed the utility executive to give renewed emphasis to managing the size and diversity of the energy load. The example of the elevator pointed him toward power consumers as the logical way to counterbalance a heavy demand during the evening with a sizable load during the day.

Insull did not have to look far to find such a prospective group of customers, whose use of electricity already far surpassed all others'. By 1902 the street and elevated railways in Chicago already used more than three times as much energy as the combined total of Edison's light and power customers. In the near future the potential of the transit business would be even greater, since its conversion from horse and cable to motor traction was not yet complete. Of the nearly 1,400 cars entering the Loop during the rush hour, 772 were drawn by cables, 97 were electric cars attached to cable trains, and 510 were powered by electricity. In the neighborhoods of the North and West sides, the horsecar remained an all-too-familiar sight. When Charles Yerkes left Chicago the year before, his Union Traction Company spent $240,000 to operate its cable system as well as $322,000 on electric traction, but it still spent $562,000 on horses. The completion of transit modernization by the city's fifteen franchise holders would boost their demand for electrical energy to truly enormous proportions.[44]

For Insull, however, the problem was that the companies were all generating their own power—all except one, the Lake Street Elevated Company. While the Fisk Street Station was under construction in 1902, this company approached Insull to make a deal. The elevated (el) line to suburban Oak Park had been purchasing the excess capacity of other transit companies until these sources dried up. For Insull the timing could not have been better. With an expensive new central station coming on, a large power load during the daylight hours would help keep otherwise idle equipment working. Eager to gain a foothold in this market, Insull proposed a contract with rates so low that the railway company could hardly afford to reject his offer. This important precedent used the demand model of a two-tier structure of rates in a modified form, called the Hopkinson system. The primary rate, corresponding to the utility company's fixed costs, was still based on the consumer's maximum demand as measured by a meter, but the elevated company made a flat rate payment annually for each kilowatt of maximum demand. In contrast, the Wright system of rates charged a variable amount determined by hourly use each month. The secondary rate, corresponding to the utility's operating costs, remained pegged to actual kilowatt-hour consumption. For the utility, the advantage of this rate structure was a key provision added by Insull that guaranteed a minimum amount of consumption equivalent to a load factor of 35 percent. In other words, the traction company promised to keep the machinery it used running at least 35 percent of the time on the average or to pay for an equivalent amount of energy. In exchange, it enjoyed what seemed like a ridiculously low secondary rate of less than 1¢ per kwh, one-third the usual charge for large consumers of power.[45]

Over the next five years, additional contracts with the transit companies fueled an unprecedented expansion of central station service in Chicago. In 1907 the energy consumption of these Edison customers would surpass the combined total of all the rest of the utility's light and power users (see chart 6). Why did more and more of the transit companies turn to the utility operator rather than building large-scale generating stations of their own? The answer to that question requires an analysis of utility finances and reform politics, which were reaching a crisis during the tumul-

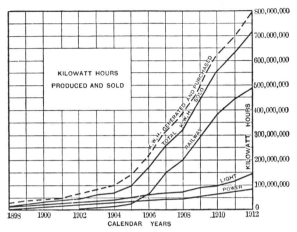

CHART 6.   Generation of electricity by Edison central stations, 1898–1912. *Source:* Keily, *Central-Station Service*, 430.

tuous administration of Mayor Edward Dunne (1905–7). Since electric, gas, and other utilities also came under the intense scrutiny of urban progressives during this period, the entire subject of government-business relations is best dealt with in a separate section below. First, however, it is important to underscore the ways the transit companies' insatiable demand for electricity triggered the cycle of falling rates and rising output.

Adding the power load of the city's transportation companies built the load of Chicago's central station system to the point where it finally realized significant economies of scale. Although the trolley and elevated companies continued to supply some of their own fast-growing needs for power, an even larger proportion of this increment was met by Insull's electric companies (see table 12). Supplying only 1.6 percent of the transportation power in 1902, within five years Insull raised this figure to almost 30 percent. The guarantee agreements with the transit companies attracted the investment capital to underwrite the utility's expansion program; their energy bills paid for most of it. With the lifting of the utility's heavy burden of finance costs, Insull was free to offer much lower rates to off-peak customers. "You are starting to get yourself in a position," he explained to fellow utility operators in 1909, "where you can afford to run your entire small-customer business at a loss, and gentlemen, you have got to do that eventually, if you expect to remain in business." By then the transit contracts had provided the funds not only to retrofit the Fisk Street

Table 12   Electrical Output of Transit Companies and Edison Central Stations in Chicago, 1902–7 (in Millions of Kilowatt-Hours)

	1902	1907
Transit companies		
Chicago Union Traction	35.3	80.5[a]
Chicago City Railway	30.3	113.5[a]
Metropolitan West Side Elevated	25.4	19.1[a]
South Side Elevated Railroad	21.5	45.8
Chicago Consolidated Traction	20.8	45.8
Ten smaller companies	27.8	47.3
Total transit company output	161.1	352.1
Edison central stations		
Transit power	2.6	148.8
Light and other power	43.1	116.2
Total Edison output	45.7	265.0
Grand total output	206.8	617.1
Edison share of transit power	1.6%	29.7%
Transit power share of total Edison output	5.7%	56.2%
Edison share of total output	21.8%	34.6%

Sources: U.S. Census, Special Report, Street Railways, 1902, 332–33, 1907, 460–61; "Kilowatt Hours Sold per Capita," manuscript, Statistical Department, Commonwealth Edison Company, 1927, in CEC-LF, F23-E3.
[a]Power also purchased.

Station with a second generation of turbogenerators, but also to build a second modern power plant. It was directly across the Chicago River at Quarry Street and contained even larger and more efficient (14,000-kw) units than its companion.[46]

The energy demands of urban transportation also provided Insull with the first significant solution to the problem of spreading the load over the entire day (see chart 7). Early-morning commuters and midday shoppers riding the transit system helped counterbalance the peak demand for light, power, and traction during the evening rush hour. Although the point of maximum demand climbed incessantly from 1898 to 1908, new demands for power during the daytime improved the utility's overall load factor from 30 to 40 percent.[47] This meant that on average the central station's equipment was in use one-third more of the time. Again the result was a steady decline in the cost of electricity, which permitted additional flexibility in setting rates.

In the decade following Insull's gospel of consumption speech

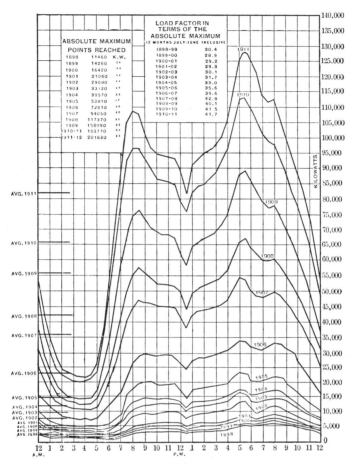

CHART 7. Central station load curves, 1898–1912. *Source:* Keily, *Central-Station Service*, 325.

to the NELA in 1898, the Chicago executive offered what appeared to be irrefutable proof of its validity. The importance of his theories cannot be overstressed, because they established a solid economic foundation for the growth of central station service in the city. The equation he drew between falling rates and rising consumption gave him key insights on how to put technology to best use to reach a goal: cheap electricity. Pursuing his vision with relentless determination, he was able to take advantage of the revolutionary pace of technical progress by following engineering trends in the United States and Europe and working closely with the equipment manufacturers. It was during this period that he became highly regarded as a system builder and financial wizard.[48] The remarkable success of his electric companies also put Insull in a strong position to weather the storms of municipal reform that were sweeping over Chicago and most other American cities at the turn of the century.

Building a Political Consensus

During the first decade of the twentieth century, the battle for municipal reform in Chicago reached a showdown. The traction baron Charles Yerkes left the legacy of an embittered and aroused community that demanded rapid improvement in service and an end to major corruption at city hall. From 1903 to 1907, public frustration reached the point where a majority of voters favored a public takeover of the city's transportation system. The clamor for municipal ownership rose in a deafening crescendo as the politicians fought among themselves for leadership of this popular crusade. With their franchises expiring in 1903, the transit companies were in deep trouble politically and financially. Investors shunned these high-risk ventures. The gas monopoly's dubious record of corporate arrogance and rate gouging also contributed to an atmosphere of hostility toward all the city's privately owned utilities. Threatened by demands for municipal ownership, the utility companies could not expect to escape a tightening net of regulation. Even conservative reformers believed the city government should set service standards and maximum rates for gas and electric lighting.[49]

In many respects the path of reform in the metropolis of the Midwest was typical of the progressive movement in American cities. But Chicago's acute ideological and ethnic conflicts gave its politics an intensity unknown elsewhere. On one hand, a tradition of violent labor clashes had branded the city's immigrants and working classes with a reputation for radicalism and anarchy. On the other, a strident posture of moral righteousness among midwestern Protestants had turned several policy questions into gut-wrenching tests of ethnic self-respect and religious fidelity. Among the issues most passionately fought over were the saloon, the regulation of the liquor trade, and the teaching of foreign languages in the schools.

In the early years of the twentieth century, a reform mandate to write a new municipal charter had the unintended effect of compounding all the various ideological, economic, and ethnocultural conflicts into an explosive mixture. Dominated by a business elite, the charter reform movement raised basic questions about the meaning of good government and about who should rule. As a special charter convention began its deliberations in November 1906, articulate Chicagoans and their special-interest groups split into two antagonistic camps at war over the future of the city.[50]

The polarizing of politics made the city council's formation of public policy on urban utility services difficult and likely to result in uncertainty or deadlock. On transportation policy, for example, the goals of the democratic idealists who advocated municipal ownership were so fundamentally different from those of the business pragmatists who favored regulation that a dialogue between the two sides was impossible. On the one side, a clear majority of Chicagoans wanted a public takeover of the transit system as the most immediate solution to their personal needs. On the other side, their political leaders were working toward private ownership under municipal regulation. Close public scrutiny and elections every other year forestalled any resolution of the issue. Instead, an ambiguous collection of reports, resolutions, and enabling acts kept piling up until Mayor Edward Dunne was elected in 1905 on a municipal ownership platform. During his two-year administration, the city council had an opportunity to make several momentous policy decisions. The new administration was asked to settle the municipal ownership of the transit system, how far the government should regulate its franchise holders, the definition of utility rate structures, and the maximum rates for gas and electric lighting. The outcome of each of these policy issues would have a major effect on the quality of daily life in the city. These decisions were also of vital concern to Insull and his electric companies.[51]

Assuming the posture of a progressive reformer, Insull was able to steer a smooth course through the troubled waters of Chicago politics, bringing him greater legal and financial strength than ever before. The contrast between his liberal approach toward government-business relations and the robber-baron style of Yerkes or the Gas Trust could not have been more complete. At the 1898 meeting of the NELA, for example, Insull bravely called for public regulation and profit limitations in front of his fellow utility managers. Two years later he joined Chicagoan Ralph M. Easley in forming the National Civic Federation to help channel the currents of reform between the shoals of radical socialism and laissez-faire capitalism.[52] At home he never seemed to tire of explaining his gospel of consumption and principles of rate making to the city council, elite men's clubs, and special-interest groups. To most middle-class Chicagoans, he appeared to be a responsible and successful businessman, a corporate officer who was a pragmatic spokesman of "reasonable" reform.

In the case of urban transportation policy, for example, he took the middle ground of acting as a neutral expert in search of the most efficient and economical supply of electrical energy for the city's trolleys and els. After 1902, of course, power contracts with the traction companies became an indispensable part of his master plan to achieve a "massing of production." But municipal ownership would probably stymie this crucial goal, especially since the city and the Metropolitan Sanitary District were already engaged in generating electricity. By keeping a low profile in this contest, he did not look like a self-serving lobbyist. Yet Insull ended up with contracts to supply almost all of the transit system's needs for more power.

At the same time the utility operator began delivering the first central station power to the Lake Street El in 1903, the reformers were laying the legal foundations in the state capital for a municipal decision on transportation policy. The enabling legislation, known as the Mueller law, was a typical by-product of Chicago's divisive politics. On the one hand, it permitted the city to take over the transit companies; but on the other, it made such a policy nearly impossible to implement. What was the real purpose of this Janus-faced act? "It was," according to historian Paul Barrett, "a tool to be used in bargaining with the companies and an attempt to control public ownership sentiment which . . . [had largely] arisen from frustration with the city's inability to get better service from the companies." Despite repeated efforts over the next two years, however, the reform faction of business pragmatists who sponsored the act failed to get the transit companies either to improve service or to accept regulation. With the election of Dunne on a wave of municipal ownership sentiment, the companies were finally forced to work out an agreement with the pragmatists on the city council.[53]

Insull was able to take advantage of his utility's strong political and economic position to influence the outcome of the transit settlement in ways highly favorable to his interests. Although his promise of cheap electricity was important, a much more powerful form of leverage was his offer to finance the construction of the power facilities for the transit system. Every proposed agreement called for vast expenditures to pay for a rapid conversion of the trolley cars to electric traction, a complete rehabilitation of the lines, and more or less ambitious expansion programs. The cost of bringing Chicago's obsolete system up to modern standards would be staggering. Conservative estimates put the figure at $70 million, including $20 million earmarked for electric generating plants.[54] For the transit companies, therefore, the need for a much larger supply of energy represented one more financial burden to be added to an already heavy load of debt. Without franchises, their futures were thrown into a legal limbo that made the cost of borrowing money prohibitive.

For Insull, on the contrary, contracts to supply electricity to the transit system represented an asset that gave easy access to the money markets. The fifty-year franchise of the Commonwealth Electric Company and a virtual monopoly of central station service in Chicago also contributed to a solid standing in the financial world. For the next decade, transit contracts would largely underwrite the rapid expansion of Insull's electric companies. As we have seen, the growth of demand for light and power after 1898 had justified investing huge sums of capital to construct the Fisk Street Station. After 1907, however, the transit contracts, with their provisions for a guaranteed amount of power consumption, ensured that Insull could borrow money at reasonable interest rates. In effect, middle-class commuters financed the growth of central station service in the metropolitan area.[55]

During the Dunne administration, the utility executive quietly negotiated to sign up the biggest transit companies. In many respects, they had few other options for meeting the skyrocketing demands of their trolley and el lines for electricity. After the final settlement of 1907, he used public platforms to argue that the contracts promoted the public interest by augmenting the modernization program and by helping to reduce the electric rates of the entire community.[56] In this case Insull presented himself as a progressive reformer who was simply striving to enhance the quality of urban life. He promised to improve the delivery of essential services to the public. In the case of government regulation of utility rates for gas and electric lighting, he attempted to project a similar image, that of a responsible civic leader.

In 1905 the Dunne administration was presented with the city's first opportunity to formulate a public policy of utility rate structures and to set their maximum levels. Like the Mueller law, an enabling act of the legislature gave the cities of Illinois new authority to regulate their franchise holders, including gas, electric, and telephone companies. The reform measure was a response to the continuing battle between Chicago and the gas monopoly. In 1900 the council had passed an ordinance to reduce the price of gas from $1 to 75¢ per thousand cubic feet (MCF). The company had shifted the controversy to the courts, which had enjoined the enforcement of the ordinance. The gas company contended that the city had no power under the general Municipal Corporation Act of 1872 to regulate utility rates. The remedy was the Reform Act of May 1905, which Chicago voters approved in a referendum during the November elections.[57]

The process of setting the gas rates warrants examination because it epitomized the self-defeating nature of a politics of confrontation in government-business relations. In the years ahead, the complete contrast between this approach and Insull's would bring major benefits to the electric company. Driven into insolvency by its obstinacy, the gas utility would have fewer and fewer resources to use in competing against the electric company for a share of the energy market.[58] For Chicagoans, the question pressing for immediate settlement by the aldermen was the price of energy. Yet more important for the future of gas service in the city was the closely related issue of the standards for measuring the quality of the fuel. At the time of the council investigation, the franchises of the gas company held it to a standard of 22 candlepower (cp). This type of measurement made sense before the electrical revolution of the 1880s, when gas was used almost exclusively for illumination. But this standard became increasingly obsolete and costly to both the company and consumers with the inexorable retreat of gas from the lighting market, the advent of the incandescent gas mantle, and the growth in demand for the energy fuel for cooking and heating.

In creating a regulatory framework for gas services, the council adhered to the tenets of urban progressivism by relying heavily on the opinions of experts. Shortly after his election in November 1905, Mayor Dunne submitted a bill to the aldermen proposing a maximum rate of 75¢ MCF of 22-cp gas in order to initiate the process of policy formation. The council's Committee on Gas, Oil and Electricity (CGOE) hired three experts to study the Chicago gas company and report their recommendations for setting a "reasonable" rate. The balanced panel of experts consisted of a past president of the gas association, E. G. Cowderly, on one side, a noted "antiutility" reformer, Edward W. Bemis, on the other side, and the president of Stevens Institute of Technology, Alexander C. Humphreys, in the middle.[59]

Although the experts disagreed about rates, they spoke with one voice in recommending a change in the standard from the old lighting qualities to a more useful measure of heating values. As one might expect, their rate proposals varied significantly, from Bemis's low figure of 84¢ MCF to Cowderly's high of 99¢, with Humphreys's estimate of 95¢ falling in between. Yet there was no discord over the need to revise the standard to reflect the transformation of gas from an illuminate to an energy fuel. "I think it is a mistake," Bemis told the council, "to require 24 c.p. gas. I think an 18 c.p. gas would be a much wiser light, providing it would give a corresponding reduction in price. It is just as good for heating and the proportion of gas used in the open burner is growing less, continuously all over the world."[60] The CGOE hesitated before making such a major concession to the gas company, given its record and its current efforts to hamstring public policy in the courts. The aldermen decided to proceed cautiously by first opening negotiations on a rate limit of 85¢ MCF for five years.

But the stubborn refusal of the gas monopoly to bargain in good faith doomed the chance of further reform, including a change in the standards. The utility was willing to accept a rate of 88¢ MCF if the aldermen agreed to a series of concessions. The aldermen accepted several of these but balked at the demand that the city drop all pending litigation against the company. In effect, the utility was asking the city to leave unsettled the very question of the constitutional foundation of the city's authority to set rates for its franchise holders. The CGOE was not fooled by this plan to overturn its policies in a future court challenge, and it reasserted city hall's determination to continue the litigation to a final ruling. In February 1906 the full council gave overwhelming approval, in a vote of fifty eight to nine, to the CGOE's proposed rate ordinance of 85¢ MCF of 22-cp gas. Mayor Dunne vetoed the measure because he believed the rate level was "unfair" and "excessive" compared with the earlier 75¢ MCF rate ordinance. Nevertheless, the council easily overrode the veto in hopes that the courts would ultimately uphold both the city's authority and the "reasonableness' of the higher rates.[61]

Over the coming years the gas monopoly would pay a mounting price for its intransigence. Its confrontational stance foreclosed the possibility of reforming the standards from light qualities to heat values. A change to a heat standard (BTUs) would have produced an immediate benefit by cutting the cost of the energy fuel. Reducing its light-bearing qualities would eliminate the need to add petroleum products to the coal gas and would open the way not only to more efficient methods of manufacture but also to the use of cheap natural gas. The new standard would also encourage the conversion of gaslight fixtures in working-class homes from crude open-flame burners to soft, glowing incandescent mantles. Looking to the future, the reform would protect consumers of all classes as they used gas less for lighting and more for cooking and heating. Instead of a settlement, however, the gas monopoly continued to battle the city in the courts, while the rising cost of petroleum eroded its profit margin to the point of economic ruin. When World War I created a national fuel shortage, Samuel Insull would have to be called in to rehabilitate the ailing utility.[62]

Insull extracted far more concessions from the city council for his electric companies by couching his objectives in the conciliatory tones of cooperation. Equally important, perhaps, the utility executive realized the value of public relations in creating a favorable political climate. To be sure, a fifty-year franchise and the recent attainment of economies of scale in generating electricity gave him strong leverage in demanding concessions from the council. Yet Insull's posture of being a public servant and his ready acceptance of city hall's rate-making authority helped shield his interests from an aroused community. Consumer-oriented reformers directed public attention to other causes, such as telephone and water rates, charter reform, and transit modernization.[63]

Most important, perhaps, Insull made well-publicized announcements of rate relief at the beginning and end of the council's investigation of his utilities. After the passage of the enabling legislation in May 1905, for example, he ordered a rate cut.[64] It is hard to imagine a better way to court public favor while sapping the strength of any political opposition. In fact the aldermen admitted that their constituents were voicing few complaints about Edison central station service. Nonetheless, Insull appeared to bend over backward to accommodate the aldermen and the reformers who found fault with the proposed settlement.

In the formation of a public policy for electrical service, Insull and his lieutenants dominated the debate in the council about the structure and maximum levels of rates. In spite of the aldermen's reliance on professional consultants, the company's highly talented lawyers and engineers overawed these experts with carefully prepared legal briefs and technical reports. By conceding the city's authority to regulate the price of utility services, the company hoped to gain official recognition of its differential rate structure. However, the company's practice of discriminating among its customers by the quantity of energy consumed and the type of use raised troubling questions of constitutional law and public policy. If the council put a stamp of legitimacy on a differential rate structure, then the levels set for each class of consumer would have major effects not only on the company's fortunes, but on the energy-use patterns of the city as well.

The real question for the aldermen to decide was who ultimately paid for Chicago's electrical energy among the community's various groups of householders, shopkeepers, manufacturers, and commuters. The Dunne administration did not shirk its duty in conducting a full and open investigation of these weighty issues. The council appointed the city electrician and its top traction adviser, Bion J. Arnold, to study the utility's operations. Arnold and other engineers helped the CGOE calculate a fair level of rates. To review the law, the aldermen turned to the corporation counsel. They also heard testimony from a long parade of reformers, including legal experts, public officials, and articulate spokesmen of civic watchdog groups such as the City Club and the Hamilton Club.

Behind the facade of this public forum, however, Insull was largely able to set the agenda of the debate and to dictate its outcome. In this case, the decision-making process exposed many of the inherent difficulties facing the government in its attempt to regulate utility enterprises in the public interest. The consulting engineers were the first to admit that they were ill equipped to act as accountants. In this highly technical field of utility finance, only a specially trained economist could arrive at an independent judgment of the maximum rates the city ought to impose on the company. After a three-month study, the engineers confessed they had to accept Insull's figures and statistics because of the sheer complexity of such an intricate technological system. To make a true accounting of the company, they reported, would require a major commitment of time, money, and personnel beyond the resources of the city government. The utility's top lawyer and lobbyist, William Beale, reinforced this imagery of insurmountable obstacles in the path of an effective regulatory apparatus. "It is obvious," he explained to the council, "that for you to make an inquiry under which you could intelligently make an extended and a reasonable classification of the basis of supplying electricity, and could fix proper rates according to such classification, would require you to sit for months and to engage in an enormous amount of work." Confronted by such a prospect, the council agreed to Beale's suggestion that it limit the scope of the city's rate-making powers to the rates paid by the ordinary incandescent lighting customer. By confining public policy in this manner, the council essentially gave Insull a free hand to apportion the cost of electrical services among the various groups constituting Chicago's community of energy consumers.[65]

The council also allowed the utility operator to set the terms of debate on what constituted "fair and reasonable rates" for lighting. The corporation counsel, James Hamilton Lewis, could not refute Beale's contention that the law permitted utility enterprises to discriminate between customers based on consumption. Some courts had even upheld the notion that competition in some markets and not in others justified rate differentials. The council, Lewis reported, could set a single rate or a "schedule of maximum rates, providing the kinds of service or classes of customers be divided according to some real differences in condition and surrounding." Although the Wright demand meter treated all customers equally, the aldermen were under no obligation to accept the company's two-tier system of rates that favored the big consumer over the small. As the experts pointed out, retail merchants did not charge big spenders less for an item than customers with more modest needs. Even accepting a two-tier system, the council still retained a broad range of policy choices in setting different ratios between the primary and the secondary rates (see table 13).[66]

To persuade the council to adopt a rate schedule favorable to the utility's interests, Insull proposed to make some enticing concessions to the city. This maneuver helped him maintain the initiative by creating a new situation in which both parties would have the right to a fair bargain. His offer required the council to transform public policy from a general ordinance into a specific contract. Insull held out two pecuniary incentives that were guaranteed to win the support of most politicians and business-oriented, pragmatic reformers. First he agreed to reduce the city's bill for each streetlight from $103 to $75 per year if a deal could be consummated.

A second incentive promised an even richer reward, but Insull shrewdly linked this prize to the city's making an important concession. Since 1898 Insull had been operating under the separate franchises of the Chicago Edison Company and the Commonwealth Electric Company. Although the two enterprises func-

Table 13   Alternative Rate Schedules Proposed for Electric Light Service, 1906

Rates (¢/kwh)		Percentage
Primary Rate	Secondary Rate	of Current Income
20	10	100.0
16	10	92.8
14	10	89.2
14	9	85.8
14	8	82.3
12	9	
10	10	
14	7	78.8
12	8	
12	6	71.6

Source: Chicago, Proceedings of the City Council, 1906, 3230.

tioned as one, they were distinct entities in law, which complicated the utility's task of raising money in the bond markets. Insull wanted to consolidate the two firms under the longer fifty-year grant of the Commonwealth company. On the one hand, it required the utility to pay 3 percent of its gross receipts from the outlying districts to city coffers. On the other hand, the franchise of the Edison company shielded its income from the CBD from this kind of tax until 1911, when the grant expired. Insull offered to end this exemption and to pay an additional amount, estimated to be $250,000 a year for the next four years, or $1 million, if the council gave official sanction to the merger. Here was a "plain business proposition" with little political risk; few politicians could afford to turn it down.[67]

For almost two years, however, a handful of critics stymied a final resolution of the controversy. The Chicago Record Herald summed up their complaints. "For the city," an editorial argued, "[the proposed settlement] means a leap in the dark, followed by a general confession of ignorance by all concerned except the company."[68] In May 1906 the CGOE presented an ordinance-contract that represented the fruits of its negotiations with the utility company. It accepted a two-tier system of rates for incandescent lighting that set out a five-year schedule of sliding, primary, and secondary charges. The following month the full council passed the proposal by a vote of fifty to sixteen. The matter seemed closed, but in a surprise move Mayor Dunne vetoed the legislation. The former judge listed several weaknesses in the provisions detailing the city's authority to regulate the firm, and he castigated the rate schedule as "unreasonably and unjustly high." The mayor was able to muster enough support to thwart a bid to override the veto, killing the bill for the rest of his administration. In November 1907, however, Dunne was himself defeated at the polls by the Republican Fred Busse, who promised a more pragmatic approach to government-business relations.[69]

In March 1908 the new administration resubmitted the settlement plan over the objections of a small but committed opposition. Led by Alderman William Dever and the *Chicago Record Herald*, the critics of the ordinance-contract uncovered the startling fact that Edison attorney Beale had been its primary author. "Is there anything wrong in that?" Beale retorted at one of several public debates between company officials and reformers. George H. Hooker, president of the City Club, and other opponents admitted there was no evidence of the old Gray Wolves type of corruption, but they were bothered by the utility's dominant influence in decision making. As the newspaper protested, "When we see the aldermen enthusiastically making a contract with the Edison company highly favorable to it, in total neglect of the powers the legislature has given them to control electric light rates on their own initiative, we may not brand them morally for what they are doing, but we must at least accuse them of hide-bound conservatism. . . . They are blind to the scope of public interests that are placed in their care." To these critics Insull replied that the city was receiving a fair compensation for the contract, "one million dollars in four years."[70]

Even some of Insull's fellow business pragmatists were troubled by the inherent difficulties of developing sources of information independent of the company itself. For example, Robert R. McCormick, businessman and president of the county's drainage board, worried about the ability of elected officials to cope with the intricacies of utility economics. To him, rate making was largely a question of "bookkeeping," beyond the grasp of all but a handful of experts. However, the precedent of the Board of Supervising Engineers for municipal regulation of the transit companies did not inspire McCormick or other reformers to propose an independent administrative mechanism.

Instead, reformers restricted their debate to improving public access to the company's records. The City Club delegated one of its directors and legal advisers, Walter Fisher, to work with the CGOE to strengthen these provisions of the proposed ordinance.

The lawyer Fisher, like the engineer Arnold, promoted "reasonable" reforms that strove to improve the workings of corporate capitalism. With Fisher's revisions incorporated in the bill, the CGOE unanimously recommended its passage. After a long debate lasting past midnight on 23 March 1908, the aldermen approved the measure fifty to sixteen.[71] The two-year delay caused by the opposition had the salutary effect of improving city hall's ability to protect the public. The reform critics gained the right to inspect the company's books annually, and they created a mechanism to arbitrate disputes between the company and the city over the selection of examiners.

The reformers also extended the rate powers of the municipal government by inserting a schedule of maximum charges for retail power consumption. The inclusion of power rates went a long way toward alleviating fears that the company was engaging in "criminal practice[s]" in fighting the small-scale block plant with "freeze-out" and "clubbing" methods of cutthroat competition.[72] Moreover, the revised settlement gave the city authority to order the utility to extend its lines into any part of the city. This expansion of the city's regulatory powers met the needs of the outlying districts, where demands for modern services were being voiced loudly by merchant groups such as the South Chicago Business Men's Association. The Fisher amendments and other minor changes allowed the critics of the bill to declare the final product a "decided improvement" over the bill vetoed by Mayor Dunne.[73]

Yet these gains were minor compared with those Insull won from the city government. The opposition helped focus attention on the key issue of utility rate structures, but the council chose to leave this crucial area of public policy in the hands of private decision makers. To be sure, Insull's demand system of rates had a certain economic logic and was not intended to benefit friends and punish enemies. At the same time, however, the aim of his discriminatory rate schedules was to enhance the company's profits, not the community's welfare. The ordinance-contract of 1908 brought Insull additional financial benefits because it established great stability in government-business relations. City hall put its imprimatur on the utility's franchises, consolidation plans, and rate schedules. In the money markets, Insull's solid legal position could be translated into lower interest charges, the largest expense of his fast-growing enterprise.

Utility rate policy in Chicago took shape in a highly charged atmosphere of political polarization. Between 1898 and 1905, the resulting stalemate on transit policy and charter reform as well as utility rates created mounting pressure on elected officials to resolve the conflict. During the Dunne administration and the charter convention from 1905 to 1907, the heated debate over the future of the city reached a climax. Older patterns of political loyalty broke down, spawning a major realignment that would gradually shift power to the ethnic leadership of Anton Cermak and the Democratic party. In the short run, however, the defeat of the reform charter and Mayor Dunne at the polls reflected an urban polity exhausted by a generation of reform agitation. Mayor Busse and the city council moved quickly to settle questions of urban utility service by adopting the progressive compromise of regulated capitalism. The battle for municipal reform was over.

The outcome of utility policy formation in Chicago put the city in the mainstream of government-business relations across the urban landscape. The city steered between what radicals called domination by robber barons and what conservatives called socialism. Elected officials created a range of regulatory mechanisms to protect the public interest while enhancing the ability of private enterprise to improve the quality of essential services to the community. In the most acrimonious case, that of the gas monopoly, the dispute wound its way up the federal judiciary for settlement by the Supreme Court. In the most favorable case, that of the electric company, city hall largely deferred to Insull and his talented assistants, who had superior access to information about a highly complex technology. In a revealing admission three years after the 1908 rate investigation, Gilchrist confirmed that "much of the consideration of rates by regulating bodies has been very superficial. . . . where such commissions have gone deeper into the subject they have, in most cases, deferred action because of the very complicated and diversified methods which they have found, and they consider it advisable to limit their decisions to the maximum rates, the schedules which affected the great majority of the customers served."[74] In the most pressing case—the transit system—the creation of a municipal commission under the direction of experts drew upon the states' experience with the railroad. Chicago's Board of Supervising Engineers also set a precedent for bringing other urban utilities under independent regulatory scrutiny.

Samuel Insull, however, had no intention of letting the public policy initiative slip into the hands of government bureaucrats. As a member of the National Civic Federation, he sponsored a highly influential study of urban public utilities in the United States and

Europe. In 1907 this blue-ribbon panel of experts—including Insull—released a massive report that recommended the creation of state-level regulatory commissions. During that year, model legislation drafted by panel members was enacted in Wisconsin and New York. Within a decade, Illinois and over thirty other states would adopt similar measures.[75]

In 1908, Insull's achievement of political stability combined with the recent attainment of true economies of scale in central station service to eliminate virtually all the remaining constraints on the supply of electrical energy. A decade of extraordinary success in finance, marketing, technology, and politics left little doubt about the validity of his gospel of consumption. Moreover, the efficient delivery of services at a reasonable price, supplemented by public relations campaigns, was building a popular consensus in favor of the performance of the Commonwealth Edison Company. Insull could feel gratified about the effects of his marketing strategy. "Daily newspaper advertising," he crowed, "has had the result of so increasing the demand for our product that in the downtown district and the thickly settled residential districts our business is obtained from people who first write us to call on them."[76]

By 1909 Insull's gospel of consumption was becoming institutionalized. At base, the electrical revolution helped transform the art of invention into the business of research and development. After 1903, for example, GE, Westinghouse, and other major manufacturers turned the technical progress seen in the invention of the turbogenerator into a steady process of improvements in efficiency and size. Engineering firms such as Sargent and Lundy also played an important role in spreading innovation in the generation and distribution of electricity from one utility to the next. Close communications and financial ties among the national giants and the local operating companies ensured a rapid diffusion of technological improvement from industry leaders to central station managers across the country.[77]

Insull played a parallel role in institutionalizing the gospel of consumption in the relationship between the utility company and the consumer. For a decade, he had been collecting statistics from the demand meter on group patterns of consumption. With the help of his brilliant researcher, Fowler, he turned the business of central station service into a science. The introduction of IBM computing machines that used Holrath cards by the chief statistician at a 1909 meeting of the industry epitomized this process. At a wide variety of association meetings, moreover, Commonwealth

Edison managers and engineers employed Fowler's studies to report on every aspect of the generation, distribution, and consumption of electricity. In the same year, the publication of an *Electrical Solicitor's Handbook* by the NELA was equally emblematic of the industry's new maturity.[78] The early emphasis on overcoming technical problems no longer seemed important. Equating technological obsolescence with progress, Insull made just this point to fellow operators. "The very best monument that any of you can erect," he bluntly declared, "is a first-class junk pile." In coming years, he urged, they should turn their attention instead to the "engineering of selling."[79]

6

The Electric City, 1902–1912

On a late afternoon in November 1903, a man boarded the Chicago, Burlington, and Quincy Railroad for a trip from the cornfields of western Illinois to the big city of Chicago. He was going to join his sister's family for the Thanksgiving holiday. As the train pulled out of the station, he was filled with excitement and anticipation. A decade had passed since his last visit to the city during the World's Fair, and he was looking forward to a day in the Loop before going to his sister's home in the wealthy suburb of Highland Park.

As the locomotive sped past harvest scenes on the prairie, the twilight faded to darkness. Although faint rays from kerosene lamps inside farmhouses occasionally penetrated the night, rural life remained locked in step with the daily rhythms of the sun, as it had been since time immemorial. Only as the train approached the metropolitan corridor did the man see the night gradually turned into day by artificial illumination. As it passed through the satellite city of Aurora, the dim glow of gaslights from homes along the tracks first became visible, then the brilliant light of arc lamps perched on high towers came into view down the middle of Main Street. The speeding train soon reentered a sea of darkness, but other islands of light became more frequent as the man rode through the suburbs of Hinsdale and Riverside and finally the outskirts of the great metropolis. He could now see the city's lights radiating from the downtown area, where the train soon delivered him.

The man was struck by the tremendous changes wrought during his ten-year absence. Traveling by cab to his hotel, he immediately noticed that the buildings were brilliantly illuminated with ornamental streetlights and electric signs of every description. Attractive window displays beckoned, though the stores had long since closed for the day. Yet the Loop was alive; the streets were teeming with people moving among the hotels, theaters, and restaurants, as well as the less refined dance halls and "gyp joints." Though drawn by the lure of nightlife in the big city, the man was tired from his journey and went straight to his room at the hotel in the Auditorium Building. He had decided to return to this grand guesthouse because its architectural beauty and its dazzling use of electric lights recalled the marvels of the Columbian Exposition[1].

The next morning the man set out to explore the Loop. Walking north on Michigan Avenue toward the heart of the central business district, he was soon struck by a strange scene at a construction site on Jackson Street. Workers were erecting yet another sky-scraper, the Railway Exchange, but gone were the "cumbersome, roaring, smoking donkey-engine [hoists], so familiar to everyone with their obstructing coal piles and ash heaps." The steam engines had been replaced by quiet electric motors that were also running cement mixers on the site.[2] The man turned west toward State Street, the main commercial thoroughfare. Only a block away, at Wabash Avenue, more unfamiliar sights and sounds greeted him. Rapid transit cars rumbled overhead on Charles T. Yerkes's elevated structure, giving new meaning to the term the "Loop." Underneath, electric trolleys now vied for space with cable cars and horse-drawn vehicles, while pedestrians used the protection of the el pillars to dodge across the rushing flow of traffic.[3]

On State Street, he observed a flurry of activity three blocks farther north at the corner of Madison Street. Moving closer, he was swept into a crowd of shoppers who were trying to get into the new Schlesinger and Mayer department store (the predecessor of Carson Pirie Scott). Designed by Louis Sullivan, the architect of the Auditorium Building, the twelve-story emporium invited shoppers inside through a grand rotunda elaborately decorated with ornamental ironwork. Passing through this portal, the man was struck by the quantity of goods on display in illuminated glass cases and by the cheery, festive atmosphere. Everything seemed bright and shiny in the glow of 23,000 incandescent lights. As he looked around this cornucopia, he was reminded of the countless displays of products at the World's Fair.[4] He could hardly restrain the impulse to buy something, but first he had to visit the other new palace of commerce, the Marshall Field's store a few blocks away.

Here the exposition's master planner, Daniel Burnham, had exceeded even Sullivan in the extravagant use of electric light and power. Having abandoned its self-contained system for Edison central station service in 1902, Field's had installed over 40,000 incandescent lights, in addition to elevators, fans, pumps, and other motors with an equivalent of about 1,000 HP. Together the two department stores set a standard of urban modernity, style, and affluence for other retailers to emulate.[5] After shopping for gifts, the man realized it was time to meet his brother-in-law for lunch under the Field's clock at the corner of State and Washington.

The two men exchanged greetings and decided to forgo the usual meal at the Union League Club in favor of a new Chinese restaurant on Clark Street near Adams. Opened just a month earlier, the Shanghai was a bustling eatery with accommodations for two hundred. Although seated in a high-backed booth near the rear, the men noticed how well the room was illuminated with electric bulbs that had been subtly worked into the Oriental motif. The Shanghai's 310 incandescent lights helped ensure that the diners would be impressed by its $50,000 worth of interior decorations.[6] The food was as good as the decor was beautiful.

While his brother-in-law returned to his law office, the man made his way to the Olympic Theater's vaudeville matinee at Randolph and Clark. Combining elements of the rural minstrel show and the legitimate stage, vaudeville represented a new form of popular entertainment for the middle class. The matinee at the Olympic offered not only traditional song-and-dance routines, but also a short movie. The man had heard about Thomas Edison's first kinetoscope motion pictures at the World's Fair, but the electric movie projector was relatively recent. Introduced in Chicago in 1896, the movie achieved such immediate success that only ten years later almost 250 movie projectors and nickelodeons were operating.[7] The man enjoyed the show and was sorry when the parade of acts finally came to an end. He returned to the hotel for his luggage and took a cab to meet his brother-in-law at the train station for the commute to the suburbs.

Like so many other aspects of his day in Chicago, the trip to Highland Park also proved a wonderfully new and exciting experience. The first part of the commute was ordinary enough on the Chicago, Milwaukee, and Saint Paul's steam-powered line to Evanston, the first suburb north of the city limits. Instead of continuing on a commuter railroad, however, the man's companion insisted they get off at the Davis Street station for a ride on the electric train. Opened in 1899, the North Shore line extended twenty-eight miles to the satellite city of Waukegan, making the interurban railway the longest in the metropolitan area. The cars were much smaller than those of the steam-powered train, but the interurban was quicker and quieter and did not belch clouds of black smoke. The man stared out the window at the houses flying past and marveled at how the future had become everyday reality within a decade.

By the time the two men arrived at their destination, it was growing dark. Usually the man's brother-in-law walked home, but tonight a carriage driven by a servant was waiting at the station. As they rode along Central Avenue toward the lawyer's lakefront home, the arc lamps began to sputter and glow. Electricity had come to Highland Park in 1890, but service was still restricted almost entirely to street lighting in the small business section. The town's primitive generating station operated only between 4:15 P.M. and 1:00 A.M. Pointing to the arc lamps, the suburbanite remarked that Samuel Insull had bought the utility the year before. A new central station was nearing completion, and Insull had promised that his North Shore Electric Company would make major improvements and extensions, including twenty-four hour service.[8]

The two men began to discuss the pros and cons of wiring the house as they pulled into the driveway. But it was not until much later in the privacy of his bedroom that the visitor had a chance to speculate about the changes electricity would bring to domestic life. His house and his sister's were similar in many respects. Neither had electric, gas, or water service. Reflecting on the evening's wonderful dinner, he realized that servants made the big difference between the household routines of the farm and of the suburbs. Whereas in a rural setting the burden of daily chores fell on the family, his sister enjoyed the leisure afforded by having a cook, maid, and handyman to do most of the work. But there was no doubt in his mind that urban amenities could improve the quality of life outside the city. This truly was an age of progress.

Although contrived, this story of a visitor from the country underscores a basic fact about city life at the turn of the century: between 1903 and 1912, electrical technology began to affect everyday life in the Chicago metropolitan area. This chapter will examine how a more intense use of energy began to change the city's physical environment and its inhabitants' daily routines. The next chapter will continue the analysis by following the spread of energy-intensive amenities to the suburbs.

The opening decade of the twentieth century was a period of incipiencies—a time when new patterns were established in subdivision planning, domestic architecture, and interior design. Spreading distributon grids and falling utility rates meant that increasing numbers of the middle class could now afford better lighting in their homes. Rapid improvements in the incandescent bulb and household appliances, moreover, made things electrical all the more desirable and attractive to an urban society fascinated with high technology. Although the electrification of domestic life had not gone far by 1912, new models of the ideal home would shape the growth of the city for the next twenty years.

Central station service was still too expensive to install in the vast majority of Chicago homes, but electricity was starting to change the physical and social fabric of the city. In the public spaces of Chicago, a more intensive use of energy was becoming commonplace. Had the visitor ventured to the West Side, for example, he might have visited one of the city's street fairs. In 1903, an estimated 100,000 people attended the one at Western Avenue and Fortieth Street. Festoons with 2,000 lights decorated neighborhood storefronts. A Ferris wheel and an automobile parade helped give a distinctly urban flavor to a traditional ritual. Electricity was also playing an integral part in the emergence of other popular entertainments such as amusement parks, nickelodeons, and penny arcades.[9] Trolley cars, miniature "White Ways" of neighborhood business strips, workplaces, downtown department stores, and a thousand other facets of the urban environment were constant reminders that technology was rapidly transforming daily life in Chicago.

Perhaps the most important influence of electricity during this period was in furthering urban deconcentration and suburbanization. With 1.7 million people in 1900, Chicago had an incredibly high density at its center while farms were still common at the periphery, in newly annexed areas such as Rogers Park, Jefferson Park, and Hyde Park. The urge to escape the crime, disease, and social disorder of the inner city stretched back to antebellum days, but only a small elite of the well-to-do could afford to live in the suburbs until the advent of electric rapid transit and modern utility services. Together, these energy-consuming technologies helped create a sprawling "flat city" of middle- and working-class residential neighborhoods. Although the new technology did not completely replace an earlier reliance on steam railroads and domestic servants, a wider use of energy allowed more and more members of the middle class to enjoy suburban life.[10]

By the turn of the century, the installation of a modern infrastructure had become an "essential" prerequisite to the development of residential subdivisions in these outlying areas. In a seminal book, *Streetcar Suburbs*, historian Sam B. Warner, Jr., shows that "utilities had to be laid before most men would be willing to build [and] adequate streetcar service also preceded homebuilders and their customers."[11] New forms of energy-intensive technology not only promoted the exodus of affluent city dwellers but also became incorporated in the very definition of their suburban ideal. Warner argues convincingly that "the rise of thousands of families of the most diverse ethnic backgrounds to a middle class competence, like their adoption of each new invention was taken as proof of the success of the society. . . . From the prosperity of the middle class and its enthusiastic acceptance of the new sanitation and transportation technology came the popular achievement of the late nineteenth century suburbs."[12]

In the late 1880s primitive electrical services had begun spreading throughout the Chicago region. The great annexation of 1889 brought many of these outlying suburbs within municipal borders and in reach of Insull's central station system. Though many other communities insisted on retaining their individual identities, their independence from the city was being undermined by the same forces of urban growth and deconcentration. The extension of rapid transit lines, improved communications, and the delivery of urban services to residential districts all contributed to blurring distinctions between the neighborhoods at the city's periphery and those beyond its borders.[13]

Electricity Comes to the Neighborhood

In the decade from 1902 to 1912, the widespread adoption of the new technology in the public and semiprivate spaces of the city affected everyday life in direct and obvious ways.[14] To be sure, those who rarely ventured beyond the end of the block in poor neighborhoods lived with few urban amenities of any kind. Dirt roads without drainage or sanitary hookups, gasoline streetlights, piles of garbage, and putrid air were all-too-common features of working-class districts in the city of the early twentieth century.[15] But for blue-collar workers and others with some intracity mobility, encounters with electrical technology were becoming unavoidable. Beyond the lights of the Loop, the modernization and extension of the transit lines and the lighting up of neighborhood business strips brought electricity to many residential areas, including working-class districts. In addition, novel institutions of popular culture were emerging that were based on an intensive use of en-

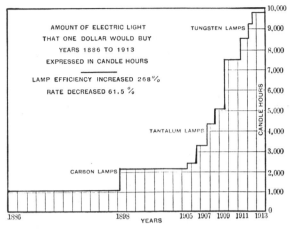

CHART 8.   Increase in lamp efficiency and decrease in rates, 1886–1913. *Source:*
William E. Keily, ed., *Central-Station Electrical Service: Its Commercial Development and
Economic Significance as Set Forth in the Public Addresses (1897–1914) of Samuel Insull*
(Chicago: privately printed, 1915), 431.

ergy. The five-cent movie house and the amusement park were
two of the most successful of these new forms of mass entertainment.

Scholars have carefully studied the transformation of the city by
the extension of the transit lines, but little attention has been given
to the neighborhood business strips they ran along.[16] Local stores
familiarized many people with the advantages of interior electric
lighting. In addition, the commercial sector continued to lead in
introducing new appliances. Success in the commercial field usually resulted in the manufacture of household models for a much
larger mass market. An example of this diffusion was the introduction of a rapid succession of improved incandescent bulbs between
1905 and 1912 that cut the cost of lighting in half (see chart 8).

At first Insull restricted the sale of the more efficient tantalum-
and tungsten-filament bulbs to commercial customers. He used
the special incentive of a fixture with a cluster of four sockets,
which he rented at low cost to encourage shopkeepers to replace
their old bulbs and arc lamps. Although the new bulbs cost more,
they emitted twice as much light as the old ones at the same cost.
The campaign was an instant success, and the number of fixture
rentals skyrocketed from 3,500 in 1908 to 25,000 just four years
later. The brighter bulbs marked the beginning of the end for gas
lighting and the arc lamp in the commercial sector. Similar "clus-

ter" fixtures of tungsten-filament bulbs also began to replace the arc lamp for street lighting. By 1911 the storekeepers' ready acceptance of the new bulbs convinced Insull to begin offering them to residential customers.[17]

Local merchants and businessmen also helped create novel physical and social environments for people in the neighborhoods. More intensive uses of energy redefined nightlife as an urban frontier, a new world of anarchy to explore and conquer. Perhaps the most obvious of the new forms of popular entertainment was the amusement park. In 1903 George Schmidt opened Riverview Park on the North Side, and two years later White City followed on the South Side. Both were modeled after the Midway of the 1893 fair, with its technological wonders and inexpensive thrills in a parklike setting. In 1907, for example, Schmidt installed a roller coaster called the Scenic Railway. "It was never intended to be scenic," cultural historian Lewis Erenberg observes; "it was a parody of Chicago's elevated railroads that gave riders a chance to take a risk under relatively safe conditions and to enjoy a minor catharsis of the noise and confusion of the industrial world. By surviving the ride, one mastered the machine." For the young at heart, the way to take full advantage of the parks was to come at night, when thousands of lights created a fantasy of romance and pleasure. White City, for instance, used so much energy to power its glittering lights, theaters, and rides that Insull had to build a separate substation to serve it.[18]

The spread of the five-cent movie houses along neighborhood business strips brought more pervasive, albeit more subtle, changes to nightlife in the city. In 1909, for instance, Jane Addams observed that "the five-cent theater is also becoming the general social center and club house for the whole family in many crowded neighborhoods. It is easy of access from the street, [and] the entire family of parents and children can attend for a comparatively small sum of money." Paradoxically, she noted that young people were attracted to these "houses of dreams" precisely because they were so dark inside. Beyond "the glamour of love making," the theaters had become a social craze among the city's teenagers. "Hundreds of young people," she complained, "attended these five-cent theaters every evening in the week, including Sunday, and what is seen and heard there becomes the sole topic of conversation, forming the ground pattern of their social life."[19]

Electric signs and streetlights also helped change the social environment of the neighborhoods. In the decade between 1902 and

1912, the number of illuminated signs mushroomed from 850 to 7,250. Although Insull did not supply power to all these advertisements, his aggressive salesmanship to build a nighttime load helped establish Chicago as the sign capital of urban America. Certainly anyone visiting the Loop could hardly miss his huge, 240-by-40-foot illuminated billboard that covered the side of the Harrison Street Station. With 2,000 lights, the sign displayed the name of the electric company and a changing list of prominent new customers.[20]

Whether in the Loop or in the neighborhoods, a greater use of energy was changing the way Chicagoans lived. Although the residential streets and homes of the working classes remained in the semidarkness of gaslight, their business strips were links to the bright city lights of the Loop. Trolleys running in the streets, el cars rushing overhead, nickelodeons, well-illuminated shops with window displays, blinking signs, and ornamental streetlights brought high technology into the daily routines of ordinary people. And for those who could afford it, electricity was becoming an integral part of family life as well.

Electricity Comes to the Urban Fringe

In the years between 1902 and 1912, electric lighting in the home began to undergo a fundamental change from a luxury of the rich to a necessity of middle-class life. At the same time, the new technology retained its cultural significance as a symbol of social status and conspicuous consumption. The easing of the constraints on central station service brought a regular succession of rate cuts in residential service, reducing the average price of a unit of electricity by 40 percent. In response, the number of households supplied with Edison service increased from fewer than 5,000 to almost 80,000, surpassing the number of Edison's commercial customers (see chart 9). This number represented approximately 16 percent of Chicago's families, or one out of every six households. In comparison, historians have found that white-collar employees composed 25–33 percent of the city's work force.[21] By 1912, then, about half of Chicago's middle-class families had installed electricity in their homes. Although central station service was still too expensive for most, new standards of lighting were rapidly becoming established in the homes of the middle class.

The diffusion of electricity in the neighborhoods of affluent Chicagoans followed two general patterns. Fulfilling residential de-

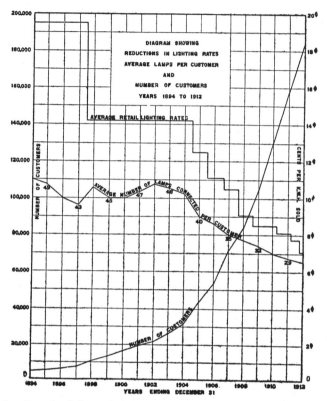

CHART 9.   Growth of electric lighting in Chicago, 1894–1913. *Source:* Common-wealth Edison Company, *Annual Report*, 1912, in CEC-HA, box 311.

mand proceeded differently in the built-up sections of the central city than in the outlying areas of new construction. Obtaining service in the settled districts was largely a matter of personal preference and family budget. By 1902, central station service was available in most middle-class neighborhoods where utility entrepreneurs, retail merchants, and home builders had strung the first overhead lines. Weaving these early power grids together, Insull created an integrated system that covered about one-third of the 170-square-mile territory outside the CBD (see map 3). Over the next five years, Insull would expand the overhead distribution network from 50 to 85 square miles.[22]

At the onset of the rapid spread of electricity to middle-class homes, differences in the availability of service reflected the existing mix of commerce, industry, and housing in the three main divisions of the city outside the Loop. Table 14 compares the distri-

Table 14   Distribution of Electrical Service in the Overhead Wire Territory of
Chicago, 1902

Division	Total Area (sq. miles)	Service Area (sq. miles)	Percentage of Area Served
North[a]	55	15	27.8
West[b]	30	11	36.7
South[c]	86	28	32.6
Total	171	54	31.8

Source: Ernest F. Smith, "The Development, Equipment, and Operation of the Substations and Distributing Systems of the Chicago Edison Co. and Commonwealth
Electric Co.," Journal of the Western Society of Engineers 8(March–April 1903): 214.
[a]From North Avenue to the city limits at Howard Street.
[b]From Ashland Avenue to the city limits between North Avenue and 39th Street.
[c]From 39th Street to the city limits at 135th Street.

bution systems in these three sections. The distribution grid
covered a larger portion of the West Side than of the other areas
because of its heavy concentration of factories and businessses. In
comparison, the South Side had long been the preserve of Chicago's affluent homeowners. Here the power grid stretched over the
largest area, although large tracts of undeveloped land at the periphery reduced the proportion of territory served to less than that
of the West Side. In contrast, the growth of the North Side had
always lagged behind that of the other sections. The Chicago River
acted as a barrier to commuters, although factories and working-
class housing hugged its north branch as far as Fullerton Avenue.
The lakefront remained largely vacant land until the great annexation of 1889, when the trolley car began to ease the flow of movement between the CBD and the North Side. The upper-middle-
class subdivision of Edgewater set the tone for the development of
residential neighborhoods in the city's most desirable lakefront locations.[23]

In these outlying areas the diffusion of electrical services went
hand in hand with the extension of the transit lines and the home
builders who followed in their wake. As one might expect, new
neighborhoods arose first around the transit stations and spread
outward from these points. The growth of the North Side provides
the best illustration of this process in the early twentieth century,
since it had been largely passed by during earlier stages of urban
growth. Land values began to climb rapidly after 1908, when the
Northwestern Elevated Railroad Company completed the final link
between the Loop and the North Shore. The rapid transit line

opened up the lakefront beyond the previous terminus at Wilson Avenue to the city limits at Howard Street.[24]

Home builders took advantage of the opportunity to apply the most up-to-date methods of subdivision planning and construction to the new land. Increasingly, they imitated the model subdivision of Edgewater. As we have seen, developer J. Lewis Cochran made extensive environmental improvements, including water, sewer, and electric connections to each house, paved and lighted streets, sidewalks, and landscaping. During the first decade of the twentieth century, subdividers who catered to the affluent began to institutionalize this type of planning. In 1909, for example, they organized the National Conference on City Planning, and they sponsored the first state law to regulate land use in residential subdivisions. Although these early efforts represented only the beginning of housing reform, they established an ideal of community building that gradually came to dominate the housing industry over the next twenty years.[25]

The trend toward planned residential neighborhoods reflected the demands of middle-class city dwellers. Real estate speculators understood that affluent families were reluctant to move into a new neighborhood unless they could enjoy the urban services and amenities they had grown to depend on in the more built-up sections. Subdividers responded by making infrastructural improvements themselves, which boosted sales and profits by eliminating the slow and uncertain business of obtaining public utilities from city hall.[26] Most Chicagoans, however, could not afford to buy a modern home on expensive land near the lakefront.

At the turn of the century, real estate speculators were providing a wide range of housing choices, from empty lots to dwellings fully equipped with the latest domestic technology. Subdivisions targeting the working class, for example, often offered homes without utility hookups or street improvements because buyers simply could not afford any extra luxuries. In these developments, new owners often went without modern amenities until they had paid the debts on their dwellings. In slightly more expensive subdivisions, homes were provided with only essential necessities, such as sewerage, running water, and bathtubs. By 1912 some blue-collar workers began to demand electric lighting in their new homes. Builders appealed to Insull to supply a cheap package deal of wiring and fixtures. He responded enthusiastically by offering to install service in each dwelling for $12, plus $2.50 for each outlet and $1.75 to $6.00 for various fixtures. In comparison, installing lighting services in the homes of the more affluent cost between $100 and $300.[27]

Population deconcentration on the urban fringe expanded home buyers' choices of the kind of community they lived in. The trend toward planning the entire neighborhood appealed to an urban society in search of order and of security from the upheavals of industrialization, immigration, and labor strife. The solution offered by the home builder—spatial separation—was a logical outgrowth of Victorian ideas about the city. "Separate spheres" for men and women, for example, had emerged early in the process of rapid urban growth and economic change. Fifty years later, the Columbian Exposition of 1893 used spatial planning to demarcate both functional and social areas of the "dream city." In 1909 the fair's master builder, Daniel Burnham, applied the same principles of land-use segregation to the real city in his comprehensive plan of the Chicago area. In both cases, the goals of creating a city both efficient and beautiful took preference over alternative models of the humane city.[28]

The desire for social segregation fueled the process of urban deconcentration and suburbanization as much as did the American dream of a house on a private plot of land. For affluent Chicagoans, especially, the new style of subdivision offered a safe haven where the price of housing restricted the neighborhood to members of the same class (and race). In many cases the urge to escape from the disease, conflict, and congestion of the central city propelled those who could afford it beyond municipal borders into the suburbs. And real estate speculators were perfectly content to enhance their profits by accommodating the separation of racial, religious, and ethnic groups in distinct neighborhoods.[29]

Electricity Comes into the Home

The introduction of electricity in the homes of middle-class Chicagoans was another sign of the emergence of the "Electric City." The electrical facilities installed in the Blackstone Hotel illustrate the coming trends in the use of electricity in the home. As in the past, the downtown hotel represented the leading edge in interior design and household technology. Constructed in 1909, the grand hotel used energy-intensive technology to achieve ultimate control over the interior environment. The Blackstone originally had 9,000 incandescent lights, over 400 telephones, and several motors totaling 450 HP. These motors ran fans for forced-air heating of the lobby and restaurant areas, and separate exhaust fans were used in the kitchen. Other motors operated refrigeration machines for ice making and food storage as well as an air-conditioning system for the dining rooms, cafés, and banquet halls. Electricity powered the water pumps, dumbwaiters, and elevators in the eighteen-story high rise; it spun revolving doors in the lobby, where it kept fountains playing under indirect lighting in the ceiling. The fountains made sure that the conspicuous consumption of energy would not go unnoticed by the guests. Behind the public spaces, the hotel also relied on electricity for an expanding array of house-keeping chores. For example, the laundry was equipped with commercial appliances, including washing, drying, and ironing machines.[30] By the time the hotel opened, the mechanism had been set in motion for the rapid diffusion of this futuristic technology into the homes of ordinary city dwellers. Chicagoans were well on their way toward an energy-intensive society.

By opening up large tracts of vacant land throughout the metropolitan area, rapid transit and modern technology encouraged the emergence of new types of housing. The first decade of the twentieth century set the patterns of housing construction for the next twenty years. Among the very well-to-do, the luxury apartment building established the social legitimacy of dwellings other than the traditional house. Patterned on the models of the fancy hotel and the men's club, the Marshall Apartments, constructed in 1906, are regarded as Chicago's first modern high rise. Among the original tenants was Samuel Insull, who was joined by his legal adviser, William G. Beale, and other members of the elite.

The luxury apartments along the Gold Coast of the Near North Side laid the groundwork for widespread acceptance of the "flat" among the middle class. The architecture of these two- and three-story buildings with one dwelling on each floor had evolved out of earlier attempts to design multifamily dwellings that were different from lower-class tenements. By the early 1900s, builders had found an efficient interior design for the long, narrow flat that remained basically unchanged throughout the great building boom of the 1910s and 1920s. The living room and bedrooms were placed in the light and air of the front, the dining room and bathroom occupied the middle, and the kitchen and servant's room were at the rear of the apartment.[31]

In addition, a new model home, the bungalow, began to replace earlier versions of the single-family dwelling at the urban fringe. The bungalow embodied the progressive ideals of efficiency and scientific management, sanitation, high technology, and renewed emphasis on the privacy of family life. "An almost austere simplicity," Gwendolyn Wright concludes, "became the basis of domestic design." The clutter and bric-a-brac of the Victorian home were replaced with clean, simple, and functional interior spaces. The growing "cult of household technology" made the kitchen the center of attention rather than the parlor. New materials such as porcelain fixtures and linoleum floors, and new appliances such as gas stoves and electric irons became important features of the "progressive" home. Since sanitary pipes and utility lines added about 25 percent to the cost of the bungalow, the house was made smaller and more compact to keep the purchase price within range of the middle class. Gradually, more and more members of the working class would also be able to buy these popular and affordable houses.[32]

The introduction of electricity in the homes of middle-class Chicagoans caused a change in domestic life, but a very limited one. In 1908, ten years after Insull instituted the demand meter system of rates, household use of the new form of energy was devoted almost entirely to better lighting. Utility salesmen confirmed that popular perceptions of electricity were narrowly defined. "The public is buying light," the *Electrical Solicitor's Handbook* of the National Electric Light Association complained, "and has no feeling for energy."[33] The typical apartment or flat, for instance, had ten to twelve light sockets, which usually hung from wires protruding from the ceiling or wall brackets. The average house had more than twice as many sockets, reflecting a larger number of rooms and, perhaps, a multilight chandelier in the parlor. To use an appliance such as an iron or a table lamp, the householder had to screw a light-socket type of plug at the end of the cord into a dangling receptacle. The two-pronged plug in universal use today would not become commonplace for another twenty years.[34]

Residential patterns of energy consumption suggest that most of the 30,000 affluent families enjoying the new technology in 1908 still regarded electric lighting as something of a luxury. The typical apartment dweller burned five to seven fifty-watt bulbs for about two hours in the evening. The light's carbon filament emitted sixteen candlepower, about one-third the illumination produced by today's standard bulb of the same wattage. Homeowners turned on about twice as many bulbs in their larger dwellings, but they used them on the average about the same length of time between 6:00 P.M. and 8:00 P.M. Actually, electric lights were normally restricted to the dining room and the parlor, or the new "living room." In the halls and living quarters, householders usually regarded the illumination supplied by gas jets or mantles as adequate.

A reluctance to use brighter illumination in the home apparently followed earlier conventions of a scarcity-minded society. The popularity of dual gas/electric chandeliers in the parlor and dining room suggests that many households made use of high technology on formal occasions but quickly reverted to gas lighting for the ordinary routines of family life. During the gaslight era of the 1890s, for example, novelist Henry B. Fuller had noted similar practices in the homes of middle-class Chicagoans. In *The Cliff-Dwellers*, the protagonist, Ogden, visited a family on the West Side. As he approached the house, the visitor observed "an excess of light [coming] through the front parlor windows, and Ogden was prepared to find that at least four of the eight burners in the big chandelier were lighted. This turned out to be the case; it was as great a tribute as the family ordinarily paid to society." The link between the level of lighting in the home and the conventions of society seems to have carried over into the early stages of electrification. In 1910, for instance, one of Insull's lieutenants explained that "one householder does not leave his lights burning while he goes to visit a neighbor who has an illumination in honor of his guest."[35] Although the prose is strained, it clearly underscores the special social meaning of burning the lights. Another solicitor comfirms that "electric lights are used in the average dwelling because of their convenience and other well-known virtues partly, but largely to be 'up-to-date' . . . [and it] costs money to be 'up-to-date' in most things." The installation of electricity in the home represented a break with the past, but its effects on domestic life were assimilated more subtly and gradually into the routines of daily life.[36]

The primitive state of homes appliances goes a long way toward explaining why little electrical energy was used for any purpose other than lighting. Household appliances such as coffee percolators and hot plates were shown at the World's Fair of 1893, but improving heating coils to a point of commercial success took another ten years. About the same time, small motors also began to be manufactured for home appliances such as vacuum cleaners and washing machines. As late as the mid-1920s, however, 80–85 percent of the electricity consumed in the home was for lighting. One device, the electric iron, accounted for almost all the remaining energy use.[37]

Although other appliances added convenience, this device eased one of the most dreaded tasks of everyday life. "Washing and ironing," a historian of domestic life points out, "was considered back-breaking labor. Ironing in particular required the constant lifting and replacement of the irons, a strenuous weight-lifting chore." Before the perfection of the electric iron about 1905, the heavy metal wedge used for this task was appropriately called the "sadiron." It had to be heated on the stove, making pressing clothes pure drudgery, especially during the summer. As with the light bulb, demand for a practical substitute for the sadiron was virtually universal.[38]

Yet even the electric iron caused only gradual changes in the age-old practices of washday. Women tended to use the new technology during the summer to avoid working over a hot gas or coal stove, but they often resorted to the old sadirons during the winter when cookstoves were in more constant use. This pattern of limited use would continue into the 1920s. The cost of buying and operating an electric iron only partly explains the persistence of traditional household routines. In the first decade of the 1900s, for example, an iron cost $3 to $5, and it took about 20¢ a week for five hours' use. Contemporaries did not consider this an extravagant amount. On the contrary, solicitors found that selling irons was the best "opening wedge" in signing up new customers. "They will not kick on minimum charge bills," a salesman reasoned, "because they will appreciate [that] they get something for their money."[39]

In 1909 Insull opened the first store in Chicago devoted to electrical appliances. Significantly, however, he designed the Electric Shop for the rich, not the middle class. "The store," one of his chief publicity men recalled, "was one of the finest in Chicago and was conducted with something of the exclusive air of a high-class shop."[40] Indeed, it looked more like a fine jewelry salon for an elite clientele than a department store for the growing number of white-collar city dwellers. Located on the first floor of the Railway Exchange Building at Michigan and Jackson, the Electric Shop featured table and floor lamps, which could cost up to $500 for a Tiffany-style art-glass model. The store also carried an amazing variety of appliances, from toasters and corn poppers to curling irons and heating pads to vibrators and other exotic gadgets claiming to have medicinal powers.[41] However, the sale of these products was negligible compared with the growing middle-class demand for better lighting. During the opening decade of the new century, household appliances remained luxury items, except for the iron.

The tone of the Electric Shop and the restricted use of energy in middle-class households cast doubt on the importance of advertising during the early spread of electricity to the homes of the middle class. Chicago presents a classic test case because of Insull's early and persistent faith in the power of advertising to influence consumers' behavior. In 1910, for example, the utility executive explained that "the increase in income in the flats and houses is simply the result of advertising, canvassing, educational work of all kinds. Our experience is that the lower we set the price per unit of energy, if we get at our customers and educate them to the uses of electricity, the greater is that use, within certain limits."[42] Yet even Insull had to admit he had no way to measure the effect of his marketing campaigns. His demand meter studies did not isolate the use of electricity for household appliances as opposed to lighting.[43]

A lack of concrete data on household routines did not stop Insull from mounting a full-blown advertising campaign. In 1898 General Electric began to promote appliances in the women's monthly magazines of the urban middle class, such as the *Ladies' Home Journal*, *Good Housekeeping*, and *House Beautiful*. Three years later the Chicago utility created a separate "business getting department" to coordinate daily newspaper advertisements, a free monthly magazine called *Electric City*, and door-to-door solicitations by a sales force, which by 1911 had grown to about ninety men.[44] In many respects though, his advertisements were more a mirror of consumer demands than their creator. For example, advertisements for appliances were limited to the Christmas season, reflecting their luxury nature. Insull's newspaper copy almost entirely emphasized the advantages of the light bulb over gas and kerosene alternatives. A 1902 advertisement stressed that "A Home Without ELECTRIC LIGHT is like a coat without a lining—unfinished, incomplete." Five years later, most newspaper copy aimed at residential customers still played on the obvious advantages of better lighting. "Make Your Wife Happy With This Gift. . . . Electric Light in the Home Means Much to Women," one proclaimed, "[because when] your home is wired for electric light there is less cleaning and dusting to do. The air is pure, the light brilliant—turned on or off instantly with the turn of a switch."[45]

In 1908 Insull began to give more attention to appliances in his advertising campaign. In perhaps his best-known publicity stunt, he had a truck laden with ten thousand irons tour the residential districts of the city. Huge signs draped on its sides announced that the mountain of appliances piled on the vehicle would be given to new customers for a free trial. A small army of salesmen followed up with a systematic canvass of the neighborhoods. In exchange for the free use of the iron for six months, the householder agreed to install central station service. As in the past, the utility company offered to defer the cost of wiring, fixtures, and appliances by spreading payments over a two-year period with no interest charges.[46]

Insull had good reason to exploit the promise of modern technology to save labor and reduce drudgery in the home as soon as practical appliances became available. Most important, perhaps, they gave him a psychological edge in his campaign to install central station service in every home in Chicago. Irons and other appliances supplied an extra incentive that helped overcome consumer resistance to the new technology. Insull's salesmen had already discovered that offering something special was an effective way to sign up businessmen for lighting services. Louis Ferguson, for example, reasoned that "you must get him [the prospective customer] to use the current first, get him into a familiarity with the advantageous points of your electrical system, and you must do it by a method that attracts him strongly." Using this marketing psychology on householders aimed for the same result. "Give service to any installation," another solicitor pleaded, "on the assumption that the use of electricity even in the most trivial manner soon becomes a habit and inevitably leads to a more extended use . . . so that almost before the householder realizes it, he is relying on electricity for his light and various other needs and wondering how he could have gone without it [for] so long."[47]

The potential of household appliances as consumers of energy gave Insull a second reason to begin promoting them at an early point. His effort to build and diversify the use of electricity brought immediate attention to these devices as a daytime power load. By 1909–10, demand meter statistics on the use of lighting in the home were confirming his theories about the importance of residential customers in spreading the electrical load over the entire day. These figures showed that the maximum use of residential lighting occurred about two hours after the peak of commercial demand between 5:00 and 5:30 P.M. during the winter. Moreover, householders did not all use their lights at the same time. This diversity of demand created significant savings in distributor equipment, which it cost the utility more to extend into low-density residential neighborhoods than into other areas. For in-

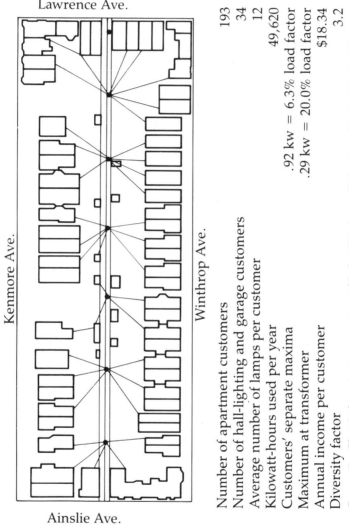

Number of apartment customers	193
Number of hall-lighting and garage customers	34
Average number of lamps per customer	12
Kilowatt-hours used per year	49,620
Customers' separate maxima	.92 kw = 6.3% load factor
Maximum at transformer	.29 kw = 20.0% load factor
Annual income per customer	$18.34
Diversity factor	3.2

CHART 10. Electrical service to an apartment block, 1910. *Source:* Keily, *Central-Station Service,* 448.

stance, Insull made a careful survey of a block of new three-flats near the Lawrence Avenue el station in the "Uptown" neighborhood directly south of Edgewater. Reflecting the trends in home building on the urban frontier, 175 of 193 apartments were served by electricity. Their aggregate demand equaled about .92 kw, but their actual maximum at any one time was only .29 kilowatts (see chart 10). The difference helped offset the cost of supplying residential districts by reducing the need for substation equipment and customer transformers. The promise of houshold appliances as consumers of energy during the utility's off-peak, daylight hours was great enough for Insull to conclude that it was well worth wiring the neighborhoods even at a short-term loss.[48]

Yet the lack of measurable results in the use of energy by home appliances cautions against placing too much importance on the power of advertising to change consumer behavior, at least until the 1920s. It was probably effective only as an "entering wedge" or educational tool. Advertising helped overcome the inertia of those who already wanted better lighting and enabled them to take the steps necessary to install service. When there was little unsatisfied demand, even the best efforts of a supersalesman like Insull had negligible effect. This was exactly the case with the battery-driven automobile, which faced insurmountable competition from the gasoline-powered vehicle. In 1898 Insull began to promote electric cars, taxis, and trucks by setting up battery-charging stations. For the next twenty years he would continue to open more of these garages, to offer special incentives, and to cut rates—but to no avail.[49]

In the home market, electric heating represents another case of the resounding failure of advertising. In Chicago and other areas where gas was available, the heavy expense of operating an electric cookstove or space heater was no match for this much more economical competitor. For example, cooking with electricity cost at least twice as much as cooking with gas or coal. The gas company, moreover, was quick to highlight the economy and convenience of its heating appliances in newspaper ads that paralleled those of Commonwealth Edison. During the Christmas season of 1909, for instance, both companies promoted room heaters. The gas company promised, "If You Had This Heater in *Your* Room you would have a hot fire the moment you applied a lighted match. The cold floor and penetrating chill which you dread in the morning would disappear in a minute." In a similar vein, Insull empha-

Table 15   Growth in Number of Residential Consumers of Central Station Service in
Chicago, 1908–16

Year	Increase in Number of Apartments	Increase in Number of Houses	Total Increase, Residential	Cumulative Total	Average per Month
1908	—	—	—	31,132	18.1
1909	12,040	1,523	13,563	44,695	18.9
1910	13,820	2,380	16,200	60,895	20.8
1911	16,046	3,777	18,582	79,477	20.4
1912	18,656	1,241	19,897	99,374	23.2
1913	24,187	3,637	27,824	127,198	26.6
1914	15,351	2,943	18,294	145,492	23.5
1915	15,620	15,264	30,884	176,376	24.2
1916	26,614	−3,263	23,351	199,727	24.7

Source: Commonwealth Edison Company, Departmental Correspondence, E. Fowler to
J. Gilchrist, 4 December 1916, in CEC-LF, F23-E4.

sized that the electric radiator "is the modern fireplace. . . . useful
for taking the chill and dampness off the room in early fall, [and]
cold winter mornings, before the furnace is fairly going." The gas
model cost $5 compared with $13.50 for the electric version.[50] In
the case of both heating devices and automobiles, no amount of
advertising could compensate for the obvious advantages of superior alternatives.

By 1912 new patterns of energy use were established in the
home, although the diffusion of electricity in the neighborhoods
was just beginning. On the one hand, 80,000 residential customers
meant that electric lighting was no longer enjoyed exclusively by a
few wealthy families that could afford any luxury. The elite found
a new symbol of social status and conspicuous consumption—the
household appliance. On the other hand, only 16–18 percent of
the city's families had electricity, indicating that central station service was still too expensive for a large proportion of the middle
class. During the prewar years, further rate cuts and rising prosperity would continue to increase the momentum toward better
lighting in the home. In 1908–12, about 15,000 households would
install central station service each year. In 1912–17 the number
would double to almost 30,000 new connections annually, the cumulative total of residential customers only nine years earlier (see
table 15). By the eve of the war, the filtering-down process would
reach a broad middle spectrum of more than one-third of Chicago
households.[51]

The opening decade of the twentieth century began the creation of an "Electric City." One indicator of the Chicago's growing dependence on an intensive application of energy to sustain everyday life was the amount of electricity used by the average city dweller. In 1902 the per capita use of central station service was 36.6 kwh, a figure that increased eight-fold over the next decade to 306.8 kwh per person. To generate this amount of electricity, Insull's powerhouses had to burn more and more fuel, surpassing a million tons of coal in 1911 for the first time. About two-thirds of this energy went to run the transit system, a quarter to provide lighting, and the remaining 10 percent to supply industrial power. These proportions would remain roughly the same until the energy crisis of World War I.[52]

7

The Suburban Matrix of Energy, 1902–1914

At the turn of the century, urban growth and technology were transforming Chicago from a city to a metropolis. In part, electric rapid transit and the automobile were responsible for opening up vast tracts of land to residential settlement on the urban fringe and beyond. To a large extent, however, faster modes of transportation were simply the latest instruments of a culture that had long ago defined the good life in terms of suburban domesticity and social segregation. In the 1900s Frank Lloyd Wright would embody this deep-seated strain of antiurbanism in a new style of suburban architecture. His "city man's country home on the prairie" epitomized America's ambiguity about the ascendancy of a distinctly urban culture and the decline of traditional rural values. Contemporaries considered the trolley car's moral influence in carrying the city's people to the open spaces of the "crabgrass frontier" just as important as its functional role in alleviating congestion and overcrowding.[1]

In a similar way, electrical energy promoted an exodus of people and industry from the central city that had been under way for at least half a century. As we have seen, utility systems fueled, and perhaps accelerated, urban deconcentration. The spread of power distribution grids to outlying areas supplied modern amenities that previously had been available only in the business districts and the residential neighborhoods of the well-to-do. Americans, moreover, had given electricity potent cultural significance as an instrument of democratic values, moral uplift, and social status. As one historian of housing reform, Gwendolyn Wright, contends, "New domestic technology was central to the aesthetic and the cultural redefinition of the model home."[2] For Frank Lloyd Wright and his clients, high technology was a conspicuous sign of progress and success.

During the opening decade of the 1900s, the combined effect of several new forms of technology was to expand the physical

Table 16   Population Growth of Chicago and Suburbs, 1890–1930

Year	Chicago	Suburban Cook County	Lake County
1890			
Population	1,099,850	92,072	24,235
1900			
Population	1,698,575	140,160	34,504
Increase	598,725	48,088	10,269
Percentage gain	54.4	52.2	42.4
1910			
Population	2,185,283	219,950	55,058
Increase	486,708	79,790	20,554
Percentage gain	28.6	56.9	59.6
1920			
Population	2,701,705	351,312	74,285
Increase	516,422	131,362	19,227
Percentage gain	23.6	59.7	34.9
1930			
Population	3,376,438	605,685	104,387
Increase	674,733	254,373	30,102
Percentage gain	25.0	72.4	40.5

Source: U.S. Census, Population, 1890, 1900, 1910, 1920, 1930.

boundaries of the community to metropolitan proportions while also weaving it more closely together. Electric service, rapid transit, automobiles, and telephones helped shrink the barriers of time and space between the center of the city and its periphery. In the case of Chicago, the suburbs' growth rate began to exceed the city's after the turn of the century (see table 16) as social, cultural, and demographic forces provided an impetus to suburbanization. The newcomers, however, were soon making demands for modern services that neither local governments nor utility companies could afford to ignore. In the period leading up to World War I, the political pressure for public works improvements grew into a popular crusade. The suburbs became involved in a massive project to construct a modern infrastructure of roads, sewers, water supplies, electrical grids, gas pipes, and telephone lines. Until these systems were in place, technology played a secondary, supportive role in the integration of the Chicago metropolitan area.

For the suburbs, too, this was a period of beginnings, a time of establishing new patterns and laying the foundations for more rapid growth. To be sure, central station electric service was only one part of this transformation of the nineteenth-century city into the metropolis of the twentieth century. Yet it was a crucial part, because more energy was integral to the emergence of a modern infrastructure, including novel forms of transportation, housing, industry, and public services such as street lighting and supplies of pumped water. New technological systems not only helped redraw the spatial relationship between city and suburbs but also helped redefine the meaning of the region. Chicago's semiautonomous satellite cities were drawn more tightly into the economic and cultural orbit of the central city. The construction of belt-line railroads, interurban railways, and high-voltage transmission towers created a web of interdependence binding these communities and the entire hinterland more closely to Chicago.

From 1902 to 1914, Samuel Insull put in place the basic components of an integrated regional power network. After the turn of the century, the social conditions and demographic trends of the suburbs established the preconditions for applying Insull's gospel of consumption. The continual extension of the electric railways meant that large blocks of energy had to be supplied at strategic intervals along the entire length of the lines. This need for power combined with steady increments in the population to boost to new levels the demand for electricity outside the central city. In 1902 Insull's associates persuaded him to expand his perspectives to a metropolitan point of view. The businessman reluctantly agreed to buy utility companies in Evanston, Highland Park, and Waukegan. Forming the North Shore Electric Company (NSEC), Insull inaugurated an experiment in electrifying low-density residential communities. The next year the success of the turbogenerator at the Fisk Street Station finally convinced him that his formula of transit contracts, large generators, low rates, and load management could be applied profitably to the outlying districts of the city.

In 1907 the first interconnection of high-voltage lines between Commonwealth Edison and the NSEC marked the beginning of an integrated metropolitan power network. The importance of this linkage cannot be overstressed, because it created a unified community of energy consumers on an unprecedented scale. The diversity of demand between the city and the suburbs was great enough to reduce significantly the unit cost of electricity for both groups of customers. Insull quickly translated the lessons of the NSEC experiments into a scheme for building a utilities empire spanning the entire Chicago region. Over the next five years, he repeated the process of acquisition, modernization, and expansion that he had used in the city to establish a virtual monopoly of electrical services in suburban Cook and Lake counties. By 1912 the NSEC was supplying electricity to 92,500 people in forty-three communities spread over a 1,250-square-mile territory.[3] Insull also set in motion an even more ambitious plan to consolidate the gas and electric utilities of the region in a single enterprise, the Public Service Company of Northern Illinois (PSCNI). This great web of power knit city dwellers, suburbanites, and farmers together in bonds of common self-interest. On the eve of World War I, the infrastructure was ready for its increasingly important role in reshaping the spatial and social patterns of Chicago's growth.

Electricity Comes to Main Street

In the late 1880s and early 1890s, local entrepreneurs had begun to supply electrical services in many of Chicago's outlying communities. In most cases these primitive utilities had been underwritten by exclusive contracts with local governments for streetlights along Main Street. Yet central station service in the suburbs had been plagued with the same crippling constraints of small scale and defective technology that had kept rates at luxury levels in the city. In addition, the cost of stringing distributor lines in sparsely settled residential areas remained prohibitive for utility operators, especially since they suffered a 30–50 percent loss of energy during transmission to their distant customers. Consequently the distribution grid was usually confined to the downtown business district and the surrounding zone of homes, and service was restricted to the evening hours defined by the street lighting contract.[4]

By the turn of the century this polygenesis had created a crazy quilt of disconnected power grids. The availability, quality, and price of utility services varied widely from community to community. Some residential suburbs, such as Evanston and Oak Park, had enough people and wealth to support both gas and electric companies. In contrast, many other well-established suburbs, including Highland Park, limped along with underfinanced systems of first-generation, small-scale equipment. Residential service was makeshift at best. A worker for the original Highland Park Electric Light Company recalled that "we usually had to take down a transformer from some vacant dwelling in order to supply a new customer."[5] At the same time, some less wealthy communities like Waukegan and Blue Island had thriving utility enterprises because industry provided a steady demand for light, heat, and power. And in many places, of course, few if any urban services were available.

The tremendous range of community types in the Chicago area makes some scheme of classification necessary. For our purposes, it will be convenient to place Chicago's outlying areas in four categories: residential suburbs, industrial suburbs, satellite cities, and agricultural farmland. The first group is characterized by low-density single-family homes occupied mainly by people who commuted to jobs offering above-average income and status. This section of the chapter will consider the effects of electrification on the residential suburbs of the North Shore, including Evanston and Highland Park. The following section will examine the industrial suburbs, where the community centered on the factory and work, not on the home and family life. Two places in the southern portion of Cook County that were dominated by heavy industry and blue-collar housing—Harvey and Blue Island—will serve as examples of this community type.[6]

A third section will take up the satellite city and the rural hinterland. Historically, these places were far enough from the central city to create balanced local economies and a full range of social classes, yet close enough to remain subject to Chicago's cultural and commercial domination. Two groups of satellite cities paralleled the growth of the central city throughout the nineteenth century. The first formed an arc about thirty-five to fifty miles outside the city. To the north was the harbor city of Waukegan, and to the west were a string of towns along the Fox River valley, including Elgin, Aurora, and Ottawa, where the Fox meets the Illinois River. Southwest of Chicago, a second group of riverfront cities formed a waterborne metropolitan corridor along the old Illinois and Michigan Canal route. This radial link between Chicago and the Illinois River valley contained a diverse range of commercial and industrial cities such as La Salle, Joliet, Kankakee, and Lockport.[7]

The North Shore from Evanston to Lake Forest provides an excellent case study of the residential suburb. These affluent lakefront communities became early champions of an antiurban, suburban ethos. In the 1850s, for example, religious men had founded the town of Evanston and Northwestern University in order to establish a unified community of absolute sobriety, moral purity, and social refinement. Forty years later the growth of the population to over 13,000 inhabitants had created social complexity as well as a crying need for better public services. Yet Evanston strove to maintain a separate identity as an asylum from the evils of the city. In 1892 its residents defeated a consolidation proposal by an overwhelming majority of 78 percent of the voters.[8]

The Evanston poll reflected the ascendancy of a suburban ethos in Chicago and similar cities of the Northeast. Reversing earlier patterns of urban expansion, the vote of Evanston and other suburbs adjacent to the city meant that municipal borders would no longer follow and absorb the outward thrust of settlement. New infrastructural technology, more adequate tax resources, and special-purpose tax districts eliminated the previous incentives for annexation to the central city. In 1889, for example, the elite Citizens Association of Chicago sponsored state legislation to establish an independent metropolitan sanitary district. Reflecting a broader vision of the city, the sanitary district supplied essential services that only central cities had previously been able to afford. The fateful decision against annexation marked the beginning of a new era of fragmented government in the metropolitan area and political conflict between the city and the suburbs in state and national capitals.[9]

The revolt against the city by Evanston and similar residential communities complemented the ideas of a rising generation of housing reformers and community builders. In the opening decade of the century, the suburbs of the well-to-do offered an inviting social environment for giving full expression to new concepts of subdivision planning, domestic architecture, and interior design. Jane Addams's Hull House became a center for a far-flung network of progressives who gathered regularly to debate and exchange ideas. Among the members of this group was Frank Lloyd Wright, who contemporaries and later historians agree was the father of modern suburban architecture. The success of the prairie house was built on Wright's sympathetic understanding of the expectations and desires of his middle-class clients. More than other reformers, he was able to translate their aspirations into a model of the ideal family home, especially during his most influential period, from 1901 to 1909, when he lived in Oak Park, a suburb on the western border of the city that was similar to Evanston.[10]

Contemporaries immediately recognized Wright's prairie houses as the perfect embodiment of a suburban ethos. His innovative plans were part of the widespread reaction of progressives against the Victorian home, with its formality, clutter, and depressing tones of heavy upholstery and dark furnishings. Like the simple new bungalows, Wright's one-story structures emphasized clean exterior lines, light and open interiors and new household technology. Of course Wright added his own special genius to these general trends in housing fashion, striking a responsive chord in his affluent clients.[11]

Wright rejected not only the spirit of Victorianism but also its most important by-product, the industrial city itself. Raised on a Wisconsin farm amid a tightly knit extended family, he hoped to return America to its agrarian roots. For Wright, the family hearth was a mystical symbol of domestic intimacy, self-reliance, and privacy from the outside world. The city represented the antithesis of these traditional values. In 1901 the architect delivered an influential speech at Hull House that expressed his generation's ambiguity about modernity. The address, "The Art and Craft of the Machine," reveals the contradictory attitudes of the progressive generation toward the city, technology, and the role of the artist in the coming era. According to Wright, a passing age of "greed" had given rise to highly concentrated and centralized cities, which he depicted as "monstrous things"—great soulless machines destructive of human values. In contrast, the reformer announced the dawn of a new age of the "creative artist" who would help dismantle the city and lead society back to the transcendental agrarianism of Thomas Jefferson and Ralph Waldo Emerson. The task ahead was to create an organic relationship between the individual, society, and nature. Paradoxically, this leader of reform was an antimodernist, or as one urban scholar put it, a minister of "innovative nostalgia."[12] And this was exactly the source of his success among Chicago's middle-class home buyers. "In a period of 'progressive' reform," architectural historian Robert C. Twombly suggests, "they clung to traditional values, and like others in the emerging metropolis felt themselves engulfed by sweeping changes not entirely to their liking."[13]

Although the architect condemned the age of steam, he took a more positive approach toward the machine in the new age of the creative artist and high technology. Always the heretic, Wright shocked the Hull House meeting of the Arts and Crafts Society by offering a defense of mass production. He called upon artists to turn the factory into an instrument of democratic values. In the right hands, technology would play a liberating role, creating new abundance and security for all members of society. Only then, he argued, would Americans be truly free to restore self-reliant individualism, family intimacy, and spiritual unity. To Wright, "the city man's country home on the prairie" was a step in this direction. It represented an antidote to urban life, a haven where people could "bring up their children in comparative peace, safe from the poisons of the great city." Over the next decade, his architectural statement of a suburban ethos attracted clients throughout the fringe areas of Chicago, including Hyde Park and Rogers Park as well as Oak Park and Highland Park.[14]

Wright's defense of the machine was intended to provoke the audience at Hull House, but he was not alone in enlisting technological innovation in the cause of tradition. With the advent of electric rapid transit and the automobile, reformers increasingly tended to link high technology to a visionary world, a nation of suburbs where consumption and leisure would become more and more important parts of daily life. Decentralization was also seen as a catalyst in the revival of republican values. For example, Henry Demarest Lloyd, another well-known progressive, predicted that greater application of electrical energy would bring about a suburban utopia of democratic equality and personal fulfillment. "Equal industrial power," the muckraking journalist asserted from his home in the North Shore community of Winnetka,

> will be as invariable a function of citizenship as the equal franchise. Power will flow in every house and shop as freely as water. . . . Women, released from the economic pressures which forced them to deny their best nature and compete in unnatural industry with men, will be re-sexed. . . . The new rapid transit, making it possible for cities to be four or five hundred miles in diameter and yet keep the farthest point within an hour of the center, will complete the suburbanization of every metropolis. Every house will be a center of sunshine and scenery.[15]

For Wright, Lloyd, and other champions of a suburban ethos, a more intense use of energy was an integral part of the plan for reconstructing urban society and the family home. Wright, like his mentor Louis Sullivan, became a master of light. His treatment of sunshine—ribbons of windowpanes encircling the house, the use of art glass and huge picture windows—was received enthusiastically by clients in search of more light and air. One of the architect's most important innovations and early "trademarks" was the casement window that opened outward to the environment. Despite clients' resistance, the architect insisted on these windows because they helped dissolve the walls between the inhabitants inside the house and the benevolent influence of nature outside. In addition,

Wright took advantage of the nonflammable incandescent bulb to devise novel indirect lighting effects. In his own home and studio, for instance, he installed electric fixtures behind a stained-glass panel in the ceiling above his dining room table. The panel produced a warm, glowing illumination like the skylights and atria in the lobbies of commercial buildings that were then reaching a peak of fashionable popularity.[16]

In a similar way, Wright incorporated the currents of progressive reform in his kitchen designs. Hull House brought him in touch with the leaders of the home economics movement. These middle-class women were driven by the same ideals of simplicity and efficiency that fueled the general revolt against the Victorian home. Focusing on domestic routines, the reformers were attracted to Frederick Taylor's scientific management and attempted to translate its principles from the factory to the home. In this context, mechanical substitutes for manual work were highly valued, especially "laborsaving" devices. Home economists believed this type of modernization would permanently solve the chronic "servant problem" by eliminating the drudgery of housework and hence the need for extra help. This approach to the reform of domestic life reinforced the image of electrical appliances as agents of national progress and personal liberation that had emerged out of the American fascination with high technology.[17]

During the 1900s, rising demand for electricity along the North Shore reflected the cultural and social redefinition of the suburban ethos that Wright embodied in the prairie house. Ten years earlier, both Evanston and Highland Park had initiated arc light service on Main Street. At the time of the Great Chicago Fire of 1871, moreover, Evanston and its university had been provided with gaslights from the Northwestern Gas Light and Coke Company (NGLCC). However, the quality of service supplied by these underfinanced utilities remained primitive until the turn of the century. For example, the NGLCC initially used crude wooden pipes to distribute gas. In the 1880s, a single worker could make the rounds reading all the town's meters and leaving monthly bills in less than a day. Similar makeshift arrangements were typical of early electric service to residential customers. John Wynn, an employee of the Evanston Electric Illuminating Company during the 1890s, recalled that his job was "to read meters, serve as lineman, troubleman, and storekeeper; and, in case of emergency, to act as the whole force at once." He also pointed out that most residential meters were in attics because this was the first and cheapest place for the company's expensive copper distribution lines to enter the home. A co-worker confirmed that meter readers had to become gymnasts and scale the rafters to the top of the attic.[18]

In the early 1900s, utility operators began to upgrade the quality of energy services in the suburbs to modern standards on a par with those in the central city. More than a dramatic rise in residential demand, the primary reason for this timing was a steady improvement in economic conditions and the availability of venture capital. A decade earlier, entrepreneurs had taken advantage of ground-floor opportunities to obtain franchises and public subsidies in the form of street lighting contracts. The electrical manufacturers had frequently helped underwrite these suburban companies, but the manufacturers were interested in selling generating equipment, not in guaranteeing the long-term success of operating utilities. For example, Bernard E. Sunny, the Chicago representative of the Thomson-Houston Company and a resident of Highland Park, helped form that suburb's electric company. In a similar way, Westinghouse encouraged the organization of a utility in Evanston. When the Panic of 1893 set off five years of economic depression, the financial foundations of these small enterprises collapsed.[19]

By the end of the 1890s, however, the opportunity to exploit accumulated demand for light and power was renewed by a gradual easing of hard times and tight money. During the decade following the panic, suburban utilities frequently changed hands. In some cases they passed from one shoestring operator to the next without adding any new capital for upgrading equipment or making extensions. This pattern helps explain the makeshift practices of early utilities like the electric companies in Highland Park and Evanston. In a few other cases, businessmen in possession of substantial capital stepped in to buy promising enterprises at bargain prices. In 1895, for example, the NGLCC was snapped up by a group of local and out-of-state investors led by Charles G. Dawes, a financier who eventually became vice president under Calvin Coolidge (1925–29).[20]

In addition, new competitors entered the suburban market to tap existing and future demands for energy. In 1900 an inventor from Oak Park, H. G. Yaryan, obtained a franchise to establish an electric and heat plant in Evanston similar to the one operating in his own community and in neighboring Cicero. A year earlier, Yaryan had begun serving these western suburbs using a patented process of recycling exhaust steam from the generating equipment to underground hot-water pipes that supplied heat to surrounding buildings. The inventor's dual-purpose system allowed him to undercut the rates charged for electricity by the competition in both sections of the metropolitan area. Yaryan was not alone in building multicommunity utility companies in the suburbs. To meet the inventor's challenge in the western suburbs, for instance, the owners of the Cicero Water, Gas, and Electric Light Company acquired utilities in Oak Park and Berwyn. Led by John R. Walsh, a powerful banker, utility financier, and political kingpin, the renamed Chicago Suburban Light and Power Company was typical of early efforts to consolidate the utility companies in outlying residential districts. These larger systems were more attractive to investors because of promised savings from unified management economies and increased revenues from integrated distribution grids.[21]

In 1901 Samuel Insull became a reluctant investor in this highly fluid market of suburban utility properties. In an unusual confession, Insull later admitted that he had not originated the idea. "At that particular time," he remembered, "my mind was running upon supplying all the energy in centers of population. I doubt very much whether I then visualized the possible influence of low cost and mass production on countryside distribution of energy." Preoccupied with the Fisk Street Station, he had paid little attention to the revival of profitable opportunities to supply central station service to the suburbs. A year earlier Emil Rathenau, his German counterpart in Berlin, first suggested to Insull that he and other electric utility operators could create unified distribution networks within a fifty-mile radius. Apparently, though, Insull paid little attention until a promoter of suburban utilities, Frank J. Baker, finally convinced him to give Rathenau's proposal more serious thought. Baker was a resident of Highland Park who had earned college degrees in engineering and law and had worked on some of the first electric railway systems. With this background, he built a successful law practice as a broker of suburban utility enterprises. In this instance he persuaded Insull to buy two capital-starved electric companies, one in Highland Park and the other in Evanston.[22]

It took Insull six years, from 1901 to 1907, to modernize and upgrade the supply of electricity in the suburbs to the standards of modern central station service in the city. Buttressed by the success of the turbogenerators at the Fisk Street Station, he formed the North Shore Electric Company. In 1902 Insull immediately sold his two companies to the new corporation at a handsome profit.

The NSEC also began purchasing all the remaining electric utilities between Evanston and Waukegan, including Yaryan's company. Two years later, an integrated hierarchy of large generating stations and distribution substations began to take shape. Insull also laid a high-voltage transmission line from Evanston to Waukegan that was the first in the Chicago region to interconnect several distinct communities. Between generating stations in the two communities were a string of new substations. The one in Highland Park, for instance, distributed light and power twenty-four hours a day to an area stretching from Glencoe on the south to Lake Forest on the north. In 1907 the interconnection of Commonwealth Edison and the NSEC by a 20,000-volt line represented the realization of Rathenau's vision of a metropolitan web of energy. In the same year the NSEC opened appliance stores on Main Street, putting the capstone on the institutional structure of Insull's gospel of consumption.[23]

Metropolitan Connections

For Insull, 1907 was a year of decision. His doubts about the economic viability of supplying electric service to suburban districts of low-density housing had been laid to rest. Even before his program for modernizing and expanding the NSEC produced significant results, the utility executive made a series of aggressive moves designed to fasten a monopoly grip on the supply of electricity throughout metropolitan Chicago. He went on a buying spree to gain control of all the energy utilities in suburban Cook and Lake counties as well as the few remaining electric franchises in Chicago (see table 17 and map 4). But more important, Insull now saw the large power consumer as the key to attaining true economies of scale. By building and diversifying the electrical load, the transportation and industrial sectors could trigger a self-perpetuating cycle of falling rates and rising use. He signed the interurban lines to the same kind of long-term contracts with minimum load guarantees that he was negotiating with the financially hard-pressed trolley and elevated companies in the city.

In this lucrative market, Insull demanded an absolute monopoly. In 1909 he complained bitterly to the president of General Electric, Charles Coffin, about recent offers by the Westinghouse Company to build powerhouses for the city's transit companies. "In my opinion," Insull explained, "it is simply a question of the Westinghouse Company 'butting in.' " Coffin tried to reassure him that the

Table 17   Growth of the North Shore Electric Company, 1902–11

Year	Company Acquisitions
1902	Highland Park Electric Co.
1902	Waukegan Electric Light Co.
1903	Evanston Electric Illuminating Co.
1904	Maywood and Proviso Electric Light and Power Co.
1904	Elmhurst Electric Light and Power Co.
1904	Freeman Power Supply Co. of Libertyville
1904	Park Ridge Electric Light Co.
1905	Calumet Electric Light Co. (Harvey and Chicago Heights)
1905	Evanston (Yaryan) Heating Co.
1906	La Grange Service Co.
1907	Leyden Light and Power Co.
1910	Illinois Lakes Light and Power Co.
	Union Light and Power Co. (Crystal Lake)
	Citizens Light, Heat and Power Co. of Dundee
	Antioch Electric Co.
1910	McHenry Electric Service Co.
1911	Kenilworth Lighting Co.

Source: Important Papers of the North Shore Electric Company, vol. 3 (Chicago: North Shore Electric Company, 1911).

transit companies would find it "extremely difficult" to finance such a project, but Insull was incensed. "I am still of the opinion," he replied, "that the proper thing for me to do is to serve the Westinghouse Company notice that they should keep their hands off our business. I am inclined to think that it has got beyond the diplomatic stage. It has got to a place where we are suffering from the disadvantage of a kind of guerrilla warfare without the advantage of a plain spoken or actual declaration of hostilities." However, Coffin's diplomatic intervention was apparently sufficient to keep Westinghouse out of the Chicago street railway business.[24]

In the five years following the pivotal decisions of 1907, Insull strove to diversify energy consumption by signing up not only the interurban lines but also the large factories in the industrial suburbs. Like Highland Park and Evanston, some of these work-centered communities were as old as Chicago itself. South of the city, for instance, German immigrants in 1835 had settled the high ground known as Blue Island. The discovery of huge clay and limestone deposits nearby soon marked the settlement as an industrial as well as an ethnic community. More often, however, the industrial suburb emerged more gradually out of the movement of city-based factories to the urban fringe. The removal of the stockyards to open prairie on the South Side during the Civil War years was an early example of this industrial deconcentration.

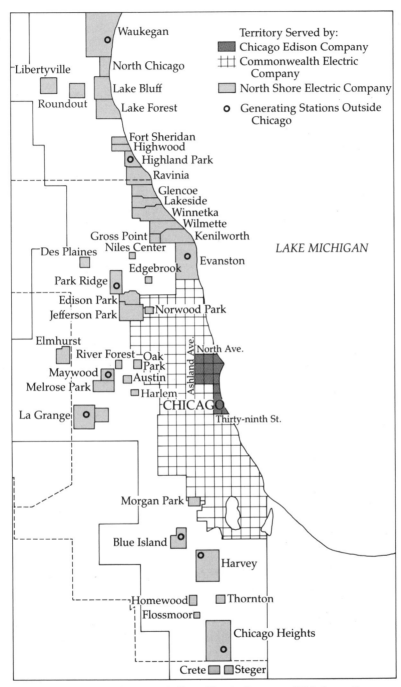

MAP 4. Service area of the North Shore Electric Company, 1906. *Source:* Imogene Whetstone, "Historical Factors in the Development of Northern Illinois and Its Utilities," typescript, Chicago, Public Service Company of Northern Illinois, 1928, 123, in CEC-Library.

After the war, many other large enterprises followed this exodus out of the congestion of the central city. Cheap land, better transportation, and isolation from the moral evils of the city were some of the attractions drawing big business to the urban fringe. In Chicago, the rise of manufacturing on a new scale encouraged this outward movement, including the iron and steel mills, the McCormick reaper works, and the railroad car shops built in 1880 by George Pullman as the centerpiece of his completely self-contained factory town.[25] In 1889, of course, many of these factory zones, including Pullman, were absorbed by the great annexation. Nonetheless, industrial deconcentration continued to accelerate, especially as engineers and fire insurance underwriters redesigned the factory as a one-story structure. Novel technology furthered the reorganization of work inside and outside the factory. Central station electrical service, for example, allowed industry to locate virtually anywhere while eliminating the heavy expense of building steam powerhouses, coal yards, and railroad sidings. Electricity also eased the problems facing scientific managers who were striving to speed the flow of work through the new one-story plants. The electric motor freed them from the rigid constraints of the steam engine's transmission system of shafts, pulleys and belts. Electrical technology, like trucks and highways, was ready-made to redefine the model factory in suburban terms.[26]

After 1889, new industrial suburbs continued to spring up just beyond Chicago's expanded boundaries. But now developers used village and municipal incorporation statutes as bulwarks against the further expansion of the central city. The origins of Harvey, Illinois, present a clear illustration of this type of community building. Shortly after the 1889 referendum on annexation, Thomas Harvey bought a subdivision known as South Lawn along the Calumet River, two and a half miles south of Chicago's new boundary and eighteen miles south of the Loop. The following year, the lumber merchant formed the Harvey Land Association to erect a self-contained "manufacturing suburb." Following in the footsteps of the founders of Evanston and Pullman, Harvey and his religious partners were determined to ban the saloon from the community. At the same time, these town planners made provisions to ensure that "every advantage to be enjoyed in the city either by the manufacturer or employée, is within reach at Harvey, to which are added more room, purer air, better drainage, more shade, and water from artesian wells." From the start, the Harvey Land Association undertook the task of supplying "modern conveniences" such as good schools, sanitary hookups, electric lighting, and even a trolley to carry workers to their jobs along landscaped boulevards. Employers, moreover, promised to give preference in hiring to residents. Local financing was made available to blue-collar workers to encourage them to purchase land and a cottage for about $1,000. Harvey became an instant success, a "magic city" containing 5,000 people and ten major industries only three years after its founding.[27]

Most industrial suburbs in the Chicago area emerged from more informal growth processes. Blue Island, about four miles northwest of Harvey, presents a typical example. By 1900 its population had reached 6,100. This ethnic community of farmers, brewers, brickmakers, and stonecutters was attractive to industry because several major railroad lines crisscrossed its territory and cheap factory sites were available along the Little Calumet River. During the years leading up to World War I, a mixture of industries located in Blue Island, such as wire mills, smelters, railroad repair shops, cigar factories, and manufacturers of medical supplies. In 1902 the NGLCC also built a major coal gas plant on the waterfront, followed five years later by an NSEC generating station that served as the southern anchor of Insull's triad of new large-scale suburban energy plants.[28]

Blue Island took on not only an increasingly industrial cast, but also the look of an up-to-date suburban community. In fact, the growth of urban public services in Chicago's industrial suburbs paralleled the construction of a modern infrastructure in their more affluent residential counterparts. Beginning in the 1880s with street grading, a municipal waterworks, and kerosene streetlights, Blue Island launched a public works boom during the 1890s, including macadam street paving, cement sidewalks, a sewer system, a municipal electric plant, an Illinois Central Railroad commuter station, and street and interurban railways. By the turn of the century, Blue Island, Harvey, and similar places on the urban fringe were distinct from the bedroom suburbs of the North Shore. The industrial suburb contained a fuller range of social classes, and community life was dominated by the work routines of the factory. Yet the two types of suburbs shared many of the same aspirations for improving domestic life. The differences between them were primarily economic, not cultural or technological. The diffusion of modern utility services into the cottages of blue-collar workers simply lagged behind the pace in more affluent neighborhoods.[29]

The deconcentration of industry along with the population

meant that the diversity of energy consumers in the urban fringe began to resemble the broad range of Insull's customers in the central city. By 1907 he had modernized the facilities of the NSEC enough to begin applying his system of large generators, low rates, and load management in suburban Cook and Lake counties. New stations on the lakefront in Waukegan on the north and on the Des Plaines River in Maywood on the west joined with the one at Blue Island to create a crescent of modern plants around Chicago. The NSEC supplied electrical energy to a mixture of households and retail shops, interurban railways, and industries and powered public services such as street lighting and waterworks. Over the next six years, the NSEC attained a load factor that exceeded Commonwealth Edison's—the suburban system's generators were in use an average of 65–70 percent of the time compared with a 40–42 percent average use of the city's generating capacity.[30]

This remarkable achievement reflected a shift in Insull's marketing strategy in 1907 from signing up commercial and residential customers to attracting large power consumers. His utility's structure of rates strongly favored these wholesale purchasers, who paid an average of 6.4¢ per kwh compared with 12.4¢ per kwh for retail lighting customers. Five years later, big power consumers accounted for a larger proportion of central station service in the suburbs than in the city. The low-density sprawl of the suburbs, moreover, continued to make wiring its neighborhoods and subdivisions much costlier than in the city. Besides larger capital outlays for stringing the lines, the NSEC suffered major losses in distributing electricity from the station to the customers' meters. Between 1907 and 1912, Insull's more efficient system cut these losses from an average of 38 percent to 27 percent, but they remained a major drain on the utility's resources. Low-density suburban areas made it even more important to reduce the operating expenses of the central station, since there was no way to recover the energy lost during transmission and distribution.[31]

Insull's consolidation of a metropolitan power network effected tremendous savings that significantly reduced the unit cost of electricity in the suburbs. The dramatic results of converting many small generating stations into a single unified system are summarized in table 18. In 1908 householders and shopkeepers benefited from these gains through a rate cut in the primary charge from 20¢ to 16¢ per kwh. In comparison, Chicago home consumers were paying only 12¢ per kwh for the primary charge under the 1907 rate ordinance. This contrast highlights the fact that cheaper electricity alone does not adequately account for mushrooming energy

Table 18 Comparison of the Operating Costs of the North Shore Electric Company, 1907 and 1909

	1907	1909	Savings (%)
Average fuel consumption, old versus new stations (lbs. of coal per kwh)	12.1	5.7	52.9
Average fuel consumption, all stations combined (lbs. of coal per kwh)	9.4	6.3	33.0
Average fuel costs (per ton of coal)	$2.24	$1.37	38.8
Average fuel costs (per kwh)	1.27¢	0.43¢	66.1
Average labor costs (per kwh)	0.47¢	0.29¢	38.3
Average operating costs (per kwh)	1.90¢	0.84¢	55.8
Average cost of electricity (per kwh)	6.62¢	4.89¢	26.1

Sources: "A History of the Public Service Company of Northern Illinois," Chicago, typescript ca. 1928, 32–37, in CEC-HA, box 4025; "Station and Distribution Output [of the North Shore Electric Company, 1907–11]," in CEC-HA, box 4227.

use. In just three years between 1906 and 1909, for example, the output of the NSEC jumped by over 400 percent, from 5 million to over 21 million kwh.[32]

This rapid rise in energy use reflected and reinforced the physical metamorphosis of Chicago into a fragmented metropolis of segregated communities. A careful analysis of variations in suburban patterns of consumption confirms that a greater use of energy in daily life was becoming an integral part of the cultural, social, and economic redefinition of the suburban ethos. In the 1900s, pent-up demand and the outward flow of people and industry sharply boosted the consumption of electricity, which grew as fast as Insull could expand generating capacity. In 1907, in fact, he apologized to the stockholders of the NSEC for not reporting a greater increase in business because of delays in receiving turbo-generators for the new Waukegan station.[33] Tables 19 and 20 present changing patterns of energy use for three typical suburban districts. Highland Park and Harvey–Chicago Heights offer case studies in residential and industrial areas, respectively. Evanston is an example of a built-up mixed suburb that remained primarily residential but contained some industries and a downtown section of shops and offices.

Even during the brief formative period from 1907 to 1911, the statistics reveal trends toward a widening divergence between the suburbs. In all three places, the importance of public street lighting declined compared with sharp rises in private consumption, especially for power. Distribution grids were expanding beyond Main Street in suburban areas throughout the metropolitan re-

Table 19   Retail Energy Consumption in Three Suburban Districts, 1907–11

		Lighting		Power		Street Lights		
Year	Population	000 kwh	%	000 kwh	%	000 kwh	%	Total
Highland Park–North Shore Residential District 1[a]								
1907	9,758	261.0	70.5	23.3	6.3	85.9	23.2	370.2
1911	10,984	935.8	73.6	168.7	13.2	168.7	13.3	1,272.2
Harvey–South Suburban Industrial District 2[b]								
1907	20,476	269.0	45.7	136.5	23.2	183.6	31.2	589.1
1911	25,165	843.1	21.9	2,572.2	66.8	432.1	11.2	3,847.4
Evanston–Mixed Suburban District 3[c]								
1907	27,412	519.2	58.0	116.5	13.0	259.4	29.0	895.1
1911	31,434	1,790.9	46.9	1,281.9	33.5	748.4	19.6	3,821.2

Source: "Station and Distribution Output [of the North Shore Electric Company, 1907–11]," in CEC-HA, box 4227.
[a]District includes Highland Park, Lake Forest, Highwood, and Glencoe.
[b]District includes Harvey and Chicago Heights.
[c]District includes Evanston and Wilmette.

Table 20   Per Capita Retail Energy Consumption in Three Suburban Districts, 1907–11 (Kilowatt-Hours)

Year	Lighting	Power	Streetlights	Total
Highland Park–North Shore Residential District 1[a]				
1907	26.7	2.4	8.8	37.9
1911	85.2	15.2	15.4	115.8
Harvey–South Suburban Industrial District 2[b]				
1907	13.1	6.7	9.0	22.7
1911	33.5	102.2	17.2	152.9
Evanston–Mixed Suburban District 3[c]				
1907	18.9	4.2	9.5	32.6
1911	57.0	40.8	23.8	121.6

Source: "Station and Distribution Output [of the North Shore Electric Company, 1907–11]," in CEC-HA, box 4227.
[a]District includes Highland Park, Lake Forest, Highwood, and Glencoe.
[b]District includes Harvey and Chicago Heights.
[c]District includes Evanston and Wilmette.

gion. In each of the case studies, moreover, the proportioning of energy use between small consumers and power customers corresponds well with community type. In Highland Park, for example, householders (and shopkeepers) strengthened their predominant use of electricity from 70.5 to 73.6 percent of total demand. In contrast, the factory quickly became the major source of demand in

Harvey, reducing the residential customer to a smaller and smaller proportion of the total demand for electricity. Evanston falls somewhere between these two extremes.

Examining the amounts of electricity used by each person in the three suburban districts supports the conclusion that energy helped further the spatial transformation of Chicago into a fragmented metropolis. Sharp increases in per capita demand for electricity reveal that consumption was rising faster than population throughout the suburbs. At the same time, however, the gap between community types widened. On the one hand, in 1911 Highland Parkers used more than two and a half times as much electricity in their homes as householders in Harvey. Only four years earlier, there was only a twofold difference. On the other hand, Harvey consumed almost seven times as much energy per person for industrial purposes as Highland Park. Earlier this gap was less than threefold. Evanston again fell between the two community types in the per capita use of residential lighting and industrial power. This suburb's downtown business zone and university campus account for its leadership in the amount of energy used for street lighting.

While Chicagoans used technology to insulate themselves in suburban enclaves, it was also binding them more closely together. Electric lines were weaving office workers, commuters, housewives, and factory workers throughout the Chicago area into a single community of energy consumers.[34] After the interconnection of Commonwealth Edison and the NSEC in 1907, electricity flowed freely between city and suburb to meet various needs of this inadvertent community. This exchange of energy helped build and diversify the electrical load of Insull's metropolitan web of power, reducing costs for both utilities (see map 5). By 1911, statistical reports on suburban patterns of energy use further verified his gospel of consumption. The results of linking the city and the suburbs seemed irrefutable proof that his method of generating and marketing energy set in motion a self-perpetuating cycle of rising consumption and falling rates. Now convinced beyond any doubt about the brilliance of Rathenau's vision of gigantic metropolitan power grids, Insull began plotting the creation of an even larger regional system.

Regional Networks

In the early twentieth century, the forces of economics, culture, and technology that were tying Chicago and its suburbs closer to-

MAP 5. Service area of the North Shore Electric Company, 1910. *Source:* Whetstone, "Historical Factors."

gether were also locking the satellite cities into the city's orbit. Symbolic of this regional integration was the completion of an outer belt freight railroad in 1889, followed by a parallel system of interurban passenger lines in the opening decade of the new century. Creating an arc with a thirty-five mile radius around the central city, the Elgin, Joliet, and Eastern Railroad triggered industrial and building booms not only throughout the Fox River valley but also in a string of towns from Waukegan on the north to Chicago Heights on the south. In 1900 the opening of the Sanitary and Ship Canal helped stimulate a similar economic revival in the communities stretching down the metropolitan corridor of the Illinois River valley. Modern communications complemented these transportation improvements, shrinking time and distance between Chicago and its formerly semiautonomous satellite cities.[35]

In 1910–11 Insull began weaving this rich hinterland of waterfront cities, extractive industries, and farms into his network of power. In another of his bold moves, he brought five large utilities, including the NSEC, under a single holding company, the Public Service Company of Northern Illinois (see table 21). The new enterprise expanded the orbit of the Insull system from 1,250 to 4,300 square miles (see map 6) and embraced a wide assortment of public utilities, including electric, gas, water, and interurban railway services. Insull quickly took advantage of a new diversity in demand to further reduce the unit cost of electricity. In a pioneering experiment in rural electrification, for example, he showed that the energy demands of agriculture, drainage projects, mines, and quarries corresponded nicely with off-peak hours of the day and seasons of the year in the city and suburbs. By the time the war erupted in 1914, Insull had put a modern infrastructure in place. The regional integration of electrical grids in Chicago would spur even more rapid urban deconcentration during the postwar decade of the twenties.[36]

In the 1900s, the appearance of large, sprawling utility enterprises similar to Insull's NSEC suggests that the demand for modern urban services was becoming universal. The formation of several of these systems in the Chicago region also shows that the financial community increasingly favored the diversified holding company over the single-utility enterprise. Closer examination of these large organizations reveals that Insull was not the only one building efficient and profitable utility services in the Chicago area. For instance, the Chicago Suburban Light and Power Company (CSLPC) was supplying electricity to a larger proportion of

Table 21 Communities Served by the Public Service Company of
Northern Illinois, 1911

Company	Population	Electric Customers
North Shore Electric Company		
Northern division (43 communities)	92,545	12,146 (13.1%)
Western division (13 communities)	54,581	7,748 (14.2%)
Southern division (14 communities)	39,911	3,002 (7.5%)
Subtotal	187,037	22,896 (12.2%)
Chicago Suburban Light and Power Company		
Oak Park	19,444	5,079
River Forest	2,456	462
Subtotal	21,900	5,541 (25.3%)
Economy Light and Power Company		
Joliet	34,670	2,527
Lemont	2,284	174
Plainfield	1,019	67
Rockdale	1,101	38
Subtotal	39,074	2,806 (7.2%)
Illinois Valley Gas and Electric Company		
Braidwood	1,958	131
Coal City	2,667	338
Cornell	536	65
Dwight	2,156	324
Gardner	946	108
Grand Ridge	403	55
Henry	1,687	270
Lacon	1,495	212
Mason	471	41
Morris	4,563	464
Odell	1,035	207
Ransom	370	36
Seneca	1,005	159
South Wilmington	2,403	61
Sparland	461	37
Streator	14,253	568
Wilmington	1,450	90
Subtotal	37,859	3,166 (8.4%)
Kankakee Gas and Electric Company		
Bradley	1,942	81
Bourbonnais	611	69
Kankakee	13,986	1,481
Subtotal	16,539	1,631 (9.9%)
Grand total	302,409	36,040 (11.9%)

Source: "Study of the Properties . . . of the Public Service Company of Northern
Illinois . . . ," Illinois Commerce Commission Case 22353 (1933), Company Exhibit 2, in CEC-HA, box 4093.

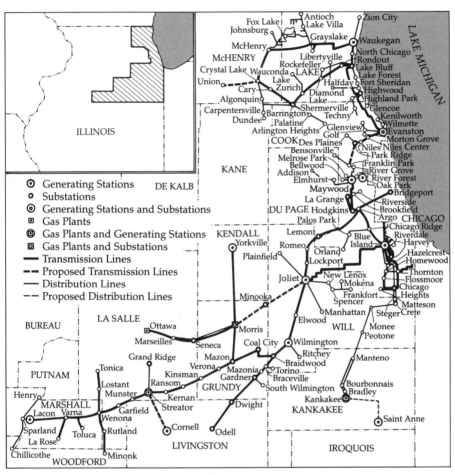

MAP 6. Service area of the Public Service Company of Northern Illinois, 1912. *Source:* Public Service Company of Northern Illinois, *Annual Report*, 1912, in CEC-HA, box 4094.

the residents of Oak Park and surrounding communities at lower rates than the NSEC. Charles Dawes's Northwestern Gas Light and Coke Company was another example of a fast-growing, successful utility in the suburbs.[37]

The growth of electric service in Joliet before its takeover by In-sull's corporate empire illustrates the building of a modern infra-structure in the satellite cities of the Chicago region. Before the turn of the century, Joliet presents a typical case of demand out-stripping the supply of electrical energy. On 21 February 1880, local entrepreneurs had been the first in the state to incorporate an electric utility company. But the riverfront community of about 11,000 inhabitants had to wait three years before the brothers A. C. and S. S. Badger from Chicago actually began to illuminate Main Street with fifty arc lights perched on 200-foot-high towers. Technological and financial constraints hamstrung the efforts of the Badger Electric Company to meet the need for better lighting, encouraging new competitors to enter the field. Over the next decade, the city experienced a pattern of interfirm rivalry, mergers, and new ventures vying for the municipal street lighting contract. In 1890, for example, local utility operators incorporated Joliet's sixth electric utility, the Economy Light and Power Company (ELPC) in order to buy out the latest recipient of a city contract, the Thomson-Houston Company, and combine it with the losing bidder, the Joliet Electric Light Company.

The ELPC started a new stage of utility competition because it was the first to take advantage of Joliet's waterpower to generate comparatively cheap electricity. For over thirty years, gristmills had been an important part of the local economy in the farm-based towns along the Illinois and Michigan Canal. Now the ELPC hooked up a DC dynamo at Hyde's Mill, though the miller retained the right to process 225 barrels of flour a day. Ice in the winter and low water in the summer necessitated the construction of a steam plant to supplement the mill's five water wheels, yet waterpower still held the potential to supply the cheapest form of electricity. A unit of electricity generated by a coal-fired steam plant cost five times as much as one produced by waterpower. This cost advantage helped the ELPC struggle through the depression of the nineties and accumulate the financial reserves to purchase potential rivals before they could become operational utilities. In 1898 the coming of the Chicago Sanitary and Ship Canal opened an opportunity for the ELPC to build a much more modern waterpower plant.[38]

By the turn of the century, Joliet had grown almost threefold to nearly 30,000 people, and its economy was thriving on new business brought by the outer belt line. The most direct benefit was the establishment of the railroad's repair shops in the city. In addition, nearby stone quarries were exploited to supply building materials for a wide range of private developments, rail lines and paved roads, and other public works projects throughout the region. Joliet's location and natural resources also attracted manufacturers such as Quaker Oats and the National Match Company.[39] Local prosperity and a rising standard of living created new demands for energy to power street railways, waterworks, streetlights, commercial shops, and private homes. For the ELPC, Joliet's prospects left little doubt about the decision to build a dam across the Des Plaines River. The new facility would contain thirty-three water wheels with the equivalent of 10,000 HP.

To raise $300,000 to pay for the project, Joliet's central station operators turned to Samuel Insull. Already regarded as a financial wizard in the utility industry, he had carefully cultivated a network of friendships with money men in Chicago, New York, and London. In this case the utility executive joined with J. J. Mitchell of the Illinois Trust and Savings Bank (the forerunner of Continental Illinois Bank) to help broker the sale of the ELPC's bonds. In exchange, Insull became a major stockholder and vice president of the firm. However, he did not take an active role in the operations of the Joliet utility, with one exception. True to the gospel of consumption, he insisted the company hire solicitors to market energy in the community. For the most part, though, electric service in Joliet was left in the hands of an extremely imaginative local manager, Charles A. Monroe.[40]

Under Monroe's direction, the ELPC became a fast-growing, innovative, and profitable enterprise. Most important, Monroe was the first central station operator in the Chicago region to sign a long-term power contract with a railway company. In October 1901, the manager proposed to supply electricity to a new interurban company that was in the process of merging three lines in the Des Plaines River valley to provide continuous service between Joliet and Chicago. The pioneering contract did not include a minimum load guarantee, but it did contain a fifteen-year monopoly clause and a two-tier rate structure to encourage consumption. The railway contract soon produced a related series of lucrative deals for large blocks of energy. At the time the new dam complex was completed in 1903, Monroe signed a contract with the canal commissioners to supply power to a pumping station in Bridgeport, a neighborhood in southwestern Chicago. In spite of the expense of building a 33,000-volt transmission line fifty miles to the metropolis, the low cost of electricity generated by the company's water wheels gave it a competitive edge over the turbogenerators of the Commonwealth Edison Company. In fact, the Chicago-based utility became one of the ELPC's customers. Towns along the transmission line also purchased electricity from the Joliet firm, including Lemont, Plainsfield, and Rockdale. By 1910 Monroe had

Table 22　Growth of the Economy Light and Power Company, 1896–1911 (Dollars)

	1896	1906	1911
Light income	43,106	101,400	167,309
Power income	2,151	102,892	253,728
Merchandise sales	—	5,238	99,867
Total income	45,257	209,530	520,904
Operational expenses	28,209	94,780	216,231
Interest charges	953	39,000	79,813
Total expenses	29,162	133,780	296,044
Net profit	16,095	75,750	224,860

Sources: Economy Light and Power Company, "Trial Balances, 1903–1911," in CEC-HA, box 4223; Economy Light and Power Company, "Corporate Record Book, 1890–1906," 129, in CEC-HA, box 4210.

built up not only a diversified list of customers but an enviable record of profits (see table 22).[41]

Up and down the river valley, the demand for more energy was fueling programs of utility modernization and expansion similar to Joliet's. In Kankakee, waterpower sites lured a succession of local and out-of-state investors to exploit the promise of cheap electricity. By 1910 the city of 14,000 people supported a unified system of utility services, including electric, gas, and street and interurban railways. Farther down the valley, rich deposits of coal served an analogous function in stimulating the growth of energy-consuming industries. Streator, for instance, was established in the Civil War era by the Vermillion Coal Company and named after its first president. As a mining center and rail hub, Streator was an obvious site to convert cheap fuel into a full range of up-to-date urban services for its 14,000 inhabitants as well as the surrounding farm villages. In these smaller communities, of course, the pace of upgrading the infrastructure was more uneven, depending from place to place on a wide array of locational, economic, and social factors. Some towns had no electric services, some were supplied with arc lamps on Main Street, and others enjoyed the benefits of being a part of larger distribution grids.[42]

Rural electrification remained the least promising aspect of the central station business. The costs of stringing wires and transmitting electricity in low-density districts were difficult to justify, especially since the farm population was declining in northern Illinois. Few utility companies were willing to invest in rural

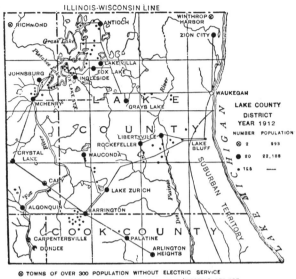

ⓧ TOWNS OF OVER 300 POPULATION WITHOUT ELECTRIC SERVICE
● TOWNS OF OVER 300 POPULATION WITH 24 HOUR ELECTRIC SERVICE
• CUSTOMERS OUTSIDE OF TOWNS WITH 24 HOUR ELECTRIC SERVICE

MAP 7.   Rural electrification in Lake County, Illinois, 1912. *Source:* William E. Keily, ed., *Central-Station Electrical Service: Its Commercial Development and Economic Significance as Set Forth in the Public Addresses (1897–1914) of Samuel Insull* (Chicago: privately printed, 1915), 360.

distribution grids while much more profitable and fast-growing markets still awaited full exploitation. Only a utility leader of Insull's stature could find a rationale even to experiment with farm electric service. Insull's interest in the project was personal; in 1906 he purchased a country home in Libertyville, just west of the exclusive North Shore suburbs. Hawthorne Farm, as it was known, reflected the immigrant's upbringing in England, where a country estate was regarded as the ultimate sign of success. Electricity was available from the local utility company only during the evening, so with his usual flair Insull set about supplying his farm with modern central station service.[43]

By 1910 Insull had upgraded utility services throughout Lake County to the point of creating the first working model of rural electrification in the United States (see map 7). Most important, Insull found new diversity of demand in rural districts that complemented the load characteristics of city and suburban consumers. Farmers, for example, used electricity most during the summer, the season of least demand in urban areas. In addition, rural districts had unique needs for large amounts of energy that could be met conveniently during the off-peak hours of the urban central station. For example, the slow, steady pumps used to drain sub-merged farmlands over long periods could be operated late at night and at other hours of slack demand. In a similar way, special low rates provided an incentive for waterworks in rural villages to fill their storage tanks during off-peak hours. Insull's experiment firmly established the potential of the agricultural sector to contrib-ute to a regional community of energy consumers. Unfortunately, financing rural electrification would remain a problem for another twenty years.[44]

In contrast, Insull and other utility operators found investors eager to underwrite holding companies that embraced the satellite cities of the great urban centers. In April 1910, Charles Monroe led the creation of the Illinois Valley Gas and Electric Company (IVGEC) to gain control of utility services in this thriving metro-politan corridor to Chicago. In less than a year, the tireless Monroe purchased fourteen existing systems and obtained franchises in fifty-six other communities, including twenty-two electric, twenty-one gas, thirty-one combined gas and electric, two water supply, and one street railway grant.[45] Although the IVGEC covered a large territory, it was merely preliminary to Insull's consolidation of a regional network of power under the PSCNI.

In the years following, up until the emergency of World War I, the main task facing Insull and his lieutenants was to modernize and integrate the resulting collection of systems. They junked ob-solete generating equipment, installed new large-scale turboge-nerators, strung high-voltage transmission lines, built substations, and extended distribution grids. At the same time, Insull contin-ued his relentless drive to establish a complete monopoly of utility services in the suburbs. In 1913 the acquisition of the Northwest-ern Gas Light and Coke Company ended the independence of the last major energy supplier in the metropolitan area. The consoli-dation boosted the number of PSCNI gas customers from 16,000 to 60,000, briefly outnumbering the consumers of central station elec-tric service.[46]

A Regional Community of Energy Consumers

Insull's integration of a regional power system had an immediate effect on Chicago and its people. By 1912 the gap between the sup-ply of energy in the city and outside it had narrowed in important ways. The ready availability of electricity (and gas) at low rates

Table 23   Comparison of Electricity Consumption in Chicago and the
Metropolitan Area, 1914 (Kilowatt-Hours)

	Chicago		Metropolitan Area[a]	
	Amount	Per Capita	Amount	Per Capita
Population	2,391,000		677,000	
Light	193,998,000	81.1	22,257,000	32.9
Power	160,829,000	67.3	36,140,000	53.4
Railway	627,538,000	262.4	29,209,000	43.1
Utilities	9,376,000	3.9	36,061,000	53.2
Total	991,741,000	414.7	123,667,000	182.6

Sources: U.S. Census, Population 1920, 1:393–405; H. A. Seymour, "History of Commonwealth Edison Company," 1934, Appendix D, in CEC-HA, box 9001; Public
Service Company of Northern Illinois, "Reference Statistics," ca. 1935, (in my
possession).
[a]For our purposes here, the metropolitan area consists of the following counties:
suburban Cook, Lake, McHenry, Will, Kankakee, Grundy, Kendall, La Salle, Marshall, and Livingston.

throughout the region fostered suburbanization and deconcentration. This outward movement was matched by a steady improvement in the flow of people and information among the residential,
industrial, and satellite communities of the region. Modernizing
the energy infrastructure was only one part of a massive public
works campaign to upgrade essential services to the standards of
the central city. But the use of more and more energy was becoming a systemic part of the delivery of a widening array of urban
services, including city and interurban transportation, water and
gas supplies, sewerage, sanitation, and communications.

Comparing the consumption of electricity in the urban core and
on the periphery reveals how energy promoted the spatial transformation of the nineteenth-century city into the metropolis of the
twentieth century (see table 23). By the end of this period, the
average person inside the city used twice as much electricity as
someone outside it. This gap corresponds well with differences in
the two areas' population density and land values. For example,
Chicago's trolley and elevated railways served vast numbers of
daily commuters whereas the typical interurban railroad handled
far fewer riders. In a similar way, the amount of artificial illumination used in the center outpaced use on the periphery. Metropolitan integration contributed to the growth of the central busi-

ness district, with its lights beaming from office towers, shops, hotels, night spots, and advertising signs. Higher densities in Chicago also meant a more rapid spread of the distribution grid in the residential neighborhoods and along their business strips.

In contrast, the growing availability of energy on the urban fringe was already furthering the deconcentration of industry in highly visible ways. In the per capita use of electricity for power, the gap between the center and the periphery had narrowed significantly. By 1914 the PSCNI's list of industrial customers was impressive for the variety of manufacturing activities as well as their territorial distribution (see table 24). The fast-growing energy business was a part of this shift to large tracts of land in the industrial suburbs and satellite cities. The sheer size of modern electric and gas plants, their need for vast quantities of water, elaborate railroad sidings, huge coal storage areas, and room for expansion, as well as their pollution of the surrounding air, all pointed the utility operator away from downtown. In the case of Chicago, water-power from the Illinois River valley provided the region with a cheap base load twenty-four hours a day, while the PSCNI's stations in Blue Island, Maywood, and Waukegan carried an increasing share of Chicago's demand for more electricity.[47]

From 1898 to 1914, the search for better lighting became something more; it was evolving into a fundamental dependence on energy to sustain modern urban life. The roots of an energy-intensive society were firmly planted during this period. It was still a formative era, however, characterized by a highly uneven pace of spatial diffusion and social change. Energy technology made its first great impact in the public and semiprivate sectors of the metropolis. For both rich and poor, electric lighting and rapid transit were inexorably altering daily life in the city streets and the Main Streets of surrounding communities. Commercial shops, service industries, and popular amusements also contributed to the establishment of new cultural standards of modernity and urban styles of life. By contrast, energy technology seems to have had the least effect in the private realm of the family. New models of ideal homes and domestic relationships appeared full blown, but traditional values retained a tenacious hold on normal household routines.

In a second important way, this period saw the emergence of the modern metropolis of the twentieth century. The revolt against the city reinforced traditional American ideals of the single-family homestead to produce a suburban ethos. New energy-consuming technology provided increasing numbers of the middle class an

Table 24 Industrial Customers of Central Station Service in the Chicago Region, 1913

Company	Type	Connected Load (HP)		
		Firm	Town	Area
Northern Area				
Algonquin				
Reinert Brothers	Sand/gravel pit	140	140	
Arlington Heights				
Bray and Kates Co.	Manufacturing	165	165	
Carpentersville				
American Sand and Gravel Co.	Sand/gravel pit	140		
Illinois Iron and Bolt Works	Manufacturing	150	290	
Evanston				
Mark Manufacturing Co.	Foundry	1,500		
Mears-Slayton Lumber Co.	Manufacturing	120	1,620	
Fox Lake				
Knickerbocker Ice Co.	Sand/gravel pit	180	180	
Libertyville				
Lake County Gravel Co.	Sand/gravel pit	130	130	
North Chicago				
Chicago Hardware Foundry Co.	Foundry	500		
Cyclone Fence Co.	Manufacturing	125		
Vulcan Louisville Smelting	Foundry	155	780	
Techny				
Society of the Divine Word	Religious order	206	206	
Waukegan				
Wilder-Manning Tanning Co.	Manufacturing	382	382	3,893
Southern Area				
Blue Island				
Chicago, Rock Island and Pacific Railroad Shops	Transportation	600	600	
Bridgeport				
Illinois and Michigan Canal	Public utility	600	600	
Chicago Heights				
Acme Brick Co.	Brick/clay	250		
Central Locomotive and Car Works	Manufacturing	500		
Edgar Allen American Manganese	Foundry	1,500		
J. Joseph and Brothers Co.	—	150		
King and Andrews Co.	Foundry	140		
United Car Co.	Transportation	130		
Whiteacre Fireproofing Co.	Brick/clay	300		
Victor Chemical Co.	Manufacturing	600	3,570	
(*Continued*)				

Table 24 *(Continued)*

Company	Type	Connected Load (HP)		
		Firm	Town	Area
Harvey				
Bliss and Laughlin	Foundry	190		
Buda Co.	Manufacturing	300		
Forsyth Brothers Co.	Manufacturing	300		
Ingalls-Sheppard Forging Co.	Manufacturing	155	945	
Hodgkins				
Dolese and Shepard	Stone quarries	2,660	2,660	
Midlothian				
Midlothian Country Club	Amusement	150	150	
Palos Park				
Green and Sons	—	500	500	9,025
Western Area				
Bellwood				
O'Laughlin Stone Co.	Stone quarry	435	435	
Franklin Park				
Franklin Park Foundry Co.	Foundry	266	266	
Forest Park				
Forest Park Amusement Co.	Amusement	134	134	
La Grange				
Illinois Stone Co.	Stone quarry	335	335	
McCook				
Chicago Union Lime Works Co.	Stone quarry	480		
U.S. Equipment Co.	—	100	580	
Melrose Park				
American Brake Shoe and Foundry	Foundry	250	250	2,000
Illinois River Valley				
Coal City				
Illinois Shale Tie Co.	Brick/clay	175	175	
Joliet				
American McKenna Process Co.	Foundry	348		
Chicago Contractors Supply	Stone quarry	250		
Elgin, Joliet, and Eastern Railway Shops	Transportation	450		
Gerlach-Baklow Co.	Manufacturing	171		
A. Groth and Co.	—	157		
Joliet Bridge and Iron Co.	Foundry	580		
Joliet Rolling Mill Co.	Foundry	297		
Joliet Sand and Gravel Co.	Sand/gravel pit	515		
Michigan Central Railroad	Stone quarry	335		
National Match Co.	Manufacturing	228		

Table 24   *(Continued)*

| Company | Type | Connected Load (HP) | | |
		Firm	Town	Area
W. E. Proatt Manufacturing Co.	Brick/clay	280		
Quaker Oaks Co.	Manufacturing	136	3,747	
Lemont				
Illinois Stone Co.	Stone quarry	345	345	
Lockport				
Barrow Lock Co.	Manufacturing	150	150	
Richie				
Wabash Railroad Co.	Stone quarry	175	175	
Sparland				
Illinois Valley Coal Co.	Mining	160	160	
Streator				
Streator Metal Stamp Co.	Manufacturing	164	164	
Wilmington				
American Straw Board Co.	Manufacturing	450	450	5,366
Grand total				20,284

Source: "Electric Service in Chicago Suburbs," *Electrical World* 61(7 June 1913): 1249–51.

affordable means of escape from city neighborhoods to the relative safety of the suburbs. One indicator of growing energy use in these residential enclaves was the wiring of houses built before electric services were available. This retrofitting quickened from 600 installations in 1910 to 10,200 in 1915, when the conversion was about half finished.[48] The movement of industry away from the congestion and the social evils of the city speeded urban deconcentration. As Chicagoans used more and more electricity, they encouraged this spreading out of the city into a web of communities segregated by class, race, and land use.

At the same time, the application of energy helped weave this sprawling region into a more or less unified economic matrix. In effect, an expanding infrastructure of automobile highways, rapid transit, and instant communication brought Chicago's hinterland much closer to downtown. The new technology helped expand the physical boundaries of the city to regional proportions while binding its economy more tightly to a centralized administration. Insull's use of the holding company to consolidate a utility empire represents a classic case of the trend toward concentration of economic power. The contrasting fragmentation of local government in the metropolitan area reflected the triumph of an antiurban, suburban ethos. This divergence between economic integration and political fragmentation would become pivotal in forming public policy on providing essential utilities.

Between 1911 and 1913 local and state governments began responding to the growing importance of electrical energy in everyday life, but Samuel Insull continued to take the initiative in the political arena. In the absence of effective political institutions on a metropolitan level, he was able to finesse a decisive shift of power from city hall to the state capital. By taking an early stand in favor of state regulation, Insull put himself in a strong position to influence the legislators' creation of a public utility commission. In 1911 the lawmakers created a study group to report on a regulatory agency modeled along the lines drawn by the National Civic Federation and already enacted in Wisconsin and New York. However, factions in both parties from Chicago demanded modifications to give the city a measure of "home rule." For two years these local forces were able to stall Insull's drive to escape from the scrutiny of elected officials in Chicago and the suburbs and substitute a remote state agency appointed by the governor.

In January 1913 the outgoing Republican governor, Charles Deneen, recommended a separate commission for Chicago. A month later the new Democratic chief executive, former mayor Edward Dunne, echoed these sentiments. A bill embodying a mechanism for the regulating of local utilities by city officials passed the Senate, but the House rejected the proposal seventy-five to forty-three. In the final bill, the cities retained only the power to grant franchises, while rate-making authority was transferred to the state commission. To what extent Insull influenced the legislature is difficult to determine, but the reform was highly favorable to his interests.[49]

In 1913, however, considerable confusion remained in Chicago over whether the city council could still set the rates of its franchised utilities. Since the act of 1905 granting this authority had not been repealed, the aldermen presumed that it was valid until superseded by specific rulings of the state commission. The expiration of ordinances of 1908 setting maximum rates for five-year terms presented the opportunity to negotiate with the electric, gas, and telephone companies. A pressing need to adjust the financial arrangements of the ailing transit companies also thrust the city council to the forefront of policy formation. In the case of the Commonwealth Edison Company, Insull's image as the preeminent leader of the industry helped him line up fifty-two of the sixty-three aldermen behind his proposal to change the formula from the mandated rates of 13¢ per kwh for the first thirty hours of maximum demand and 7¢ per kwh for additional use to the current charges of 10¢ and 5¢ per kwh. The council added a third tier of 4¢ per kwh for use beyond sixty hours.[50]

Insull's achievement of political stability combined with the recent attainment of true economies of scale in central station service to eliminate virtually all the remaining constraints on the supply of electrical energy. A decade of extraordinary success in finance, marketing, technology, and politics left little doubt about the validity of his gospel of consumption. The industry's leading trade journal, the *Electric World*, agreed that "considered from the viewpoint of men interested in the manufacture, distribution and sale of electricity, Chicago stands in a class by itself. . . . it is perhaps the most interesting large city in the world—the one where the methods employed in electrical merchandising can be studied with the greatest profit. The truth of these statements is widely recognized."[51]

Part III

The Integration of a Region, 1914–1932

8

The Energy Crisis and the Birth of the Machine Age, 1914–1919

World War I caused an energy crisis, the first in the nation's history. At the depths of the coal shortage during the fearsome winter of 1917–18, half-frozen mobs stormed the gates of fuel yards in Philadelphia and New York City. There were no riots in Chicago, but people were found frozen to death, and cases of pneumonia and related diseases soared into the thousands. "Send us aid at once, or we will perish," an apartment dweller pleaded in a letter to the *Chicago Tribune*, "we are again out of coal. . . . The windows are frozen over worse than in the slums."[1] Without fuel, milk, or food, entire communities faced starvation and exposure. In the suburbs, families were forced to huddle together in the houses of neighbors who had wood-burning fireplaces. The crisis became so grave that President Woodrow Wilson had to issue draconian orders to shut down all industry and business east of the Mississippi River for several days. A series of heatless and lightless days followed as federal officials made desperate efforts to stem the slide toward the breakdown of the nation's economy. The energy crisis and widespread food shortages brought the war home to every family in urban America.[2]

In spite of the magnitude of this unprecedented national emergency, however, American historians remain sharply divided over the meaning of the World War I for the home front. On one side, social and economic historians usually downplay the domestic impact of the war. Since the late 1920s they have consistently rejected the contention that the conflict of 1914–19 was a historical turning point or a catalyst of rapid change in America. Instead, they have stressed the continuity of change by emphasizing the temporary nature of the American intervention and the quick return to "normalcy" after the armistice of November 1918. Scholars usually underscore the contrast between the effects of the war on the United States and on Europe. Over here there was no massive destruction, heavy loss of life, chronic malnutrition, political revolution, or even painful taxation. Within a transatlantic context, the United States seems to have occupied a minor position on the periphery of sweeping change.[3]

On the other side, students of American culture and art consider World War I a historical crossroads, giving credence to contemporary descriptions of the conflict as the birth of the machine age. For example, Gilman M. Ostrander contends that "World War I came as the catalyst which separated American materialism from American idealism and opened the way for the Roaring Twenties. . . . Both for better and for worse, modern culture *is* mass culture, and the American Jazz Age was nothing less than the world premier of modern culture."[4] For American artists and writers, the war marked a great divide. Their work reflects impressions of a society undergoing rapid changes in perception of an environment that was becoming increasingly technological and urban. During the postwar decade, the avant-garde of art and literature continued to explore the dynamics of this new relationship among the individual, the city, and the machine. Could the war have changed so much and so little at the same time?

Perhaps it takes the broad perspective of a European scholar to resolve this apparent contradiction in American historiography. For example, the Englishman Neil A. Wynn observes that wars produce changes in perception that are just as important as changes in a nation's material conditions. The new directions in art and literature that appeared during the period support this interpretation of the war as an important crossroads in American history. Artists are often the first to reflect upon society's changing sensibilities toward its environment. To Wynn, the new art's central motif of the machine was an appropriate symbol of the changes wrought by the upheaval. "By the end of the war," he reasons, "modern industrial methods had become widely established, and America was well on her way to becoming an organizational and managerial society."[5]

The Chicago case suggests that World War I was a pivotal experience that gave birth to a machine age. Decisions made during the war-caused emergency set the city decisively on a new course. The war marked a turning toward an intensive use of electrical energy in manufacturing, food processing, and the home. In each case this historic shift had profound and long-lasting ramifications that transformed the very nature of urban life. Where people lived, what kind of neighborhoods they lived in, how they got to their jobs, what kind of environment they worked in, how much they were paid, and how they spent their leisure time were all affected by the creation of a society dependent on energy. This chapter will focus on the most direct effects of the war by examining the revolution in factory production. Following chapters will consider the more subtle effects of the creation of a ubiquitous world of energy. New patterns of energy consumption would promote industrial deconcentration, residential suburbanization, and the economic integration of the Chicago region.

At base, the postwar decade in the United States was marked by the emergence of a materialist society that intellectuals have never accepted—nor have they given sympathetic treatment to this emerging culture of consumption. Negative perceptions of the changes wrought during 1914–19 may have turned American historians away from a more thorough examination of their society at war. For example, economic historian Ellis Hawley accurately describes the postwar era as follows: "In essence, [the American economy] . . . became the first industrial economy geared to the production of consumer durables and cultural fare for the masses, and as such was widely hailed as a new kind of economy, a 'people's capitalism' seeking profits through volume production and in the process healing social divisions and creating uplifting abundance." But Hawley quickly denies the validity of the new consumerism. In the 1920s, he argues, the new economy represented only a "premature spring," a full thirty years early. Hawley would like to push this transformation of American cultural values into the 1950s in spite of his own evidence to the contrary.[6]

In Chicago, the food and fuel crises accelerated the transition from old to new. For better or for worse, a consumer-oriented society would emerge full blown in the city during the postwar decade. The machine age was marked by an intensive use of energy, an "ethos of mass production," a reliance on installment buying, a new style of advertising, and a sharp rise in popular entertainment and leisure pursuits.[7] The decade preceding the outbreak of war in 1914 had seen the use of more and more energy to sustain urban life. The commercial and public utility sectors had led the way, while industries and households adhered to older technology and traditional patterns of energy use. Yet though the pace of change was uneven, it was gathering force in virtually every area of urban life.

In manufacturing the war triggered radical change. Coal shortages and government policies persuaded industrialists to abandon their steam engines and self-contained electric generators in favor of central station service. More important, the electric motor spawned a complete reorganization of the factory, its work force, and its consumer products. "Much of this innovation," economic historian John L'Brant states, "took the form of 'full mechanization,' identifiable primarily by the installation of mass-production methods. Frequently, innovations were tied to the beginnings of industrial electrification." In just a decade, from 1919 to 1929, the conversion of American industry to the electric-powered motor would create a new economy based on assembly-line methods, standardization, specialization, large-scale operations, and higher wages. A generation of influential business leaders had become apostles of this historic shift from scarcity to abundance, including Thomas Edison, Henry Ford, Herbert Hoover, Edward Filene, and Samuel Insull. In the postwar decade, America would embrace the ethos of mass consumption, which had already became known as "Fordism."[8]

The transformation of the food industry by an intensive use of electrical energy also had a major impact on urban America's standard of living and daily routines. Again the war marked an important watershed in the eclipse of older patterns and the rapid adoption of new ones. The first careful study of the nutritional history of the United States concludes that "if in 1930 pollsters had asked adult Americans what had been the most important factors in improving their eating habits, it is likely that a large proportion of middle-class Americans would have cited their experience during World War I." The lessons learned from the food shortages of the war years led to rapid changes in virtually every step of the food chain, from the use of artificial fertilizers on crops to the processing and storage of fresh meat and produce, the acceptance of the new supermarkets, and the spread of iceboxes in the home. Over the course of the twenties, Americans would significantly alter their diets and standards of nutrition.[9]

"This Is War"

"This is War," United States fuel administrator Harry A. Garfield retorted in response to the howls of protest against his decision to shut down the nation's factories, offices, and stores.[10] Although Garfield's decree was limited to a five-day period during January 1918, the energy crisis spanned the entire length of the American intervention in the war. During the twenty months from April 1917 to November 1918, acute shortages of fuel (and food) made an indelible impression on decision makers in government and business as well as on the popular imagination. Virtually no one in the urban centers of the United States escaped the hardships. As one of the country's leading industrial cities, Chicago provides an instructive illustration of this national emergency. Chicago's strategic location as the transportation hub of a vast coal-producing region should have softened the shocks of any wartime dislocations, yet the metropolis of the Midwest faced the same severe shortages as cities on the eastern seaboard. How could a country with a chronic oversupply of coal suddenly run out of such a basic necessity?

Famine in the midst of glut developed for two reasons, both related to the coming of the war. First, skyrocketing demand outstripped the ability of the nation's miners to increase the supply, especially after United States entry into the European conflict. Beginning in August 1914, the miners responded admirably, increasing coal production by about a third over the next two and a half years. In 1917, for example, they extracted an extra 42 million tons of bituminous coal over the previous year to boost output to a record 544 million tons. Yet the needs of industrial, military, merchant marine, and railroad consumers were growing twice as fast. By the middle of the year the result was sharply rising prices that threatened to trigger runaway inflation and speculative chaos in the fuel markets.

A second, more damaging problem arose from the railroad's inability to handle an explosion of freight traffic. Chicago alone required the delivery of two thousand cars of coal a day to maintain its economy. Yet fuel, in increasing amounts, was only one component of a war economy moving almost entirely over a single transportation system of railroads. The lines soon suffered from this tremendous overload, which was exacerbated by a lack of coordination among the various companies. An insufficient number of coal cars soon became hopelessly entangled in monumental traffic jams, especially as men and supplies were funneled to the seaport cities of the East Coast for shipment abroad. Snarled in these blockades, the cars could neither deliver coal to consumers nor remove it from the mines. As the traffic situation deteriorated in the summer and fall of 1917, some mines were forced to close simply because there was no place left to store the bulky commodity.[11]

Perceptions of an impending crisis were so widespread by the time America entered the conflict in April 1917 that fuel consumers demanded immediate government action. Each state, including Illinois, set up a council of defense to help coordinate the war effort. Governor Frank O. Lowden appointed Samuel Insull to chair the prestigious panel of businessmen, civic leaders, and politicians. Hundreds of letters were waiting for Insull from consumers who pleaded for the government to stem the rising price of coal and restore order to the transportation system. With typical enthusiasm, Insull mounted a publicity campaign. He encouraged Chicagoans to "buy and store" during the summer in anticipation of expected shortages during the winter. However, his call for cooperation went unheeded by fuel dealers, who held out for a government-ordered reduction of prices. In July, Insull brought together representatives of the fuel business to break the deadlock. But this effort at voluntarism also ended in failure as both the mine operators and the mine workers balked at a rollback of the "excessively high" prices. Concluding that "a coal shortage was inevitable," the Illinois state council issued a call for federal seizure of the nation's mines and railroads.[12]

On 10 August 1917, Congress responded to the mounting problems facing Illinois and other states by passing the Lever Food and Fuel Act. Ten days later, President Wilson set the price of bituminous coal at $2 a ton. The creation of a Food Administration is well known, because its manager was Herbert Hoover, whose leadership would soon make him the most popular man in America since Ulysses S. Grant. In comparison, historians have virtually forgotten the Fuel Administration and its boss, Harry A. Garfield. The son of a president, Garfield was an academic colleague and close friend of Wilson's.[13] Although a fixed price helped restore stability to the coal market, the worst winter storms in fifty years frustrated every attempt to untangle the gridlock of the nation's railroads.

In Chicago, the first period of subzero weather turned the early warning signs of impending famine into a full-scale public health crisis. In the fall, Hoover asked every American to endure one "meatless" and one "wheatless" day each week. To conserve fuel, Garfield ordered advertising signs darkened after 11:00 P.M. Insull believed that dimming the city's lights had a beneficial impact on the popular imagination, preparing people throughout the metropolis for the hardships to come. And come they did, in December, when a cold snap brought the gravity of the fuel shortage to the forefront of daily life. The homeless and the poor suffered first; the fuel administrator of Cook County, Raymond E. Durham, began adding 250 to 300 names a day to a list that already included 2,000 needy families without heat.[14]

On 9 December 1917, Durham assured the public that plenty of coal was on its way to Chicago, but the disruption of normal routines became widespread less than a week later. Schools began to close because of lack of coal, office buildings were "hardly habitable," the Wisconsin Steel Company shut down, and the U.S. Steel Company laid off four hundred workers. By 17 December, Durham admitted that coal supplies were down to "dangerous levels" as the winter weather slowed the movement of trains into the city. To alleviate the worst suffering, Insull opened the coal yards of the Commonwealth Edison Company to home consumers. The situation became truly grave farther east; New York City alone reported 263 deaths from pneumonia and over 1,000 new cases during the week. On 27 December 1917, Wilson instructed his secretary of the treasury, William McAdoo, to put the railroads under government control.[15]

Two days later, however, a devastating winter storm stretching from New Orleans to New York brought the transportation system to a standstill. As local supplies of coal were exhausted, yet another storm with winds of 50–60 mph buried Chicago under fourteen inches of snow. Called "the worst blizzard in the city's history," the bad weather completely stymied movement to and from the city as well as in the streets. Rail passengers were marooned along the lines, streetcars ground to a half, and delivery vehicles could not get through the snowdrifts. "God help the poor if it turns colder," the *Chicago Tribune* prayed.[16]

But more bad weather followed, frustrating efforts to dig out and to resume the delivery of fuel and food. By 13 January 1918, the situation was desperate. "Chicago is a blockaded city this morning," the newspaper reported, "its railroads practically useless, its milk supply exhausted, its coal famine growing more serious hourly." The plight of the suburbs and satellite cities was the same, since they too were isolated by mountains of drifting snow. The schools and most factories, including the steel mills and packinghouses, were forced to shut down. Over 150,000 idled workers joined 60,000 students in the battle to clear the city's streets and railroad tracks. While the war machine ran out of fuel in the industrial heartland of the Midwest, a fleet of ships loaded with men and supplies sat with empty coal bunkers at the docks in the seaport cities of the East.[17]

The magnitude of the crisis facing the home front of a nation at war demanded a response of equally grand proportions. On 17 January, Garfield's decree of a shutdown east of the Mississippi

River was a drastic reaction to a desperate situation. The five-day holiday allowed government officials to clear the tracks, to order priorities in the delivery of essential commodities, and to get 480 ships laden with vital materials on their way to the battlefront. In Chicago the order gave official sanction to the forced idleness of students and workers, whose ranks now swelled to include 350,000 factory hands, 50,000 retail employees, and 100,000 office clerks.

While continuing bad weather hampered efforts to break the transportation logjam, Chicagoans adjusted to the emergency. There was no escape; the famines of World War I touched the life of virtually every city dweller. Women social workers from society's upper crust learned to drive trucks laden with coal to settlement houses in the slums. Saloon keepers learned to evade "heatless" and "lightless" orders by using candlelight to serve thirsty patrons bundled up in overcoats. The emergency gradually dissipated with the coming of spring, though daily life in Chicago did not return to "normalcy" for another two years. Occasional shortages of food and fuel would reappear during this wrenching period of postwar adjustment.[18]

"Mechanization Takes Command"

If the Chicago experience was typical, the nation's first energy crisis made a lasting impression on the home front. Under the stress of war, the coal shortage posed a grave national challenge that triggered major changes in perception among the managers of the American economy. Before 1914, businessmen had given the subject of fuel conservation little if any serious thought. In meeting the needs of a nation at war, however, they were suddenly confronted with a whole new set of questions about saving fuel as well as speeding up production. Fighting the war on the home front shifted orientations from the top policymakers in Washington down to the factory foremen in Chicago who faced empty coal bins. The new consensus that emerged out of this common experience led to a rapid transition away from the use of coal, steam engines, and other self-contained energy systems and toward a reliance on central station service and electric motors. "Mechanization took command," Sigfried Giedion declared in a pathbreaking study of technology and society.[19]

Beginning in 1914, the American economy underwent a second industrial revolution. ("Revolution" is used here to emphasize the rapid acceleration of change rather than to imply a complete break with the past.) Like most other major turning points in history, the birth of a machine age had its roots in the preceding period. In the 1900s, the apostles of progress had worked out the technology, economics, and philosophy of a new order. Men like Insull, Ford, Hoover, and Filene had outlined a social vision that drew its inspiration from engineering principles of scientific management, assembly-line production, and progressive ideals of a rising standard of living. Their business careers provided a model of success that encouraged others to depart from traditional practices. Their leadership in winning the battle of the home front would elevate them to the nation's highest positions of political and financial trust. The war-caused fuel crisis, then, brought old ways of manufacturing to an abrupt end while speeding up the adoption of more modern techniques of factory organization.

By the 1910s, engineers had completed a lengthy process of designing a new style of mechanized factory, but tradition and inertia slowed the pace of industrial reform. As we have seen, for example, Insull largely failed to convince factory managers to junk their steam engines in favor of central station service. Even a schedule of rates that discriminated in favor of industrial consumers at the expense of residential and commercial customers had little effect. By 1914 the Commonwealth Edison Company supplied only 17 percent of the primary power used by Chicago industries. Consuming less than a fifth of its total output, the electric utility listed just 490 large power customers and approximately 1,500 smaller companies in a city that boasted over 10,000 manufacturing firms.[20]

Before the war, factory managers generally considered energy a negligible, virtually "free" cost of production. At base, fuel used to power machinery constituted a very minor part of the total cost of making most goods. In contrast, large amounts of capital were sunk in power equipment, including steam boilers plants, mechanical shafts, belts, and pulleys, and factory design specifications to support this cumbersome transmission system. However, the fuel to run the factory's furnaces and boilers was usually written off to the heating account rather than calculated as a separate amount for power. "The most specious argument," a utility advertisement carped, "is that electricity can be made for almost nothing as the by-product of a steam-heating plant that must be operated in any event." As long as the factory needed heating equipment, few managers worried about the cost of energy for power. Insull's sales pitch fell on deaf ears because it asked factory managers to junk their investment in steam power equipment, install a costly set of new wires and motors, and begin paying a monthly power bill.[21]

Between 1900 and 1914, many manufacturers instead pursued a middle path between a traditional reliance on the steam engine and a complete conversion to central station service. A self-contained electric generator offered many of the advantages of the new technology while preserving the illusion of "free" power. Simply attaching one of these devices to the old steam engine was also relatively cheap and posed little risk of upsetting daily work routines. At first factory managers installed self-contained systems to fulfill pressing needs for artificial illumination that was brighter and safer than gaslighting.[22] But they soon discovered how to use electrical wiring to transmit power from the steam plant to the machines on the work floor, thus breaking loose from the straitjacket imposed on factory layouts by mechanical shafts and pulleys. These were superseded by electric motors, which now drove groups of previously arranged machines. Reflecting the transitional nature of this "group drive" system, the motors were commonly placed on the ceilings and attached to belts, just like the shafts they replaced.

As new plants were built, however, group drive gave manufacturers greater freedom to redesign work spaces into specialized units and to reorder the flow of work into an efficient, assembly-line operation. No longer needing heavy ceiling supports, industrial engineers created spaces with higher walls, skylights, better ventilation, and overhead power systems for moving materials. Some manufacturers, moreover, took full advantage of the flexibility of electric technology by employing a system of "unit drive," coupling each machine with its own motor. Early applications often harnessed powerful motors to large machines such as printing presses, ice makers, and overhead cranes. In the 1900s, however, the electrical industry quickly expanded their product lines to offer manufactures a full range of motor sizes, speeds, and types.[23]

In 1907–8, the establishment of the Clearing Industrial District in a suburb southwest of Chicago marked the final step in the transformation of the factory. The 3,000-acre setting for the new workplace featured "green lawns, flowers around the buildings, boulevarded streets, attractive buildings and healthful surroundings," besides central station electric (and gas) service, transportation facilities, and one-story factories. "[Their] advantages are at once apparent," the district's promotional literature trumpeted: "Perfect light, especially where overhead lighting is used; closer supervision; easier handling of materials; no elevator expense and delay; better shipping facilities; cheaper and better foundations for machinery, etc.; less wear and tear on the structure."[24]

In the prewar period, Henry Ford's Model T car plant was perhaps the best-known example of a factory powered by a self-contained group-drive system. Opened in 1910, the Highland Park, Michigan, works was built on a sixty-acre suburban site. Ford, of course, had worked as a young man for the Detroit Edison Company and was one of the wizard's greatest admirers. The centerpiece of the new factory was an 8,000-HP electric plant. Group drive gave the automaker the flexibility to keep experimenting with the arrangement of the machinery in the search for faster and cheaper methods of production. In 1913 he installed a moving assembly line, quickly boosting the plant's annual output over the next four years from 82,000 to 585,000 cars.[25] Ford's achievement demonstrates that a new age of factory mechanization, energy use, and mass production was dawning at the beginning of the First World War.

The Chicago experience suggests that the war ushered in a second industrial revolution by speeding up the conversion of American industry from steam- to electricity-driven machinery. The coal famine shattered the myth of "free" power and jarred factory managers into undertaking their first real examination of energy costs and fuel conservation. And for the first time, a federal agency made conserving coal a national priority as well, as a patriotic duty. Sharply rising fuel prices highlighted energy-related issues, but the sudden unreliability of coal supplies caused the greatest changes in perception and policy. For factory owners in Chicago and other cities, the winter famine of 1917–18 hammered home the lesson that central station service guaranteed a reliable supply of power compared with erratic fuel markets. A series of bitter strikes in the coalfields following the armistice would accentuate the insecurity about the ability of individual businesses to secure adequate coal for self-contained power plants.[26]

In 1917–18 manufacturers began a sustained and accelerating movement toward complete reliance on central station service to supply power in the factory. The Fuel Administration reinforced this decision by giving public utilities priority over private companies in the delivery of coal. In Chicago, Commonwealth Edison

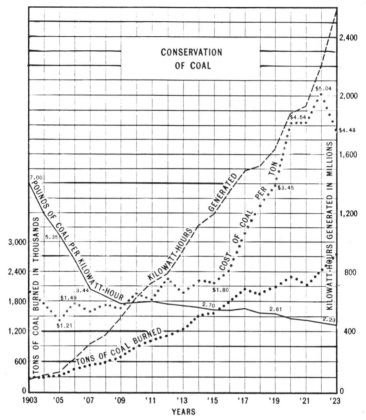

CHART 11.   Conservation of coal, Commonwealth Edison Company, 1903–23.
Source: William E. Keily, ed. *Public Utilities in Modern Life* (Chicago: privately printed, 1924), 50.

gained from the abandonment of steam-based technology, the addition of more and more electrical machinery in existing plants, and the construction of more fully mechanized factories. The large-scale turbogenerators of the utility companies burned significantly less coal to generate a unit of energy than the smaller self-contained systems of the manufacturers (see chart 11). For example, Commonwealth Edison consumed about 2.5 pounds of coal to generate 1 kwh of electricity compared with 8–9 pounds per kwh for privately owned machinery. Insull took great personal satisfaction from government policies that seemed to give official sanction to his long crusade against the "isolated" electric plant.[27]

The coming of war sounded the death knell of self-contained

electric power technology. The energy crisis persuaded factory managers to purchase electric power (and gas) from central station utilities. In Chicago, the installation of new electric generators dwindled to an average of less than twelve per year, prompting city inspectors to discontinue counting them after 1919. In an analogous manner, 1917 was the high-water mark for the number of central station generators in the United States, which dropped from over 6,500 to fewer than 4,000 over the next dozen years. During the same period, the annual output of the average utility turbogenerator rose more than six times, from 3.9 million to 24.7 million kwh. In other words, the war marked a new trend toward fewer, larger generators. Insull's biggest turbogenerator grew from 30,000 to 208,000 kw, for instance. These economies of scale allowed electric rates in the United States to fall by one-half between 1900 and 1929, in spite of a threefold increase in the price of fuel and labor (see chart 12). For most manufacturers, the decision to purchase electric power rather than generate it themselves made good economic sense.[28]

The coal famine was a pivotal experience in Chicago, turning factory managers from several alternative energy supplies onto a single path of central station service. Beginning in 1918, the demand for industrial power jumped significantly as manufacturers abandoned their self-contained plants while adding more and more machines (see chart 13). In the prewar years, the utility gained about thirty major converts annually, rising to a peak of sixty-eight in 1922 and then leveling off at twenty-five to thirty for the rest of the decade. After 1916, no major office building or manufacturing concern installed a self-contained electric plant. In 1922 Insull estimated that electricity supplied about two-thirds of the city's demand for power, while the remaining third was provided by steam engines. Commonwealth Edison supplied 70 percent of the electrical load, which represented about 45 percent of Chicago's total demand for power. Just four years later, he could report that central station service was supplying 87 percent of the city's fast-growing demand. More important, perhaps, than the narrowing of electric supply alternatives to a single form of technology was the emergence of a new industrial economy characterized by an intensive application of energy, the mass production of consumer goods, and a corresponding culture of affluence, youth, and leisure.[29]

Unprecedented labor shortages and record demands for goods combined to convince manufacturers to mechanize their opera-

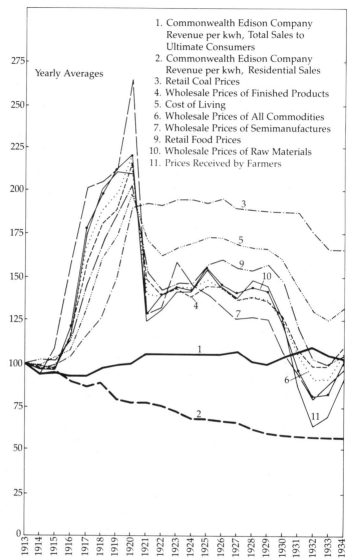

1. Commonwealth Edison Company Revenue per kwh, Total Sales to Ultimate Consumers
2. Commonwealth Edison Company Revenue per kwh, Residential Sales
3. Retail Coal Prices
4. Wholesale Prices of Finished Products
5. Cost of Living
6. Wholesale Prices of All Commodities
7. Wholesale Prices of Semimanufactures
9. Retail Food Prices
10. Wholesale Prices of Raw Materials
11. Prices Received by Farmers

Yearly Averages

CHART 12.   Comparative price trends, 1913–34. *Source:* Illinois Commerce Commission, Case 22337 (1934), Commonwealth Edison Company, Exhibit 176, in CEC-HA, boxes 42–47.

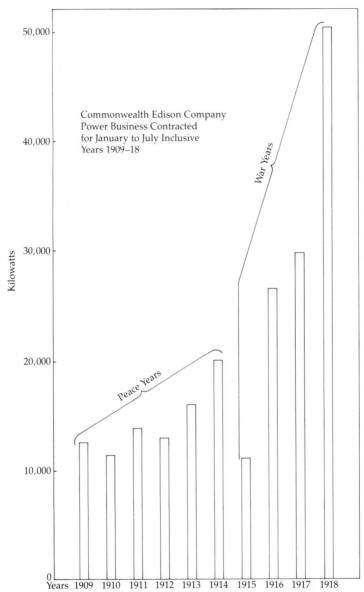

Commonwealth Edison Company
Power Business Contracted
for January to July Inclusive
Years 1909–18

War Years

Peace Years

Kilowatts

50,000

40,000

30,000

20,000

10,000

0

Years 1909 1910 1911 1912 1913 1914 1915 1916 1917 1918

CHART 13.   Growth of industrial power demand, 1909–18, in CEC-LF, F4-E2.

tions. The steel and metals business illustrates how the war spurred industry to make more intensive use of electrical energy to boost the pace and volume of production. According to the War Industries Board, iron and steel represented a classic case of an industry unprepared to meet the exigencies of a world at war. Orders poured in during the second half of 1915, prodding steel mills in the Chicago region to begin using the electric furnace. The installation of these energy-intensive devices not only increased the mills' output but allowed them to improve quality. The companies also built their first mechanized rolling mills to speed up the transformation of molten metal into a variety of useful products. The elaboration of machine systems for handling materials also helped turn steelmaking into a continuous, semiautomatic operation. The war, moreover, largely paid for the modernization of the mills. For example, the profits of the U.S. Steel Company jumped from an average of $76 million annually in the prewar period to $478 million in 1917 alone. In this case an absence of physical destruction from the fighting helped boost the pace of industrial reform in the United States ahead of that in Europe, which was occupied with rebuilding its economic infrastructure. In a similar manner, war profits underwrote the mechanization of many other leading industries of the 1920s, including electrochemicals, electrical appliances, rubber, garments, publishing, and automobiles.[30]

The results of industry's wholesale conversion to electric power supplied by central station service were revolutionary. "Acceleration rather than structural change is the key to understanding our recent economic developments," a blue-ribbon presidential panel wrote in 1929, "[including] the increased supply of power and its wider uses; the multiplication by man of his strength and skill through machinery, [and] the expert division and arrangement of work in mines and factories, on the farms, and in the trades." Looking back on ten years of phenomenal economic growth, the report of the Hoover commission is a remarkably perceptive and detailed testimonial to the birth of a machine age. Hoover's experts properly placed the origins of industrial mechanization in the prewar period, but recognizing the war's impact on perceptions, they concluded that "the war . . . did focus attention upon the advantages of mass production."[31]

A comparison of energy-use patterns in Chicago before and after the national emergency supports this conclusion. A useful way to measure the mechanization of industry is simply to plot the rise in the amount of electricity supplied by central station service to run machine motors. Table 25 lists consumption statistics

Table 25   Growth of Electrical Demand, 1914–29 (Thousands of Kilowatt-Hours)

Group	1914	1919	1924	1929
Large light and	160,829	415,948	892,883	1,802,432
power consumers	(16.4%)	(29.4%)	(37.4%)	(47.2%)
Small light and	193,998	305,072	664,441	1,054,552
power consumers	(19.7%)	(21.5%)	(27.8%)	(27.5%)
Railway	627,538	695,014	831,497	964,749
consumers	(63.9%)	(49.1%)	(34.8%)	(25.2%)
Total	982,365	1,416,034	2,388,821	3,821,733

Source: H. A. Seymour, "History of Commonwealth Edison," 3 vols., manuscript, 1935, Appendix, 176, in CEC-HA, box 9001.

for three groups of customers: large wholesale power consumers, small retail light and power consumers, and street railways. From the eve of the war to the onset of the Great Depression, industrial power use increased tenfold, or a spectacular 68 percent annually over the fifteen-year period. Energy consumption by commercial and residential customers also grew at a vigorous rate of almost 30 percent a year, while public transportation lagged behind with an anemic annual rate of 3.5 percent.

Dramatic shifts in the three groups' patterns of energy demand reflected the mounting importance of industrial mechanization in the ascendancy of a consumer-oriented society. In 1919 the explosive growth of the factory's use of central station service reduced public transportation's share of the electric supply to less than half of the total for the first time since Insull signed the original contracts with the transit companies in 1907. Four years later in 1923, the manufacturing sector pushed past the streetcars in the use of central station service. And by 1929 the positions of the two were practically reversed. The slow but steady increment in the small consumer's proportional share of the electric supply primarily represented the completion of residential electrification in the city.[32]

Another important indicator of the inauguration of a machine age was a sharp rise in the amount of electricity the average city dweller used. Table 26 summarizes the per capita use of central station service for power, light, and transportation. Industry's skyrocketing demand for power during the war years was the largest increase recorded during the entire period. The war, moreover, set the stage for a decade of sustained growth, resulting in a 700 percent gain in the use of industrial power for every person in Chicago. An extraordinary jump in energy use by small consumers

Table 26   Per Capita Consumption of Electricity for Power, Light, and
Transportation, 1914–29 (Kilowatt-Hours)

Year	Power	Light	Transportation	Total
1914	67.4	81.2	267.2	415.8
1919	156.3	115.9	274.7	546.9
1924	300.0	223.2	279.4	802.6
1929	544.8	318.7	291.6	1,155.3
Increase (%)				
1914–24	345.1	123.6	4.6	93.0
1919–29	248.6	175.0	6.2	111.2
1914–29	708.3	292.5	8.9	177.8

Sources: H. A. Seymour, "History of Commonwealth Edison," 3 vols., manuscript,
1935, Appendix, 176, in CEC-HA, box 9001; U.S. Census, Population, 1910, 1920,
1930.

immediately after the conflict was also directly related to the war's
impact on society. The conflict generated tremendous wealth, but
many purchases of goods and services had to be deferred until
after the armistice. In 1919 Chicagoans went on a spending spree,
especially for electrical appliances.

In contrast, the war worsened the economic plight of the street
railways. Still hamstrung by the political legacy of Charles Yerkes,
Chicago's transit companies had trouble convincing policymakers
to keep fares in step with steep rises in the cost of labor and materials. For the straphanger, the political stalemate meant deteriorating equipment and reduced service. The economic stagnation
of the city's railways was eventually reflected in the relative decline
of the per capita energy-consumption figures for the 1920s. Nevertheless, the transportation sector conforms to emerging patterns of
an energy-intensive society. After the war, urban Americans expressed a strong preference for the gas-guzzling private automobile. In 1915 Chicago had 40,000 cars, or one for every sixty persons. Over the next fifteen years, the number of private vehicles
multiplied tenfold, to put one of every eight city dwellers in the
driver's seat.[33]

A closer look at the changing relationship between machines
and workers highlights the role of electrical technology in the second industrial revolution. Between 1914 and 1929, manufacturers
added machine power at a rate three and a half times as great as
for manpower. In just fifteen years, the average worker was using
almost 60 percent more horsepower on the job. Mechanization and
assembly-line methods meant that factories employed fewer, more

highly skilled workers to operate high-speed machines. The resulting fourfold increase in labor productivity made the postwar decade unique in the economic history of the United States during the twentieth century. The 1920s was the only period recording an increase in the ratio of capital to labor. The year 1919 also marked a historic reversal in the ratio of capital to output. The decision to invest in laborsaving machinery and factory reorganization paid off by boosting output while reducing the cost of each unit. In other words, the new type of factory was able to make more goods with less labor and capital per unit of output. The achievement of mass production provided the economic foundations for the rise of a consumer-oriented society.[34]

During the 1920s, factory workers generally enjoyed a rising standard of living, but as producers they paid a heavy price for these gains. The story of workers' responses to the mechanization of the factory falls largely beyond this book, but it is important to note briefly that the advent of the machine age was a painful process that brought a denigration of the dignity of labor and a loss of self-respect for large numbers of Americans. Ford's institution of the five-dollar day in 1914 to keep the men "on the line" was a clear recognition of their subordination to its monotonous routines and constant demands. Ford's machinery eliminated some of the factory's backbreaking drudgery, one critic observed, but it also created the "fascinating and horrible spectacle of the jobs along the famous assembly line at Highland Park out of which the last residue of workmanlike interest has been relentlessly squeezed." The spread of "Fordism" during the twenties would give rise to labor strife and the industrial union movement during the following decade.[35]

By the mid-twenties, the composition of the city's electric power consumers closely resembled the makeup of its industrial base (see table 27). Without question, food processing had become Chicago's most energy-intensive industry. In 1917 the meat-packing firm of Swift and Company had been Insull's prize convert from a self-contained system to central station service. Over the next eight years, the company's reliance on electricity grew at an astonishing rate, from 750,000 kwh to over 41 million kwh a year, making it the utility's largest nontransit customer. As we will see in the next section, the war initiated the massive use of artificial refrigeration by the entire food industry, including the meat packers. Other major consumers of electricity represented the mainstays of the city's economy, such as the steel and metals trade, printing, railroads,

Table 27   Composition of Industrial Customers of Central Station Service, 1925
(Customers Using 500,000 Kilowatt-Hours or More)

Customer Group	Number	Consumption (000 kwh)	Maximum Load (kw)
Food			
Refrigeration[a]	79	108,728	29,120
Meat packing	10	101,695	21,630
Grain elevators and mills	5	10,728	2,990
Food processing[b]	14	9,718	2,920
Subtotal	108	230,869	56,810
Steel and metal products	63	88,997	23,810
Chemicals[c]	15	30,865	8,220
Printing	20	24,911	8,440
Railroads	12	19,670	5,770
Electrical products	12	16,715	6,770
Automotive	5	9,167	3,780
Rubber	3	9,033	3,340
Garment	5	4,779	1,860
Woodworking	3	3,870	1,510
Leather	4	2,518	1,350
Other manufacturing	22	21,627	8,120
Total	272	463,021	129,630

Source: Commonwealth Edison Company, Statistical Department, "Output . . . 1925
and 1926 . . . Customers Having an Annual Consumption of 500,000 K.W.H. or
More, Grouped as to Classes of Business," in CEC-LF, F23-E3.
[a]Includes ice manufacturers, ice cream makers, cold storage facilities.
[b]Includes bakeries, groceries, dairies, candy makers.
[c]Includes paint manufacturers, yeast and malting makers, drug suppliers.

garments, leather, and general manufacturing. Chicago also benefited from several of the new high-growth industries of the twenties, such as chemicals, automobiles, rubber goods, and electrical
equipment. In many cases the war required specialized assembly-
line factories for these new products. By the end of the postwar
decade, the mechanization of Chicago's major industries was virtually complete.[36]

"Diversify, Fertilize, Motorize, Specialize"

A case study of the transformation of the food business illustrates
the profound effect the second industrial revolution had on daily
life in urban America. Again the war triggered rapid systemic
change. In a nation graced by natural abundance, the sudden
prospect of "wheatless" and "meatless" days came as a shock that

altered the diet of Americans. In the quotation above, the meat packer J. Ogden Armour outlined a program of reform shortly after the war began. Significantly, Armour's prescription for improving the food supply did not appear in a farm journal but was in the *Saturday Evening Post*, a glossy magazine marketed to the urban middle class.[37]

Armour was representative of the progressives who had formulated far-reaching programs for upgrading America's nutritional standards and eliminating its chronic waste of huge quantities of food. The war presented reformers with a unique opportunity to convince producers and consumers alike to break the habits of a lifetime. The distressing revelation that the army had to reject one of every three young men as unfit raised the question of nutritional standards to a level of sustained public debate in the popular press and mass-circulation magazines. In the final months of the war, moreover, the worst epidemic in modern times swept the nation (and the world). Reaching frightening proportions, "the 1918 flu was unique," Alfred Crosby has written, because "no other influenza before or since has had such a propensity for pneumonia complications and pneumonia kills." In Chicago alone, over 14,000 persons died in just twenty-seven weeks between September 1918 and March 1919. The terrifying scourge hammered home the domestic reformers' lessons about the need to raise the nutritional value of foods and improve their freshness.[38]

The application of massive amounts of energy to every step in the commercial food chain was chiefly responsible for the revolution in what Americans ate. The war brought recent innovations in the manufacture of artificial fertilizers to technological maturity, helped ice makers kill off the natural ice business, turned shoppers toward the new cash-and-carry supermarkets, and made processed foods socially acceptable among the middle classes. During the 1920s, the food industry made intensive use of heat and refrigeration to offer a wider variety of better-tasting canned and baked goods as well as fresh fruits, dairy products, vegetables, and meats year round. "Foods formerly limited to the well-to-do," Hoover's economic experts noted in 1929, "have come more and more within the reach of the masses." Robert and Helen Lynd, the keen observers of daily life in "Middletown," concurred. They recorded a "marked spread in the variety and healthfulness of the diet of medium and low-income families throughout the bulk of the year when fresh garden products are expensive."[39]

Like several industries, in 1914 food processing was poised on the threshold of a major transformation. Refrigeration technology, for example, stood between the traditional practice of harvesting natural ice and the innovative technique of making a substitute using electric-powered machinery. As early as the 1860s, the growth of urban populations had encouraged the improvement of ice harvesting and storage to a peak of economic efficiency. In the post–Civil War era, Chicago's meat packers and brewers had become the largest consumers of ice, and the use of iceboxes in the home remained confined to the wealthy. Moreover, Gustavus Swift's practical perfection of the railroad refrigerator car in the 1870s had been the key to the national concentration of the meat-packing industry in Chicago. By the 1880s several companies supplied ice to the city from as far north as Green Bay, Wisconsin, to Wolf Lake, Indiana, in the south.[40]

But refrigeration by natural ice imposed severe bottlenecks on the food processors, initiating a search in the 1860s for a mechanically made substitute. As the brewers and packers became big businessmen, they had to build expensive icehouses and hire extra men to handle the heavy, cumbersome blocks. The meat packers, for instance, used ice to cure the beef as well as to store it. The temperature of the ice fixed the curing time; there was no way to shorten it without achieving colder temperatures. In addition, ice was too moist for the efficient storage of meat and produce. High rates of spoilage and quick deterioration of taste helped account for the widespread unavailability of fresh foods during the winter, especially in the urban areas of the North. The Lynds found that before the 1920s, most families in Muncie, Indiana, had a diet rigidly defined by seasonal gluts and famines. In "Middletown," the typical pattern included meat at three meals a day during the fall and winter packing season, followed by a period of "spring sickness" from a long absence of green vegetables and fresh fruits, and finally, a summer of recovery when garden produce briefly became more abundant.[41]

In the 1900s, steady advances in mechanical refrigeration began to offer food processors a superior alternative to natural ice. However, the new ice-making machines had to be coupled with the old steam technology because clear ice could be made only from distilled water. Only large firms could afford huge capital outlays for coal-fired boilers, steam engines, and a powerhouse. This limitation effectively excluded the use of electric power for ice making until 1910, when a novel method of blowing air through the chilling liquid eliminated the need for distilled water. The air-blowing method allowed the manufacturers of electric refrigerators and air conditioners to miniaturize their equipment so it was practical in commercial establishments such as hotels, restaurants, dairies, bakeries, and candy and ice cream factories. Yet the new machines made little headway in gaining acceptance as long as electric power remained expensive compared with the price of natural ice.[42]

Insull was among the first electric utility managers in the country to compete successfully against the natural ice business. In 1909 his statistical studies of group patterns of energy use revealed that the ice cream makers had one of the highest load factors—the ratio of average use to maximum use. In addition, the demand for refrigeration peaked just as the need for artificial lighting was bottoming out during the summer. By offering ice makers special low rates to use their equipment during the evening, Insull found an opportune way to build and diversify the utility's off-peak load. In 1911, a year after the new tap-water ice machines appeared, he secured his first contract for ice making with the Anderson and Goodman Ice Cream Company. Five years later, the horsepower of refrigeration machines connected to Insull's generators equaled the motor capacity of his steel and metals customers. On the eve of the American entry into the war, 20 percent of the 2.1 million tons of ice used in Chicago was made by electric power, 26 percent by steam engines and the remaining 54 percent by natural means.[43]

The war tipped the scales of economic competition against natural ice and steam power and in favor of artificial refrigeration and central station service. Soaring labor costs and freight rates combined to deliver a deathblow to ice harvesting. And as we have seen, the coal famine turned businessmen away from self-contained steam plants to purchased electric power. Across the nation, the proportion of ice made by using steam engines fell drastically during the five war years, dropping twenty points to 71 percent. In Chicago, Insull's aggressive salesmanship and special rate incentives further accelerated the switch to electric power. By 1922, central station service supplied the electricity to make 70 percent of the ice used in the city, a figure that increased to give Insull a virtual monopoly within the next few years. In addition, the conversion of some of the meat packers and cold-storage companies from self-contained systems to purchased power boosted the food industry to the top of the list of large energy consumers (see table 27).[44] This incredible rise in refrigeration reflected new directions in both the technology of food processing and the eating habits of city dwellers.

Food shortages during the war gave businessmen new incentives to speed up the processing and distribution of fresh meats and produce. As Armour pointed out, the food problem was ultimately one of distribution, not supply. To end the age-old cycle of glut and famine, he argued that "the cold-storage plant and various preserving processes must be amplified until they represent a great reservoir." Using energy to cool and heat foods to more extreme temperatures became one of the most important elements in the industry's transformation. The food industry not only introduced more extreme temperatures in established procedures but also discovered entirely novel applications of energy. The meat packers, for example, found that flash cooling shortly after slaughter made cutting the beef easier, and hence faster. Likewise, vegetable canners started prechilling fresh produce before shipment and using ultrahigh temperatures during cooking and packaging to preserve taste while ensuring proper sterilization. New refrigeration technology also fostered a rapid expansion of cold-storage facilities, which were redesigned to incorporate better mechanical control of temperature, humidity, and air circulation. By the mid-twenties, electrical energy helped bring much greater quantities of appealing fresh foods to the urban market at prices that broad segments of the population could pay.[45]

To aid in distributing an expanding array of prepared and fresh products, the war ushered in a new type of food store, the "supermarket." In 1916 Clarence Saunders opened the first of these cash-and-carry, self-service groceries, a Piggly Wiggly store in Memphis, Tennessee. Imitators soon followed, especially other chain stores such as the Great Atlantic and Pacific Tea Company. Between 1914 and 1919, A & P alone built 2,200 of the new economy stores. Self-service and a turnstile operation helped the chain stores offer shoppers lower prices than independent retailers. More important, however, the refrigerator display case transformed the old-fashioned grocery store. Now consumers were offered the convenience of doing all their shopping in one place rather than having to go to several specialty stores. The supermarket's refrigerator units contained fresh dairy products, fruits, vegetables, and meats, while its open shelves carried an expanding selection of canned and baked goods. Public approval of the new grocery stores was reflected in an astonishing rate of growth from 10,000 in 1920 to 53,000 by the end of the decade.[46]

Wartime shortages helped create new habits of shopping and eating by making food reform a national patriotic crusade. Food administrator Herbert Hoover became a household name, in part because he mounted a massive propaganda campaign to convince American women to conserve food and change the family diet. Hoover commanded 1,400 government workers who bombarded the public with a barrage of 43,000 posters and 2,000 press releases. Another 750,000 volunteers, mostly women, took up Hoover's cause by gathering 20 million pledges of "meatless' Tuesdays. Hoover also found support among business and civic leaders. In July 1917, for example, Illinois Council of Defense official J. Ogden Armour enlisted middle-class housewives across the country in the battle of the home front. In a *Ladies' Home Journal* article entitled "Lest Women Realize," Chicago's meatpacker declared that "the woman who is handling the food supply in the home is equal in importance to the man who handles the gun on the battlefield. Food shortages are imminent unless averted by efficiency in the kitchen." Illinois followed up the propaganda offensive with educational bulletins on such subjects as planting "war gardens," canning vegetables, and eliminating waste in the kitchen. Armour estimated that the average family spent 43 percent of its income on food and half of that on meat—figures confirmed by later studies—suggesting that the "meatless" day once a week as every good citizen's patriotic duty meant a break with the habits of a lifetime.[47]

The need for sacrifices on the home front became an opening for reformers to press for further departures from tradition. Armour called upon housewives to accept the new "cash-and-carry" system of shopping as the only way to hold down the fast-rising price of food. Storekeepers could not "reduce their overhead expenses while the customers demand four or five deliveries a day and extra fine wrapping paper and colored string and all that sort of thing." Fewer, larger stores, he promised, would mean smaller markups and faster turnover of fresh foods. Changing social conventions in the food store paralleled a new ethic of conversation in the kitchen. The "skilled housewife," for instance, did not buy unneeded items, spoil food by careless cooking, overload guests' plates as a gesture of hospitality, or throw away leftovers. Appeals for eliminating waste at the grocery store and at the dining room table easily spilled over into discussions of nutrition.[48]

During and after the war home economists mounted a sustained campaign to improve the American diet, and they targeted the housewife, especially in the mass-circulation women's magazines. "Whole industries," the Lynds observed, "mobilize to impress a new dietary habit upon her." In Muncie, Indiana, and many other places, home economics courses became a regular part of the school curriculum. Moreover, the electric and gas utilities often sponsored free "cooking schools" that advocated a scientific approach to improving the family's diet and health care. Of course the utilities also used the classes to introduce housewives to their modern appliances. In the 1920s, educational and advertising campaigns complemented rising standards of living and more widely available fresh foods to make the icebox commonplace in working-class homes. Electric refrigerators remained a luxury, though the adaptation of a temperature regulator in 1918 marked a major step toward a fully automatic, self-contained machine.[49]

By the late twenties, the application of massive amounts of energy to every step in the commercial food chain had changed the nutritional standards of the nation. Precise statistics are lacking, but there is little reason to doubt the Lynds' assessment that "new mechanical inventions as the development of refrigeration and cold storage and of increasingly rapid transportation have . . . furthered sweeping changes in the kind of food eaten." The Hoover commission on economic trends noted a marked decline in the consumption of cereals as opposed to increases in sugar, vegetable oils, fresh fruits, vegetables, and dairy products. Baked, canned, and other prepared goods also showed significant increases. The transformation of food processing shows how the war speeded up mechanization and mass production to meet the demands of rapidly expanding markets. In this respect the food industry became one of the basic underpinnings of an emerging consumer-oriented urban society.

"A Gold-Plated Anarchist"

The Chicago experience with military conflict abroad and economic crisis at home suggests that World War I marked a crossroads, one that contemporaries labeled the birth of the machine age. Rapidly changing patterns of energy use, industrial mechanization, food processing, and dietary habits in Chicago demonstrate a need to reexamine conventional interpretations of the war as a short-term interruption of long-term trends. The most obvious support for traditional approaches to the study of war and society comes in the broad area of government-business relations. Historians usually emphasize that most wartime agencies such as the Fuel Administration were disbanded in 1919, with dizzying speed. Domestic supplies of coal and oil, moreover, quickly returned to "normal" conditions of glutted markets and low prices. In Chicago, reformers mounted a major political assault against the utilities after consumers suddenly confronted unprecedented rate hikes for gas and transit services. In the heat of the battle over the price of gas, a frustrated Donald Richberg, the city's special counsel, publicly denounced Insull as "a gold-plated anarchist." The businessman replied in kind, calling the reformer a "crook." But popular interest in public utility regulation quickly waned with the return of domestic prosperity in the early twenties. In some respects, then, the nation's first energy crisis conforms to the mainstream view that the war had a negligible impact on the political economy.[50]

Yet the battle of the home front helped initiate dramatic shifts in perception among policymakers, reinforcing policy choices that were giving rise to a new civilization, an urban-centered consumer society. For example, government priorities reinforced the practical lessons of the fuel and food crises to effect a major alteration in the provision of urban energy supplies. To be sure, legislators at the state and national levels produced little change in the institutional framework of public utility regulation. Equally important, however, the wartime experience of government-business cooperation carried over into peacetime politics. New attitudes toward big business and businessmen foreclosed further consideration of several of the progressives' farthest-reaching policy goals.

In the area of urban public services, Chicagoans came to accept a political economy of regulated monopoly under state jurisdiction. Reform ideals of municipal ownership and home rule faded permanently in the wake of a stream of decisions made by wartime managers in response to the energy crisis. After the fighting stopped, this reorientation of popular opinion and expert advice contributed significantly to narrowing the policy debate over energy supplies to a single choice of giant power systems on a regional scale. Although advocates of public and private ownership continued to champion their causes, both sides now assumed that American society would become increasingly dependent on energy to sustain a new style of life.

The effects of the war on gas service warrant attention here because the fuel shortage brought to a climax the protracted conflict between the Peoples Gas Light and Coke Company and the city government. Because neither side could afford to prolong the stalemate, the way opened for negotiations that would encompass a complete review of the technological and economic foundations of gas service in Chicago. Between 1914 and 1921, the energy crisis was the overriding influence mandating a comprehensive settlement with the gas company. In contrast, Insull was able to deflect political agitation from his electric companies by avoiding the need to request higher rates. A cost-plus partnership with the Peabody Coal Company and the ownership of transportation links to downstate mines put these utilities in a strong position to endure the war's dislocations at minimal cost. In addition, the English immigrant's contacts abroad put him in the forefront of the preparedness campaign. His stockpiles of coal not only kept the lights burning and factories running but also generated invaluable goodwill for the Commonwealth Edison Company by saving many families during the depths of the fuel shortage.

In 1913 Illinois had followed the general thrust of the progressive movement by creating a state commission to regulate its public utility companies. In the final bill, the cities retained only the power to grant franchises, while rate-making authority percolated to the state commission.[51] For Chicago's gas monopoly, the meaning of the new law was problematical. Governor Edward Dunne signed the legislation to establish a precedent for the state government to set energy policy, including the rates charged by privately owned utility companies. At the same time, he renewed the call for municipal ownership of essential utilities and for home-rule powers that would allow Chicago to regulate its franchise holders.[52] The arrogance of the gas company handed Governor Dunne and other progressive reformers a ready-made cause in rallying popular support into an electoral majority.

On the eve of America's entry into the war, the Peoples company was still battling with the city over rates and standards. Although the utility had complied with the 85¢ per thousand cubic feet (MCF) rate set in the 1905 ordinance, it had refused in 1911 to accept a further reduction to 80¢ MCF. Rather than negotiate a compromise of mutual benefit with the city, the company again turned to the courts. Its appeal for an injunction against enforcing the rate ordinance now contended that the 1905 state enabling act was unconstitutional. After five years of compliance with this law, city officials were outraged at the company's new defiance. The gas company's stubborn refusal to reach a settlement with the city council soon turned into a complicated legal contest. In 1915 the city hired Donald Richberg, a brilliant young lawyer, to countersue the company on behalf of Chicago's 550,000 gas consumers for a $10 million refund of overcharges. The aldermen also refused to change the standards from light-giving to heat-bearing qualities.[53]

The coming of the war, however, radically altered the formation of energy policy in Chicago. In 1917 the skyrocketing price of oil finally forced the company to negotiate with the city council. The fuel shortage undercut the utility company's profits so badly that it had to suspend the stockholders' quarterly dividends for the first time in its history. To meet the obsolete candlelight standard in 1916, the gas plant had to use a "water gas" process that required 8 million gallons of petroleum additives, or more oil than the United States navy consumed. On behalf of the city, Mayor William Hale Thompson led the patriotic charge to reach a settlement that appeared to contribute to the war effort by saving fuel.[54] On 25 June 1917, a one-year agreement was worked out between the city council and Insull, who had been acting as the Peoples company's chief lobbyist and chairman of the board for the past three years. In exchange for a reduction in the basic rate from 80¢ MCF to 70¢ MCF, the aldermen agreed to adopt a modern heat-bearing standard of rates. A new generation of gas-making equipment eliminated the need for petroleum additives, though the new "fuel gas" had a lower heating value than the old "water gas." The politicians could take credit for a rate cut, but it largely represented a corresponding reduction in the heat value of the gas. Nonetheless, reformers considered the linkage of rates to heat values a major victory that would set an important national precedent.

In exchange for a lower base rate, the council made several concessions to the gas company that amounted to a complete departure from the past. The mass of residential customers paid dearly for a new structure of differential charges that discriminated in favor of the large consumer. Insull was again allowed to institutionalize his gospel of consumption in a regressive schedule of rates that dropped by 50 percent to only 40¢ MCF for use beyond 50,000 cubic feet per month. In effect, the small consumer subsidized the large in a second way by making up lost revenue from the gap between a flat charge and the lower rates paid by big industrial and commercial customers. The council allowed the company to impose a novel minimum charge of 30¢ per month regardless of the amount of gas used. About 65,000 Chicagoans, those who could least afford any extra household expense, became immediately subject to the service fee. Under the pressure of the national campaign to conserve fuel, Insull prevailed in setting public policy on a new course that ironically rewarded greater use of energy.[55]

For the gas company, however, reform came too late. Even before the one-year agreement expired, it had to appeal to the state commission, in January 1918, for an emergency rate increase. The utility was saddled with equipment that was at least ten years out of date, and its fast-rising fuel bills were pushing it to the brink of financial ruin. But the regulators were reluctant to grant concessions to what Insull's biographer Forrest McDonald has called "probably the least popular utility in the state."[56] The commission delayed its approval of the request until June, when the company's agreement with the city expired. The rate hike, coming on the heels of annoying service disruptions from the changeover of gas service to a lower heat value, stirred Chicagoans to seek political retribution. In most cases residential lighting fixtures and cookstoves merely needed adjustment, but the hard-pressed utility was slow to make the necessary service calls. Consumers turned their anger on both the gas company and the state commission.

After five years of little progress, municipal reformers suddenly found broad popular support for their political platform. The reformers demanded passage of laws to create a utility commission for Chicago and to authorize a constitutional convention that could give the city additional home-rule powers. Between 1918 and 1920, they seemed to have achieved both goals. In January 1920 a constitutional convention began to draw up a replacement for the state's fifty-year-old charter. A hotly contested gubernatorial race kept the question of utility regulation in the forefront of politics. Republican Len Small won the election because he pledged to abolish the state commission and shift the power to regulate utility companies to the local level. The economic and political fortunes of the gas company appeared never to have sunk so low.[57]

By 1920, however, the utility began to show a profit again. Insull applied his formula of success by matching technological innovation with aggressive marketing techniques. After his management team took over in January 1919, investors were willing to finance the construction of an ultramodern, $20 million coal-gas plant. Embodying lessons about fuel conservation learned during the war, the new technology was designed to recover a full range of coal by-products. In the process of making gas, the new plant extracted coke, ammonium sulfate, benzol, and tar. Insull could sell the gas and by-products for three times the cost of his raw material—cheap bituminous coal. He also opened promising new gas markets in industrial and home heating. Under this "Insullization" of the gas company, a sliding scale of rates went a long way in persuading larger consumers to abandon their self-contained heating systems in favor of central station service.[58]

Insull not only calmed the financial waters of the gas company, he dispersed the political thunderclouds as well. At the top of the political hierarchy, the support of powerful politicians like Governor Small for municipal regulation of utilities appears to have been a mirage conjured up to lead reformers into a political dead end. According to Forrest McDonald, the utility executive had secretly plotted with Small and other leaders in both parties to use deception to defeat the forces of reform. Although McDonald's evidence is based on off-the-record interviews, there is no doubt that sleight of hand was involved in the abolition of the Public Utility Commission. In 1921 the legislature killed the regulatory agency with great fanfare, but it was quietly resurrected later in the same session as the Illinois Commerce Commission. The following year, home rule for Chicago met a similar fate in the constitutional convention, forcing reformers to campaign against the ratification of a new state charter.

At the other end of the political hierarchy, Insull mounted a publicity campaign of his own to reverse consumers' hostile attitudes toward utility companies. To head the public relations drive, he turned the sophisticated propaganda apparatus of the State Council of Defense's Committee of Public Information into the privately funded Committee on Public Utility Information. The businessman exploited his wartime experience with innovative methods of manipulating public opinion to the advantage of his utility enterprises in a second way. Shortly after the fighting ended, he retrained his employees to sell utility securities door-to-door instead of war bonds. Insull was now convinced that giving customers a stake in their utility companies was a sure way to build political support at the grass roots.[59]

In the case of the gas company, then, a national fuel emergency changed the perceptions of policymakers and consumers by imposing a whole new set of conditions on government-business relations. The utility's well-deserved reputation for corporate arrogance made the technological transition from "water gas" to "fuel gas" and coal by-products difficult and expensive. The war, however, created a special urgency to shift public priorities almost completely around, from confrontation to accommodation. Local and state regulators went along with Insull's plans for the reform of gas service in Chicago because he offered cooperation and the promise of lower energy bills for the ordinary householder. In 1921, for example, the utility czar lectured the members of the American Gas Association on the need for enlightened approaches to government-business relations. Echoing his 1898 address to the National Electric Light Association, Insull proposed that "if the rights of the public are properly taken care of in producing lower costs and steadily improving service, the rights of the stockholder will take care of themselves."[60]

Although the war did not alter the institutional structure of utility services in Chicago, it did channel energy policy irreversibly toward state regulation of privately owned monopolies. During the postwar decade, Insull stepped up his contributions to the politicians' campaign coffers and increased his image-making effort in the daily newspapers, which also profited from advertising revenues. His well-oiled public relations machine helped ensure that the voters would no longer seriously consider the policy alternatives of the prewar years, such as the municipal regulation of Chicago's energy utilities.

The absence of any demand for a review of the decisions made about gas rates under the duress of the fuel crisis reflected Insull's success in reorienting public opinion. The Peoples company spokesman attempted to draw a direct analogy between electric and gas services to justify a sliding scale of rates, but the two forms of technology are fundamentally different. Whereas the demand system of rates bears a plausible relation to the instantaneous nature of the supply of electricity, the gas storage tank makes load and diversity factors irrelevant. In other words, the cost of making a unit of gas for the small consumer is exactly the same as for the large one. Yet the inequities of a sliding scale of gas rates for residential customers seemed unimportant compared with reassuring consumers that plans to expand utility services would keep up with their demand for more and more gas and electricity.[61] In a society that was becoming increasingly dependent on huge amounts of energy, public debate narrowed to the best way to build bigger and bigger networks of power.

Beginning in 1917, the national government strongly reinforced this shift in policy toward the regional integration of utility systems. Within six months of United States entry into the war, several cities faced critical shortages of central station capacity. The shortages posed an immediate threat to the military effort because huge quantities of electricity were needed to extract the nitrates used in making explosives. For the first time, the national government became involved in planning energy systems in such "military districts" as Niagara Falls and Buffalo, New York; Philadelphia and Pittsburgh, Pennsylvania; Akron, Ohio; northern New Jersey; and northern Georgia. The Fuel Administration ordered shipments of coal to priority locations, while Bernard Baruch, head of the War Industries Board, diverted war work to areas like Chicago that had more adequate power reserves. As one of Wall Street's financial leaders, Baruch also tried to purchase additional turbo-generators for needy districts, but he found that lead times of eighteen to twenty-four months for constructing the equipment precluded any quick solution. Lags in the manufacture and installation of new electrical equipment help account for the lack of dramatic changes in statistical indicators of industrial energy use. My argument here is that the changes caused by the war radically reoriented perceptions and policy goals rather than conditions.

Reflecting these new directions in public policy, Baruch explored interconnecting the distribution grids of utilities to create regional power pools. He directed another new institution, the Council of National Defense, to study the country's power systems. The research agency finished the survey just as the fighting stopped, and no plans were formulated to implement its recommendations.[62] Nevertheless, after the war the new alliance of government and business continued to determine national policy objectives for electric suppliers and consumers. The energy crisis generated new perspectives that set the agenda of public debate during the twenties.

Advocates of public and private ownership were equally enthusiastic about spreading the technological benefits of greater energy use throughout society, including the farm sector. The proponents of public ownership spoke about electric grids on a regional scale in terms of "giant power," while the champions of investor-owned utilities talked about "superpower" systems. In the 1920s, political debate over energy policy would center on the fate of the federal government's power dam and fertilizer plant at Muscle Shoals, Alabama, which had been built during the war. In a similar way, the war ended the prewar controversy between the conservationists and the power industry over the use of the nation's waterways. Marking "the end of the conservation era," the Water Power Act of 1920 represented a defeat of reform ideals for the multipurpose development of the nation's waterways.[63]

Official agencies also gave impetus to the drive for modernizing industry through mechanization and mass production. In the electrical appliance industry, for example, Baruch had instituted a "very comprehensive program of standardization, directed to the conservation of raw materials, labor, and transportation." In the twenties this work was carried forward by the quasi-government National Research Council (NRC) which initiated industrial research in such areas as electric arc welding, fan-motor efficiency, and artificial lighting standards in the factory. Significantly, four out of six of the members of the NRC's Industrial Relations Division were representatives of the country's largest electrical manufacturing companies.[64]

The Chicago case suggests that the reorienting of popular perceptions and public policies during the war helped usher in the machine age and its cultural ethos of consumption. The United States may have escaped the horrors of military combat on its own soil, but the battle of the home front—at least in Chicago—had a major impact on its domestic life. New directions in art and literature were not simply the statements of an alienated avant-garde, they were indigenous reflections of a society moving through a historic period. The war itself set the cultural motif of the machine age by plunging the world into a new era of mechanized death on a massive scale from such technological innovations as the machine gun, submarine, and poison gas. In the decade following the armistice of 1919, the fate of the individual in an impersonal world of assembly-line methods and mass society would remain one of the most vexing problems of contemporary life.

Changes in the popular imagination paralleled systemic mutations in the American economy. In Chicago, the industrial sector underwent a metamorphosis from coal-fired steam engines and handicraft techniques to electric-powered motors and mass production. The war accelerated the eclipse of the old and the ascendancy of the new. The triumph of "Fordism" created a need for changes in consumer behavior commensurate with higher levels of factory output and worker income. As department store magnate Edward Filene pointed out, the true meaning of mass production "is not simply large-scale production . . . It is production for the masses."[65] In the postwar decade, Chicagoans and other American city dwellers would use more and more energy in the effort to achieve Filene's vision of an urban-based society of consumption and leisure.

9

The Emergence of an Energy-Intensive Society, 1919–1928

In 1925 the Commonwealth Edison Company won the Charles A. Coffin Award, presented annually by the electrical industry to the operating utility making "the greatest contribution toward increasing the advantages of the use of electric light and power." Company president Samuel Insull undoubtedly deserved the accolade for turning Chicago into the most energy-intensive city in the world. Over the previous ten years his herculean sales efforts had more than doubled the per capita consumption of electricity to 936 kwh per year. Capped by the award-winning Christmas promotion "Give Something Electrical," the advertising campaign of 1925 blanketed the city with $7.5 million worth of newspaper copy, 25,000 ads in church bulletins and theater programs, 2,500 streetcar posters and billboards, and 1.6 million fliers inserted in customers' bills. These promotions worked; consumers bought over 110,000 household appliances from the company's stores and solicitors, spending ten times more than they had just a decade earlier. By the end of the year the average city residence had acquired three appliances: an iron, a vacuum cleaner, and perhaps a washing machine, toaster, or coffee percolator. Equally important, the electrification of Chicago's homes was virtually complete. Approximately 92–95 percent of the city's families were now supplied with central station electric service.[1]

After World War I, what is referred to as "modern society" emerged full blown in the United States. Urban based and consumer oriented, it required huge amounts of energy to sustain a new style of life. Cheap, abundant energy provided the economic and technological underpinnings of a culture that was driven by a suburban ethos of affluence, youth, and leisure. Mechanization and mass production affected every aspect of daily life, from the breakfast foods Americans ate to the leisure time they spent listening to the radio or watching a movie. Building on the foundations of the prewar years, new patterns of work and play, of domestic and public life spread rapidly among the middle and working classes. By the late twenties the use of more and more electricity, gas, and oil in everyday life had become so ubiquitous as to wrap urban America in an "invisible world" of energy. Even the shock wave of the Great Depression could not halt a steady rise in household consumption of electricity, preserving the new standard of living.

In the Chicago region, more intensive energy use furthered a dual process of suburban deconcentration and economic integration. At base, the great wartime migration from farm to factory continued to fuel the growth of the metropolitan area during the twenties. The United States census of 1920, in fact, marked the historic turning point when more than half of the population lived in the nation's urban centers. Over the decade, Chicago added 675,000 new residents to reach a population of 3,375,000. The suburbs grew even faster. The city, for example, increased by 25 percent compared with a growth rate of 72.4 percent for suburban Cook County. The attraction of rural Americans to the bright lights of the metropolis reflected the eclipse of traditional agrarian values by a vibrant urban culture. In the twenties Chicago was a good place to live, a triumph engendered by an industrial economy and by progressive reforms that had given the city vastly improved systems of sanitation and health, transportation, recreation, schools, and other public services.[2]

The war and the prosperity it generated spawned the greatest building boom in the city's history. During the 1920s, private construction and public works projects transformed the urban environment. A string of office towers and high-rise apartments along the lakefront created a new "Gold Coast" at the center, while a patchwork of residential and industrial subdivisions produced metropolitan sprawl at the fringe and beyond to the satellite cities. Only the farm sector was still excluded from the new world of energy that was enveloping the Chicago region. During the twenties, cultural conflict between city and country helped draw public attention to the crucial role that electrical technology now played in maintaining a modern style of life. Although farmers constantly deplored the eclipse of traditional values, they were eager to join their urban counterparts in enjoying the benefits of an energy-intensive consumer society.[3]

"How Long Should a Wife Live?"

"How long should a wife live?" Commonwealth Edison asked in newspaper copy written in 1925 by the advertising genius Bruce Barton. The question and Barton's answer were emblematic of the changes in postwar America's social conditions and cultural perceptions. Calling for the mechanization of the home, the adman presumed that housewives, not servants, would have to take care of domestic chores. "The home of the future," Barton suggested, "will lay all of its tiresome, routine burdens on the shoulders of electrical machines, freeing mothers for their real work, which is motherhood. The mothers of the future will live to a good old age and keep their youth and beauty to the end."[4] The advertisement reflected the new era's focus on child rearing and its related fascination with youth. Ironically, Barton's answer also revealed a common assumption about family life. The introduction of novel technology in the home would not upset traditional relationships but would serve socially conservative purposes. Electric-powered appliances would end the drudgery of housekeeping, but only to give mothers more time to meet rising standards of cleanliness, nutrition, and child care.[5]

The provocative appeal of this newspaper copy, moreover, typified a new age of advertising that grew out of the wartime experiences of Barton and others in mass communications. Emerging as "apostles of modernity," they learned how to raise the art of persuasion to new levels of subtlety and sophistication. Most important, perhaps, the content of advertisements changed from objective information about a product to subjective appeals to the emotions. In contrast to Insull's earlier stress on the educational mission of advertising, the new copy "brought good news about progress," according to a recent assessment. Admen now contended that "modern technologies needed their heralds . . . [and] modern styles and ways of life needed their missionaries."[6] They were especially successful in linking one of society's most prominent icons of progress and modernity—electrical technology—with new symbols of the machine age such as youth, leisure, and laborsaving efficiency. Barton's advertisement neatly tied together in a single question all these expressions of a culture undergoing rapid change.

The war stimulated not only modern techniques of mass persuasion and mass production but mass consumer markets as well.

Two basic economic changes helped reorient consumer attitudes toward the use of electrical technology in the home. For more and more Chicago families, the question shifted from whether they could meet the expense of replacing gaslight with electric lighting to how many additional fixtures and appliances the household budget could handle. First, wage and price levels rose dramatically relative to steady or falling rates for electrical energy (see chart 12). Insul's demand system of rates, moreover, further encouraged energy use by reducing the marginal cost of each additional block of electricity. The more you consumed, the lower the average unit price. From 1919 to 1929, for instance, the average Chicago family increased its use of electricity in the home by 63 percent, from 386.5 kwh to 629.2 kwh per year, but its bills rose only 20 percent, from $2.01 to $2.42 per month. A declining unit price accounted for the difference. Several minor rate cuts dropped the average unit price about one-quarter, from 6.27¢ to 4.62¢ per kwh. In other words, central station service was getting cheaper and cheaper compared with other basic consumer items.[7]

Second, the war permanently dried up the supply of domestic servants. The flow of cheap labor from Europe was cut off, making factory wages attractive to working-class women. After the red scare of 1919, the United States Congress closed the door to foreign immigrants, though a "Great Migration" of African-Americans from the South to the urban centers of the North became a new source of domestic help. Nonetheless, labor historians regard World War I as a watershed. The number of household servants dropped for the first time, and the workers changed from white live-ins to black day helpers who commuted back to the ghetto. Reflecting these new conditions, advertisements like Barton's appeal to motherhood now showed electrical devices in the hands of middle-class housewives, not their servants. For these housewives, appliances were supposed to become the "modern servants," agents of feminine liberation and progressive democratic values. For women in need of cash income, laborsaving technology aided the transition out of the home and into the workplace.[8]

After the armistice of 1919, working-class and middle-class families came to regard electric service as a necessity of domestic life. In the year after the war, the utility company's retail stores were "oversold," unable to keep up with the demand of better-paid workers for lamps and appliances. Sales in these "Electric Shops" jumped by 140 percent over 1918 figures, to $2.4 million, beginning a sustained period of mounting consumer demand.[9] Contemporaries noticed how the war affected consumers' attitudes. "There is no question . . . that [this] is the age of electric power," a company store manager reported in 1919 to a symposium of Illinois electric utility men. "The public," he realized, "is already 'sold' to the electric idea and is predisposed towards things electric; in fact, 'Do It Electrically' is in reality the slogan of the hour." The company's historian also noticed this reorientation of attitudes, which he attributed to the coming-of-age of a new generation. In 1922 he reflected back: "Electricity was something remote, strange and mysterious in 1890; now a child pushes a button and electricity does the rest." Seven years later, President Herbert Hoover's Committee on Recent Economic Changes confirmed that "as a consequence of the World War and the boom preceding the crisis of 1920, various cakes-of-custom were broken. Old buying habits changed and consumers subsequently manifest receptivity to new types of merchandise without all the customary delay and diffidence."[10]

Beginning in 1919, an energy-intensive society emerged out of the wartime changes in consumers' attitudes, economic conditions, and advertising techniques. A larger and larger proportion of the working class could afford central station electric service and a few basic appliances, while the more affluent bought all the latest gadgets. As table 28 shows, the spending spree in Chicago exemplified emerging consumer habits across the country. In just five years, Americans doubled the amount they spent on electrical appliances. By the end of the decade, President Hoover's blue-ribbon panel of economic experts reported that "the growing use of electricity and of electrical appliances is one of the most typical characteristics of our contemporary industrial civilization, with its tendency toward increasing mechanization and the use of labor- and time-saving devices, not only in the factory but in the home as well." The committee also found that the promotion of electrical appliances in national magazines was significant. It made special note of large increases in advertising copy for new products such as refrigerators and radios.[11]

Methods of mass production introduced during the war helped the appliance manufacturers offer a widening array of products at reasonable prices. In 1920, for example, Insull's Electric Shops offered a broad range of these devices, from irons costing $7.50 to $10 to vacuum cleaners for about $50 and from coffee percolators for $25 to washing machines priced at $165 to $190. Householders could make their purchases in cash or by installment payments

Table 28 Growth in Number of Household Electrical Appliances in the
United States, 1922–28 (in Millions)

Appliance	Sales, 1922 ($)	Number in Use January 1923	Sales, 1927 ($)	Number in Use January 1928
Iron	11.0	7.000	11.2	15.300
Vacuum cleaner	40.0	3.850	49.3	6.828
Clothes washer	65.0	2.915	118.5	5.000
Fan	11.0	3.500	11.2	4.900
Heater	3.3	1.260	2.5	2.600
Toaster	2.0	1.000	2.3	4.540
Ironing machine	6.0	0.116	10.9	0.348
Refrigerator	4.0	0.027	82.1	0.755

Source: Committee on Recent Economic Changes of President's Conference on Unemployment, *Recent Economic Changes in the United States*, 2 vols. (New York:
McGraw-Hill, 1929), 1: 325.

added painlessly to their monthly bills. Five years after the war, a
survey of the utility's residential customers showed that 85–90 percent of working-class homes had at least one appliance, usually an
iron. At the other end of the economic spectrum, about three out
of four affluent city dwellers had a vacuum cleaner, and over half
had a washing machine (see table 29).[12]

In Chicago the postwar surge in consumers' demand for modern technology brought rapid completion of home electrification.
From 1919 to 1926, Commonwealth Edison gained a phenomenal
206,000 residential customers, setting a record in 1922 of over
72,000 additional connections. The "true engineering methods of
selling," as Insull put it the following year, had become an institutionalized routine, an orchestrated campaign that included constant advertising in the mass media, special incentives, installment
plans, and door-to-door solicitation. New business within the city,
he proudly pointed out, was generating a demand greater than the
total amount of electricity consumed in several of the utility's
neighboring states. In 1927 Commonwealth Edison achieved "saturation" of the residential market. For the rest of the decade, the
number of additional customers paralleled the growth of the city's
population and the construction of new housing.[13]

Barton and other "apostles of modernity" helped stimulate desire for things electrical, but their influence on the rise of an
energy-intensive society should not be pushed too far. On the contrary, the Chicago experience cautions against a facile acceptance
of the proposition that advertising during the 1920s became a pri-

Table 29   Household Electrical Appliances in Use in Chicago, 1926

Appliance	Number in Use	Percentage of Central Station Customers
Iron	595,800	85.2
Vacuum cleaner	495,300	70.8
Washing machine	293,800	42.0
Toaster	213,700	30.5
Percolator	127,900	18.3
Sewing machine	92,500	13.2
Heater	80,900	11.6
Fan	73,100	10.5
Ironing machine	26,100	3.7
Refrigerator	8,900	1.3

Source: Commonwealth Edison Company, Statistical Department, "Survey of Electrical Household Appliances in Use in Chicago," manuscript, 10 April 1927, in CEC-LF, F23-E3.

Note: Based on a census of 15,629 households, or 2.24 percent of Commonwealth Edison's 699,715 residential customers. The figures above are prorated from the survey data.

mary agent of social change. The new advertising was probably effective to a degree, but consumers were not brainwashed. At one extreme, some historians claim to have uncovered a gigantic conspiracy on behalf of the electrical industry. The mass media, according to these critics of our culture of consumption, used hidden persuaders to get housewives to buy appliances that they neither needed nor wanted. However, the best historical research suggests that media imagery was as much a mirror of popular trends as an engine of material desire. In fact, the more closely scholars examine this interaction between consumers and advertising, the more it eludes description in terms other than the proverbial chicken and egg.[14]

In Chicago, electrical technology in the home continued to serve essentially conservative purposes. Better lighting remained the primary reason for installing central station service (see chart 14). A 1923 survey of residential customers showed that lighting accounted for 75–85 percent of electrical use in the home. The minor role that appliances played was confirmed by the finding that on average only 5 percent of the sockets were devoted to their use. Promotion of the "convenience outlet" (and the adaptor plug) was just beginning. Its relatively late introduction reflected how slowly household routines changed.

The survey also revealed that the iron remained the only home

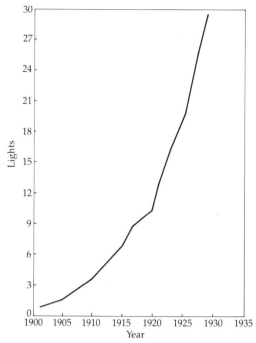

CHART 14.   Incandescent lights connected to central stations in Chicago, 1900–1930 (millions). *Source:* City of Chicago, Department of Gas and Electricity, *Annual Report,* 1900–1930.

appliance in common use and accounted for almost all the electricity consumed apart from lighting. Although market penetration of the iron reached 80–90 percent in working-class homes, the next most popular device, the vacuum cleaner, was found in fewer than a quarter. About half of middle-class dwellings had one of these useful devices, but only about one in four contained a kitchen convenience such as a toaster or percolator. The surveyors noted that even after twenty years in the home, the iron had a limited impact, since it was used largely in the hot summer months. Housewives often returned to their old sadirons during the wintertime when stoves and furnaces were burning.[15] These household patterns suggest that the power of advertising to influence consumers' habits was tightly constrained by traditional family and gender roles. A recent sociological study of ninety families in Boston comes to a similar conclusion: "It appears clear," the researchers found, "that modern household equipment is designed and marketed to reinforce rather than challenge the household-family pattern in contemporary North American society."[16]

For the most part, consumers used a rational process to decide which products to buy. A comparison of product successes and failures illustrates how city dwellers exercised discretion in making choices from the cornucopia of electrical devices. On the one hand, no amount of promotion could get Chicagoans to buy electric cars, despite Insull's best efforts. From the early days of the automobile, consumers preferred the peppy gasoline engine over the heavy storage battery, which weighed 2,000 pounds or more. In 1897 Insull had been involved in the original plan to monopolize the taxi business in Chicago, the "lead cab trust." Although that scheme had collapsed only three years later, Insull remained optimistic about the electric car and soon sank money in a new venture to manufacture commercial trucks. The utility executive believed that electric vehicles had a tremendous potential to build demand for central station service during the highly profitable off-peak hours late at night.[17] Yet Insull's campaign made little headway until America's entry into World War I.

The war seemed to open a bright new opportunity for the electric car. Food, fuel, and labor shortages caused a sudden jump in the cost of both horse-drawn and gasoline-powered transportation. In contrast, the price of electricity remained the same. Between 1917 and 1922, Insull mounted a new publicity campaign, and he invested a large amount of money to build a series of garages where the battery-powered vehicles could be conveniently recharged and repaired. Insull wanted this business badly, because an electric car consumed twenty times as much energy as the average residential customer and four times as much as the typical retail store. Chicagoans, however, continued to reject the electric car in favor of gasoline-powered alternatives. In this case Insull's well-coordinated and persistent sales effort was a complete bust.[18]

On the other hand, the radio was an instant success, the first true mass-consumption electrical product. In just ten years, the radio became more popular than the iron. Perfected during the war, it reached the mass market in 1923 with the formation of broadcast networks and the manufacture of all-electric (no batteries) home models. Two years later, Insull remarked that the radio's fantastic popularity was convincing the last residential holdouts to install central station service. By the end of the decade, the radio came to epitomize the triumph of a new ethos of mass consumption. For example, President Hoover's commission stated that "[the] radio sets forth the outstanding example of a new type of merchandise placed on the market with an almost phenomenally rapid increase in demand." Between 1923 and 1930, more than 60 percent of Chicago families purchased sets, a figure that climbed to 94 percent over the next five years.[19]

The contrast between the success of the radio and the failure of the electric car shows that, by and large, consumers bought only the products they really wanted. An immediate and unprecedented demand for the new form of home entertainment also suggests that city dwellers were not barred from buying more appliances by a lack of discretionary income. Consumers chose to own some products and not others. The three other applicances that Chicagoans most commonly purchased in the twenties—irons, vacuum cleaners, and washing machines—were practical housekeeping tools. Even the most ardent critic of consumerism would find it hard to deny that these appliances saved time and eased the drudgery of housework. There is just no comparison between running a vacuum cleaner over a rug and taking the carpet outside and beating it, for instance. The influence of advertising on the popular imagination appears to have been balanced more or less by rational decision making about the utility and convenience of various products.

Ultimately, analyzing householders' motives for buying electrical appliances is less important than understanding their impact on daily life. As we have seen, Americans used the new technology to preserve traditional gender roles. At the same time, however, the mechanization of the home began to change the rhythms of family life. The radio, for example, established new patterns of nightlife by bringing entertainment from across the nation into the home. Families now stayed up later to sit around the set, enjoy their favorite shows, and hear the latest news. In their perceptive study of "Middletown," Robert and Helen Lynd found that the radio, along with two other popular energy-intensive innovations—the automobile and the motion-picture palace—were "remaking leisure." The Lynds observed a new popular culture emerging as these new forms of technology were incorporated into everyday routines. "Indeed," they contended, "at no point is one brought up more sharply against the impossibility of studying Middletown as a self-contained, self-starting community than when one watches these space-bending, leisure-time inventions . . . reshaping the city."[20] The mechanization of the home by electrical appliances represented only part of the enveloping of family life in an "invisible world" of energy. The redesign of working-class and middle-class housing also played an integral part in creating an energy-intensive society.

"Where Will You Rear Your Children?"

In 1923 Insull built a model electric house in River Forest, a western suburb next to Oak Park in the terrirory served by his Public Service Company of Northern Illinois (PSCNI). The utility executive tapped a strong popular interest in new housing. During the sixteen days the demonstration house was open, 46,000 people filed through to see the latest appliances and modern amenities of domestic life. But unlike the dream house of tomorrow that would be built for the Chicago World's Fair of 1933, Insull's model home was carefully designed to resemble the archetypal single-family dwelling of the day. He presented visitors with images of a home that was attractive and modern but at the same time comfortably familiar and conventional. A complementary campaign of newspaper advertisements stimulated thoughts about family life-styles by asking such questions as, "Where will you live in 1930?" and "Where will you rear your children?"[21]

The answer was not as obvious as it might seem, because home builders offered Chicagoans a chance to live in new and better dwellings in a wide variety of urban as well as suburban environments. More than ever before, people were asked to choose not only a housing unit but also a community and a style of living. To be sure, the choice of housing had always implied "buying into a neighborhood," but in the 1920s improvements in transportation and communications created unprecedented opportunities for residents of metropolitan Chicago to sort themselves out by class, ethnicity, and race. New land-use regulations such as zoning laws and self-perpetuating deed restrictions encouraged social segregation and suburban fragmentation.[22] The postwar building boom remade the landscape of Chicago, from the rise of culture palaces and skyscrapers in the Loop to the erection of luxury apartments along the lakefront, and from the development of new shopping and entertainment districts in the neighborhoods to the planting of residential and industrial subdivisions throughout the urban fringe.

A faster flow of people and information also fostered regional integration by boosting the patronage of the city's central institutions of high culture, popular entertainment, and retail shopping. Rapid transit lines, interurban railways, better roads, and telephones all helped make the Loop readily accessible to the suburbs and satellite cities. Drawing on a metropolitan population, downtown Chicago flourished as never before. Suburbanites as well as cliff dwellers helped underwrite the city's literary renaissance and its new temples of highbrow culture that had been established at the time of the 1893 World's Fair, such as the Art Institute, the Chicago Symphony Orchestra, and the Museum of Natural History. The patronage of those living beyond the city limits also contributed to Chicago's dubious Jazz Age reputation as a wide-open frontier of nightlife, bootleg gin, gangsters, and sin. The "record-breaking volume" of construction during the twenties, President Hoover's economic experts pointed out, was not restricted to housing. "Advances in the design and equipment of buildings . . . had an important effect," the commission remarked, making older structures seem obsolete and creating "a demand for improved dwellings and structures of various types, including places of work and recreation."[23]

By the end of the decade, Chicago had become two cities: an upscale modern habitat and a premodern milieu of dilapidated housing and blighted neighborhoods. During the building boom, new construction incorporated more and more energy-intensive technology to provide a contemporary level of comfort and convenience. In 1920, for example, several impressive structures were completed downtown. A brief survey of these buildings will illustrate how a more intensive use of energy helped give rise to an attractive, modern urban environment alongside the old. In the Loop, the opening of the Chicago Theater inaugurated a new era of the motion picture palace, furnishing amusement for the masses in an ornate setting of romantic intimacy and splendor. "Indeed, to many millions of people" historian Carl Condit explains, "the movie theater came to be regarded as the very exemplum of the architectural art: it was what the word *architecture* meant."[24]

Creating a world of fantasy inside these gargantuan spaces required elaborate mechanical systems, and it took enormous amounts of energy to maintain a comfortable temperature. The basement of a typical movie palace contained, according to a leading architectural journal, a "curtain machine, vacuum cleaners, remote control [lighting] board, organ blowers, ozone machines, electric air scenting machines and tanks, fire pumps, oil burning heating equipment, transformer vault, . . . motor generators, etc." Although the Chicago Theater initially used only fans for cooling, its owners—Abraham and Barney Balaban and Samuel Katz—and other movie operators soon began installing early forms of air-conditioning. In 1924, for example, the Capitol Theater was built in the Loop with "a complete, scientifically controlled heating and ventilation plant . . . , containing refrigeration apparatus, by use of which fresh tempered and treated air is delivered to all portions of the building—warm washed air in winter and cool, washed air in summer." By the middle of the decade, the journal reported that the demand for this energy-consuming technology was "growing by leaps and bounds." The operators clearly understood the relation between rising thermometers and falling attendance.[25] Reversing traditional attendance patterns, signs promising "Cool Inside" began to make the movie theaters havens from the sweltering summer heat.

The Wrigley Building epitomized a new mastery in the design of office and apartment skyscrapers. It was built in conjunction with another monumental construction project, the Michigan Avenue Bridge, completed in the same year. A centerpiece of Daniel Burnham's 1909 plan for Chicago, the double-decker bascule bridge redirected the growth of the central business district from south to north. The twenty-eight-story Wrigley Building set a high standard of design for the ensuing development of North Michigan Avenue, the "Magnificent Mile." Like the movie palaces, high-rise buildings incorporated more and more elaborate mechanical infrastructures. The higher the skyscraper soared, the more energy its elevators, pumps, and other machines consumed to deliver utility services to the upper floors.

To show off this shining temple of commerce, the chewing-gum magnate installed a bank of spotlights across the street to bathe its white terra-cotta skin in 25 million candlepower of electricity. Exterior lighting added a new dimension to the decorative design of buildings and became almost as important as ensuring adequate interior illumination. After World War I, standards of lighting in the office rose sharply as the principles of scientific management were applied to clerical work. Studies showed, for instance, that increasing the amount of artificial lighting in the United States Post Office from 3.3 to 5.9 footcandles boosted the productivity of letter sorters by 20 percent. New guidelines of office management also called for more laborsaving devices to supplement the typewriter. Increasingly considered the tools of women workers, these specialized machines helped complete the mechanization of the office.[26]

Facing the lakefront at the opposite end of North Michigan Avenue stood a third new structure, the Drake Hotel. Playing a role similar to that of Insull's model home, it was a harbinger of the acceptance of apartment living in the city. The fancy hotel represented the ultimate expression of a new style of urban life that sacrificed none of the modern conveniences of a private home yet enjoyed all the downtown's cultural treasures and recreational facilities as well as the late-hour entertainment of a cabaret society. Significantly, the Drake was designed by Chicago's leading architect of luxury high rises, Benjamin H. Marshall. It was built directly west of his earlier innovative apartments for the rich at 1100 Lake Shore Drive, but more than a mile north of the Loop. This beachfront location suggests that the hotel's proprietors sought to attract a clientele that could afford to mix pleasure with business. Marshall and his business associates continued to reinforce the high-class, resortlike tone of the beachfront by filling in the space between the two buildings with a row of exceptional apartment towers for the elite.[27]

The rise of the "Gold Coast" set a conspicuous standard of urban life that large numbers of the middle class sought to emulate. During the postwar decade, more than half of new housing units were apartments. Stretching along the lakefront and rapid transit lines, apartments were in such demand following the wartime shortage that 18,000 were added in 1922 alone. This number soared for the next five years, reaching a peak in 1927 of 37,000 new homes. The large U-shaped courtyard building joined the high rise and the three-flat as a commonplace feature of the urban landscape.[28]

As Insull's model home demonstrated, the demand for new housing stemmed not from the exterior design of buildings but from the modern comforts packed inside. What separated the new style from the old in housing construction were quality materials and an energy-intensive infrastructure. "Improvements in small house design," Hoover's economic panel underscored, "include better lighting, heating, insulation and ventilation, economy in the use of materials, regard for efficient construction methods, and for durability and appearance of finished structures." In addition to electrical appliances, the experts listed several features of new homes that eased the burden of housekeeping, such as hardwood floors, linoleum in kitchens, tile floors in bathrooms, and built-in bathtubs. Beginning in the mid-twenties, moreover, the electrical industry mounted a national campaign to encourage builders to install convenience outlets. Whereas the proportions of apartments became more generous, the size of the average house shrank to compensate for the increased cost of its mechanical equipment. For contemporaries as well as latter-day architectural critics, the postwar decade was the golden age of the apartment in spite of a notable lack of innovation in the design of building exteriors. Carl Condit, for example, argues that "we are forced to the same conclusion about apartment buildings that we reached about hotels: in the matter of spaciousness both of individual rooms and total floor area, soundness of construction, adequacy of insulation, reliability of mechanical equipment, and general appearance, the conservatively designed apartments of the 1920s, with few exceptions, are much superior to their counterparts in the latest of postwar booms."[29]

Behind a lakefront facade of tall buildings, Chicago remained a horizontal "flat city." The construction boom of the 1920s had the effect of creating a new city alongside the old. By the end of the decade, the stark contrast between the new energy-intensive environment and the older premodern neighborhoods had become an embarrassing social scandal. A book titled *The Gold Coast and the Slum*, by sociologist Henry Zorbaugh, aptly described this glaring gap between the surroundings of the urban rich and poor. The problem of obsolete, decaying housing was especially acute in the South Side black ghetto, where overcrowding and exploitation by landlords produced some of the worst conditions. Yet the spreading blight of the slums was not confined to any single area or ethnic/racial group. In general these deteriorating neighborhoods had been built up between the Great Fire of 1871 and the World's Fair of 1893. Except along the lakefront, they formed a belt around the downtown district, beginning a mile outside of the Loop and covering a crescent with a radius of about five miles. During the 1920s, 150,000 people deserted the mean streets of these neighborhoods, leaving them to black migrants and others who were too poor to "buy into" the energy-intensive world of the modern city.[30] Beyond this old city, a new belt of middle-income and working-class bungalows and two-flats filled in the urban fringe up to and beyond the municipal border.

The bungalow boom provided an opportunity for families of modest means to move into a quasi-suburban environment that offered many of the modern conveniences found in more affluent neighborhoods. Over the decade, builders erected about 100,000 of these affordable brick houses in Chicago and suburban Cook County. Though not unattractive from the outside, their appeal lay in their Frank Lloyd Wright style of floor plans and their incorporation of contemporary building materials such as hardwood and linoleum floors, bathroom and kitchen wall tiles, and built-in plumbing. Those who could afford to live in these pleasant domestic surroundings appear to have also enjoyed the use of several electrical appliances. In 1929, for instance, a survey of appliance ownership divided Chicagoans into four categories based on the monthly cost of their housing (see table 30). Most apartments and houses fell in the two middle categories, with the third category comprising the single largest group of lower-middle-class and working-class households. In these homes irons and vacuum cleaners were nearly universal, most already had radios, and three out of five owned washing machines and toasters.[31]

The same impulses that gave rise to a bungalow belt around the inner city also fueled the growth of a patchwork of more affluent communities throughout the metropolitan area. After the First World War, energy-intensive technology helped make the lure of the suburbs irresistible. The Hoover commission provided a succinct explanation of how Americans turned the products of the machine age into agents of urban deconcentration. The experts reasoned that "the family's enlarged radius of movement due to the automobile, together with the neighborhood movie, the radio, and shorter working hours, strengthens the call towards the suburbs." On the one hand, it is ironic that the comforts of urban life seriously undermined the ascendancy of the American city at the very point when its residents finally became a majority of the population. On the other hand, improved transportation, communications, and energy services merely aided the fulfillment of the long-term cultural process that had generated a suburban ideal. Calling the suburbs the "hope of the city," the first careful study of these communities, *The Suburban Trend*, by Harlan Paul Douglass, defined this ideal as the belief that "the way to live in an urban situation is to preserve town forms, the small community, the single-family dwelling, and as much as possible of family privacy." The war experience reinforced these orientations by reviving a focus of social attention on child rearing, youth, and family relationships.[32]

In the 1920s, the automobile took the lead over other forms of transportation technology in giving shape and direction to urban growth. Cars and highways encouraged a more sprawling, deconcentrated pattern of settlement than railroads, which has spurred development along a system of arterial lines. The fragmented metropolis of Los Angeles became the model of the automobile city.[33] In Chicago a similar suburban sprawl occurred at the fringe simultaneously with a continuing buildup of skyscrapers and apart-

Table 30 Household Electrical Appliances in Use in Chicago, 1929 (Percentage of Central Station Customers)

	Type of Housing[a]				
Appliance	Luxury	Upper Middle Class	Lower Middle/ Upper Working Class	Lower Working Class	Total
Iron	100	97	96	95	96
Vacuum cleaner	97	94	87	85	87
Radio	71	63	56	42	53
Toaster	74	60	41	17	37
Washing machine	58	50	38	25	36
Percolator	45	28	19	5	16
Refrigerator	68	32	9	1	10
Fan	61	27	10	2	10
Heater	35	20	11	5	10
Waffle iron	42	20	9	2	9
Heating pad	30	14	7	2	7
Ironing machine	29	12	5	3	5
Curling iron	13	8	5	0.5	4
(Radio) battery charger	3	6	4	2	4
Table stove	0	3	2	1	2
Ventilating fan	13	5	1	0	1
Clock	6	2	1	0	1
Vibrator	3	3	1	0	1
Forty others					−1

Source: "Wiring and Appliance Survey in Chicago Homes Reveals Extensive Future Market," *Edison Round Table* 21(31 December 1929): 10.

[a]Based on the following monthly rental costs of housing: luxury, $150+; upper middle class, $75 to $150; lower middle/upper working class, $50 to $75; lower working class, −$50.

ments at the center. For some the automobile offered a means of escape from the old neighborhood to a new style of life.

Americans loved cars, making them the supreme symbol of a mass-consumption society. "Simultaneously with the advance in the use of the automobile," Hoover's experts noted, "there has been a marked development in the purchase of many commodities that a decade ago would have been described as luxury goods, but which have since entered so universally into the average [family] budget as no longer to be regarded as such." Between 1920 and 1925 the number of cars in Chicago tripled, to almost 300,000, while another 125,000 made the daily commute from the suburbs. The novelist Sinclair Lewis also made note of the automobile as the epitome of an emerging cultural ethos of consumption in his social satire *Babbitt*. Real estate salesman George Babbitt has trouble relating emotionally to his family, but he gains deep satisfaction from the unfailing response of his car's electric cigarette lighter. The Lynds too believed that the automobile helped break the customs of a scarcity-minded society in Middletown, replacing them with a free-spending attitude and a "vacation habit."[34]

Popular preference for the private auto over mass transit translated into a public policy that pumped hundreds of millions of dollars into city, county, and state road-building projects in the metropolitan area. This massive public works program was matched by private construction of the accoutrements of a car culture, including garages, drive-ins, roadside motels, and other tourist accommodations. Public subsidy of the automobile in the form of highway construction helps account for the decline of the streetcar and elevated companies, which continued to operate under the debilitating burdens of political controversy and heavy taxes.[35]

Nonetheless, the rail lines retained an important place in the growth of the suburbs. In 1916, for example, the North Shore interurban railway came under Insull's control. Over the next six years, his efforts to rehabilitate the line increased ridership from 2.7 million to 15.2 million passengers a year. On the South Side, the electrification of the Illinois Central Railroad's suburban routes in 1926 brought a similar boost in patronage by speeding up commuter service and spurring housing construction. Part of the postwar drive to conserve coal and reduce the smoke menace, the conversion of the rail line was enthusiastically endorsed by its 90,000 daily commuters. On 7 August 1926, over 2,000 invited guests including Mayor William Hale Thompson inaugurated the new era of suburban service by taking a ride in the faster electric trains along a thirty-mile route to their terminals in Matteson, Blue Island, and South Chicago. Upon returning to the city, the dignitaries joined in a day of pageantry and celebration of the "progress of transportation" at the new Soldier Field Stadium. During the same year, Insull opened a new interurban line, the Skokie Valley route, which paralleled the North Shore to the west of its built-up lakefront communities. A real estate boom instantly developed in the wake of this transportation improvement.[36]

Like the automobile and the interurban railway, modern utility services helped open the "crabgrass frontier" to homes, businesses, and industry. Mirroring the 1920s spotlight on family life, the public relations campaign of PSCNI played on social anxieties about child rearing and youth to promote an exodus from the city to the "edge of tomorrow." "Easily accessible," a typical newspaper advertisement promised, "these beautiful places enable you to respond with confidence to the alluring call of country life and its healthful recreations. Adequate educational facilities . . . ensure proper instruction for children of all ages, under [the] most modern conditions. . . . What greater joy can your boy have than a dog trailing along faithfully at his heels? . . . So make your plans now to live out where the country begins." Of course the clincher in newspaper appeals invariably was to remind city dwellers that their suburban counterparts now enjoy all the modern conveniences of gas and electric service.[37]

Insull's creation of a 6,000-square-mile regional network of power during the prewar years speeded the buildup and expansion of Chicago's suburbs over this vast territory during the succeeding period (see map 8). In the postwar decade, the final stages of electrification in the suburbs paralleled a similar completion in the city. Between 1919 and 1929, the number of suburban families with electric service increased by about 200 percent, from 97,000 to 275,000 households. Residential consumption of light and power grew at an even faster rate of 511 percent, an average increase of over 50 percent a year. Three minor rate cuts between 1916 and 1926 reduced the average net price to residential consumers by 16 percent. The declining cost of electricity helped make more energy-intensive life-styles seem a painless addition to the family budget.[38]

The main task of Insull's PSCNI was to keep pace with the incredible demand for more energy. The utility had to add capacity in the older, more builtup communities and extend distributor lines from the transmission system to new homes and subdivisions on the frontiers of suburban settlement. The affluent and well-established North Shore, for instance, remained Insull's testing ground for innovations in energy services. In 1923 he strung a 132,000-volt line between the Chicago-Evanston border and Waukegan, establishing one of the first prototypes of a "superpower" system in the United States. The need for a new generation of high-voltage transmission technology also stemmed from deconcentration. During the postwar decade, the utility responded to the growth of outlying suburbs by extending power lines to 150 additional places. Supplying electricity to a total of 300 communities in 1928, the PSCNI initiated service in such towns as Stickney, South Holland, Western Springs, Downers Grove, Saint Charles, Elmhurst, and Lombard.[39]

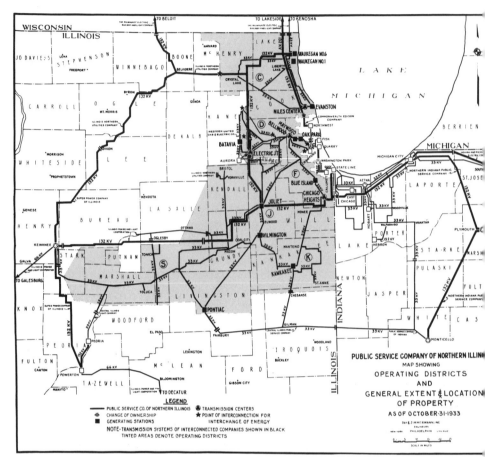

Map 8. Operating districts of the Public Service Company of Northern Illinois, 1933. *Source:* Illinois Commerce Commission, Case 22353 (1933), Public Service Company of Northern Illinois, Exhibit 1, in CEC-HA, box 4093.

Although electricity supplied the single most useful form of energy, gas became increasingly popular for cooking and home heating. Wartime coal shortages, followed by crises and labor strife at the mines, triggered a transition to oil and gas. New central heating plants using these modern fuels also provided laborsaving convenience by eliminating the drudgery of shoveling coal for the furnace. In 1923, moreover, Insull initiated a campaign to encourage residential consumers to install gas heating units by offering a special low rate equivalent to the wholesale industrial rate. Over the next five years, the PSCNI's gas customers increased by about two-thirds to approximately 167,000 households. Only 2,000 of these homes used gas heating, but they helped account for a 100 percent gain in the use of gas.[40]

Insull's energy utilities formed a crucial component of the complex infrastructure that gave rise to a modern suburban environment. Maintaining urban convenience and comfort over a metropolitan expanse of territory would not have been possible without using more and more energy to deliver public services to a dispersing population. In the twenties the PSCNI supplied electrical power for other utilities, for streetlights and other municipal services, for public transportation, and for water, gas, and sanitation pumps. Like the private side of an emerging consumer society, the public sector consumed energy at an ever-faster pace. In 1919, for example, it used 87,903,000 kwh, or 41.4 percent of the total supply. Ten years later public services consumed 502,210,000 kwh, or 47.0 percent of the total output of the PSCNI. These figures represent an average annual growth rate of 47.1 percent, only slightly less than the rate of over 50 percent for residential consumption. These statistics, moreover, reflect the major effort by government authorities not only to build roads to the suburbs but to provide all the amenities of modern life there as well. In *The Suburban Trend,* Harlan Douglass observed that the suburbs "are undergoing the vast expenditures necessary to secure the full round of standard civic conveniences in a day in which the demand for every up-to-date luxury is rapidly mounting."[41]

The spreading web of a networked metropolis helped home builders meet the demands of prospective buyers for all the amenities they could afford. In the suburbs as in the city, home buyers gave greater priority to domestic comfort and to neighborhood than to exterior facades. The floor plans and mechanical systems pioneered by Frank Lloyd Wright and other housing reformers during the Progressive Era were incorporated into a full range of architectural designs, from historical revivals to modern versions of the prairie style.

By the mid-twenties, moreover, the opening of the suburbs by the automobile and power lines resulted in a large surplus of building lots and subdivisions in the metropolitan area. A period of chaotic speculation followed as subdividers attempted to beat the competition by luring specific groups of buyers to their land. Calling for more planning, one of Chicago's leading real estate brokers, Helen Corbin Monchow, was highly critical of this rivalry because one group of subdividers will rush in to supply lots for the industrial workers who are expected to be employed in a proposed plant, while other subdividers hasten to supply lots for the white-collar and executive group. What is more, each of these subdividers has his own ideas as to where the respective groups will choose to live, or where he thinks he can induce them to live. One will supply close-in lots for the industrial workers while another will develop a subdivision farther out, using the argument that cheaper land will provide cheaper lots. When it comes to providing lots for the executives, there is no limit to the imagination of the subdividers.[43]

Faced with too many empty lots, Monchow and other real estate brokers started constructing houses on their land, and a few of the largest subdividers evolved into community builders. Following in the footsteps of Edgewater's J. Lewis Cochran, they became adept at using zoning laws and deeds with restrictive covenants to gain control over land-use planning for entire suburban environments, including various classes of residential areas, shopping centers, industrial districts, and recreational facilities. The rise of the community builders suggests that house buyers wanted all the modern amenities of suburban life outside the home as well as all the energy-intensive comforts of domestic life inside it.[43]

Comparing community patterns of electric use lends strong support to the thesis of an emerging society of consumption that became enveloped in a ubiquitous world of energy. (see charts 15–17 and tables 31–33).[44] By the 1900s, several of Chicago's suburban areas had distinct identities. The communities along the North Shore, for example, had a well-established reputation as the exclusive preserve of the upper crust of urban society. The contiguous area between Highland Park and Lake Forest exemplifies the residential suburbs of the well-to-do. Evanston represented a more balanced mix of housing and commercial enterprise. It primarily contained middle- and upper-class housing, but it had a significant shopping district. In a similar manner, Waukegan to the north of the city and Chicago Heights to the south shared a common image as working-class, industrial suburbs. Joliet provides a typical case of a satellite city with a thriving industrial sector.

Turning to the statistics, the first and perhaps most important observation is that the growth in the per capita consumption of light and power was phenomenal in all the test communities (see chart 15 and table 31). An increase in the use of electricity in the

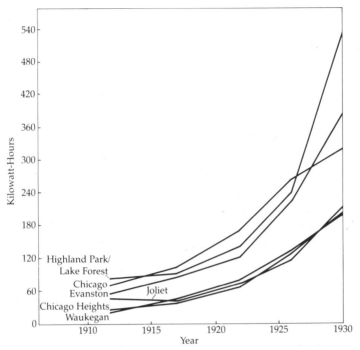

CHART 15.   Average annual per capita consumption of electric light and power by stores and homes, 1912–30 (kilowatt-hours). See note 44 for a list of townships included in each statistical district. *Sources:* Public Service Company of Northern Illinois, Statistical Department, "Average Annual Kilowatt Hours per Customer by Towns" (photocopy in my possession); H. A. Seymour, "History of Commonwealth Edison, 3 vols., manuscript, 1935, Appendix, 176, in CEC-LF, box 9001; U.S. Census, Population, 1910, 1920, 1930.

home and the retail store ranging from 19.0 to 40.8 percent a year for every person over an extended period between 1912 and 1930 offers convincing proof that an energy-intensive society developed in the Chicago region. Second, variations among the communities form a coherent pattern that generally conforms to their respective wealth, size, and social identity. A lack of perfect consistency reflects the unique historical experience of each place within larger regional and national contexts. Beginning in 1917, for example, the central city and the relatively rich suburbs of Evanston and Highland Park group together at the high end of energy usage, while the less affluent, more work-oriented communities of Chicago Heights, Waukegan, and Joliet cluster around a lower level of consumption. But the growth of electric use in this second group out-

Table 31   Average Annual Rate of Growth per Capita Consumption of Electric
Light and Power by Homes and Stores, 1912–30

Community[a]	1912–22	1917–22	1917–27	1922–30
Chicago	13.7	12.1	16.3	10.9
Mixed suburb				
(Evanston)	11.4	7.6	14.8	26.7
Residential suburb				
(Highland Park/				
Lake Forest)	6.8	10.5	17.3	33.9
Industrial suburb				
(Waukegan)	24.2	15.0	20.6	17.9
Industrial suburb				
(Chicago				
Heights)	16.6	14.6	24.2	23.7
Satellite city				
(Joliet)	5.7	15.6	19.3	23.6

Sources: Public Service Company of Northern Illinois, Statistical Department, "Average Annual Kilowatt Hours per Customer by Towns" (photocopy in my possession); H. A. Seymour, "History of Commonwealth Edison," 3 vols., manuscript, 1935, Appendix, 176, in CEC-LF, box 9001; U.S. Census, Population, 1910, 1920, 1930.
[a]See note 44 for a list of townships included in each statistical district.

stripped growth in the others between 1917 and 1922, mirroring the effects of wartime prosperity on these industrial centers.

Looking more carefully at the patterns of residential consumption, we see a direct correlation between wealth and energy use (see chart 16 and table 32). In 1912, the wide range between a high consumption level of 76.2 kwh per person in Highland Park/Lake Forest and a low of 13.9 kwh per person in Joliet resulted as much from differences in the proportions of residents with and without central station service as from variations in household usage patterns. Fifteen years later, with electrification virtually complete, the statistics show a more accurate picture of the average household in the various communities. In the domestic realm, the leadership of Highland Park/Lake Forest was unmistakable. In his pioneering study of the suburbs, Douglass argued that this type of affluent social environment typified the "decentralization of consumption." "There can hardly be a greater error," he reasoned, "than to think of the residential suburbs as mere dormitories. They are rather the realms of consumption as against production, of play in contrast to work, of leisure in exchange for business. Only incidentally are they places to sleep."[45]

A review of the growing use of light and power by retail mer-

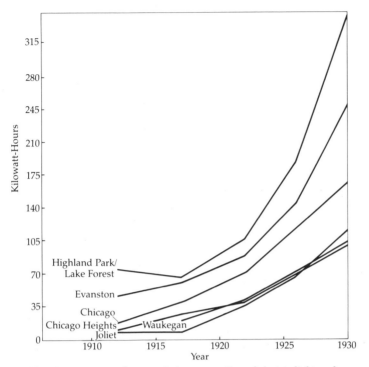

CHART 16.   Average annual per capital consumption of electric light and power by homes, 1912–30 (kilowatt-hours). See note 44 for a list of townships included in each statistical district. *Sources:* Same as for chart 15.

chants outside the central city confirms Douglass's argument about the decentralization of shopping and entertainment (see chart 17 and table 33). In 1912 the two cities, Chicago and Joliet, stood far ahead of the suburbs in use of energy for these commercial purposes. But the continuing dispersal of the population prompted merchants to follow their customers away from the downtown areas to the neighborhoods. In the 1910s, several strategic corners with heavy traffic became commercial/entertainment subcenters, complete with department stores, movie theaters, and office space for a variety of professional and consumer services. This type of shopping subcenter also arose in close-in suburbs such as Evanston and Oak Park, to serve their residents and those living farther from the Loop. By 1917 this decentralization was already evident in the convergence of per capita consumption patterns for all the communities outside Chicago. This clustering continued until 1929, when the shock waves of the stock market crash began to

Table 32   Average Annual Rate of Growth per Capita Consumption of Electric
Light and Power by Homes, 1912–30

Community[a]	1912–22	1917–22	1917–27	1922–30
Chicago	29.4	24.8	20.8	16.6
Mixed suburb				
(Evanston)	9.0	8.4	14.9	23.0
Residential suburb				
(Highland Park/				
Lake Forest)	4.0	11.5	20.1	28.6
Industrial suburb				
(Waukegan)		17.7	26.0	19.8
Industrial suburb				
(Chicago				
Heights)	12.4	8.4	16.7	19.3
Satellite city				
(Joliet)	16.8	37.8	47.5	27.3

Sources: Public Service Company of Northern Illinois, Statistical Department, "Average Annual Kilowatt Hours per Customer by Towns" (photocopy in my possession); H. A. Seymour, "History of Commonwealth Edison," 3 vols., manuscript, 1935, Appendix, 176, in CEC-LF, box 9001; U.S. Census, Population, 1910, 1920, 1930.
[a]See note 44 for a list of townships included in each statistical district.

reduce commercial activity. The 1930 figures indicate that merchants in the more affluent North Shore suburbs felt the early effects of the depression less than merchants elsewhere in the region.[46]

In 1926, downtown merchants met the challenge of the "decentralization of consumption" by erecting new streetlights along State Street. Creating the "brightest outdoor mile ever made," the 2,100-watt lamps raised the illumination level on the street to that used for indoor reading. On 14 October, President Calvin Coolidge threw a switch in the White House to inaugurate the great "Daylight Way" in Chicago. Over 200,000 people jammed the Loop to witness the historic event. Thomas Edison sent a telegram that praised the city "for such a unique and practical idea. I think it is bound to be successful." The merchants of Randolph Street agreed, and they immediately announced plans to install a similar system.[47] Yet the bright lights of the city were no longer a match for lure of the suburbs. In fact Marshall Field and Company, the ultimate bastion in the Loop, had already opened branch stores in Evanston, Oak Park, and Hyde Park.

In the 1920s, energy-intensive technology helped give Chicagoans a much wider choice of where to live. On the one hand, the postwar building boom offered a simple choice between the old

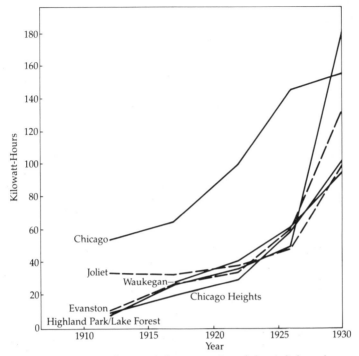

CHART 17. Average annual per capital consumption of electric light and power by commercial stores, 1912–30 (kilowatt-hours). See note 44 for a list of townships included in each statistical district. *Sources:* Same as for chart 15.

city of obsolete housing in deteriorating neighborhoods and a new metropolis that was becoming more and more dependent on massive amounts of light, heat, and power to maintain an infrastructure of modern public services. Except for middle-class blacks confined to the ghetto, every city dweller who could afford it decided to escape from the premodern "blighted areas" to this energy-intensive world.[48] On the other hand, those participating in the new consumer society were offered an unprecedented range of styles of living, including luxury high rises on the lakefront, courtyard apartments and bungalows in outlying areas, and houses in the suburbs. For growing numbers of Chicagoans, the automobile and the public utility network helped fulfill a suburban ideal.

"How You Gonna Keep 'Em Down on the Farm after They've Seen Paree?"

This popular song posed a question that struck at the heart of rural life in America during the postwar decade. Farmers too pondered

Table 33   Average Annual Rate of Growth per Capita Consumption of Electric
Light and Power by Commercial Stores, 1912–30

Community[a]	1912–22	1917–22	1917–27	1922–30
Chicago	8.5	10.5	13.5	6.8
Mixed suburb (Evanston)	21.6	5.9	14.7	36.0
Residential suburb (Highland Park/ Lake Forest)	32.8	7.8	10.2	49.5
Industrial suburb (Waukegan)		12.8	16.0	16.2
Industrial suburb (Chicago Heights)	25.2	9.6	20.5	29.4
Satellite city (Joliet)	1.2	5.9	7.0	20.0

Sources: Public Service Company of Northern Illinois, Statistical Department, "Average Annual Kilowatt Hours per Customer by Towns" (photocopy in my possession); H. A. Seymour, "History of Commonwealth Edison," 3 vols., manuscript, 1935, Appendix, 176, in CEC-LF, box 9001; U.S. Census, Population, 1910, 1920, 1930.

[a]See note 44 for a list of townships included in each statistical district.

where and how they should live. And like the city poor, they too remained locked outside the emerging consumer society. After the war, the gap between this urban milieu and the world of the farmer grew wider and wider. In the 1920s the arts and popular entertainment flourished in the city as never before, while economic distress, population decline, and psychological doubt and despair seemed to sap the lifeblood of the countryside. For farmers, the question of staying on the land or leaving for the city had long carried moral overtones, stemming from a pastoral ideal of America. The eclipse of rural life by modern urban culture made the decisions facing farm families all the more wrenching. They seemed confronted with a stark choice of either fighting to preserve the traditional values of rural life or renouncing them for a more affluent, but mechanized and alienated, life in the city. A farmer from Tennessee expressed this dilemma perfectly. "The greatest thing on earth is to have the love of God in your heart," he testified at a church meeting, "and the next greatest thing is to have electricity in your house."[49]

The war triggered this crisis of rural life by throwing the farm economy into an irreversible depression. Huge increases in production to meet the food shortages of the war years turned into

ruinous surpluses after the war. Commodity prices and farm values plummeted after 1919, leaving farmers with a heavy burden of
debt they had incurred to buy additional land and machinery during the wartime expansion. National tariff policy exacerbated their
plight by reducing agricultural exports while increasing the cost of
manufactured goods. Farmers were caught in a viselike grip between falling income and rising expenses. For many this debilitating economic squeeze left only one choice—either sell out or be
pushed out by a bank foreclosure. During the 1920s, for example,
the farm population in Illinois declined by 105,000 people, or about
10 percent. Farm values fell even more sharply, by 30 percent,
which represented a loss of $1.8 billion. The resulting migration to
the city helped swell the urban-rural ratio from 2:1 to 3:1.[50]

Although some were pushed off the land, others—like the soldiers in the song—were drawn to the city by the pull of its bright
lights and modern amenities. After the war, the daily grind of
household and farm chores changed little from earlier times, but
the drudgery became more and more unbearable as the gap widened between this traditional style of living and the new, mechanized routines of an urban-based, energy-intensive society. As
League of Women Voters activist Ann Dennis Bursch observed,
"Brilliant incandescent bulbs . . . turned the city night into another
day, lined the streets and roofs with gaudy signs, brought the moving picture industry up to its present appalling hugeness and at-
tractiveness. Since then, there has been no keeping the growing
children and the 'hands' on the farm." In 1919, for example, the
Department of Agriculture reported that the average family in the
country spent over ten hours a week pumping and hauling water
and another five hours keeping the kerosene lamps in working
order. These were perhaps the two most universally despised domestic chores on the farm.[51]

In contrast, even the working poor in Chicago generally had
water service in their buildings and gas lighting in their homes.
Living in a slum area, moreover, did not bar one from the city's
public facilities, popular entertainments, and commercial services.
Inability to purchase an icebox did not preclude putting fresh produce and meat from the neighborhood grocery on the dinner
table.[52] For farmers, however, every improvement in the urban
condition only added to their feelings of isolation from the promise
of American life.

On the surface of national politics, the countryside fought back
by mounting a vigorous protest against the city's godless materi-

alism, foreign influences, and commercialized sin. To be sure, the conflict between agrarian and urban values has been a perennial theme of American culture and politics since the founding of the nation, and scholars usually portray the 1920s in these terms. The triumph of Prohibition, the execution of Nicola Sacco and Bartolomeo Vanzetti, the Scopes monkey trial, the revival of the Ku Klux Klan, and the struggle over the 1924 presidential bid of Al Smith were all prominent signs of a vicious clash of cultures.[53]

But below the surface of these symbolic crusades against modernity, the farmers were making a decision to enjoy the material comforts of urban life. "In the consumer-oriented ethos of the postwar world," historian Don Kirschner argues, "technology promised finally to transform the very substance of country living." A growing convergence of reform thought centered on the electrification of the countryside as the single most important antidote to the exodus of youth to the city. "Farmers were no longer walking the line between their values and their material commitments," Kirschner contends, "they had crossed it, they had made their choice. Every time they bought an urban-manufactured radio, they made their choice, and every time they tuned in to a broadcast from Chicago, they confirmed it; . . . and whenever they left the humdrum of the farm for a weekend of excitement in Chicago, St. Louis, or Kansas City—or even permitted themselves such a dream—they made the choice."[54]

The farmers' eager adoption of the radio illustrates how modern technology acted as a savior of rural life and an instrument of urban culture at the same time. With the first broadcasts after the war, Henry C. Wallace and other champions of the farmer immediately grasped the significance of the radio as a tool that could break the isolation of the countryside. Two months after the first broadcast from Chicago by station KYW from the top of the Edison Building on 11 November 1921, farmers began receiving reports on market and weather conditions. And two years later, 131 stations were operating within a 380-mile radius of the center of the state. For example, the philanthropist Julius Rosenberg and his Sears Roebuck Agricultural Foundation started a station with the call letters WLS (World's Largest Store) specifically to provide farmers with business news and family entertainment. Thousands rushed out to buy battery-operated sets for their homes, and others gathered in banks, churches, and community centers to tune in such instant hits as Big Ford and Little Glen, National Barn Dance, and other favorite programs. Perhaps the best measure of the farmers' acceptance of the new technology was the rapid growth in the number of sets in rural Illinois. In just five years between 1925 and 1930, the proportion of farm families with radios increased from one-eighth to one-half.[55]

By the mid-twenties, the antimodernist crusaders had to share the public stage with those who saw electrical technology as a solution to the crisis of country life. In the national forum of the United States Congress, reformers turned the Muscle Shoals controversy over the disposition of the federal power dam and artificial fertilizer plant into a cause célèbre, which kept the question of rural electrification before the general public. The most important public statement on technology and society, however, was sponsored at the state level by Pennsylvania governor Gifford Pinchot. A veteran reformer, he had been a leader of the conservation battle as chief forester of the United States under President Theodore Roosevelt. Concerned about the decline of rural life, Governor Pinchot couched his bold proposal for public initiatives in setting energy policy in traditional democratic terms. He reasoned that "only electric service can put the farmer on an equality with the townsman and preserve the farm as the nursery of men and leaders of men that it has been in the past."[56]

In 1923 he chose another seasoned reformer of the Progressive Era, Morris L. Cooke, to conduct a study of a regional "giant power" system. Cooke supported the governor's call for long-term planning by the public sector because "we believe that electrical technology has advanced to a point where the use of current can be made so inexpensive as to revitalize the whole social fabric [of country life]." His report provided reformers with a comprehensive model of social planning on a regional scale and an attractive vision of the American dream in a rural setting.[57]

The Giant Power Survey also outlined the major obstacles to bringing modern technology to the farm. The biggest stumbling block was the cost of extending the power grid over a vast expanse of territory with a sparse population. The cost of stringing a distributor line was about $2,000 a mile in an area with two to five farms per mile. In the twenties, utility companies generally refused to finance these rural lines, for two related reasons. First, the rural population was declining while suburbia was booming. The simple logic of maximizing profits directed the investor-owned utilities to use their venture capital to keep pace with this fast-expanding, lucrative market. Second, the depression of the agricultural economy hamstrung the farmers' ability to pay for line

extensions to their homes, either individually or through a cooperative arrangement. In addition to this heavy expense, each
household faced the costs of installing service wires and purchasing essential appliances such as lighting fixtures, water pumps,
machine motors, housecleaning tools, and radios.[58]

For the great majority of the already hard-pressed farmers,
modern technology remained an impossible dream. In the 1920s,
rural America stood outside the emerging consumer society. In
1925–26, for example, Insull's PSCNI served only 1,000 out of almost 20,000 farm families in the Chicago region.[59] The nonfarm
use of electricity by the townspeople of these areas also lagged far
behind that of people in the city's residential suburbs. Energy consumption in rural households was roughly half the amount used
in the homes of affluent suburbanites. The gap between the two
worlds looms even larger when we consider commercial and industrial uses of energy. Table 34 compares average per capita electrical use for several of the PSCNI's service districts in the Chicago
region (see map 8). The relatively low figure for the rural areas
underscore the rapid growth of energy use and the mechanization
of daily life in the metropolis.[60]

On the eve of the Great Depression, the moral crusade of the
farmers against urban life had been largely superseded by the campaign to bring rural America into the modern energy-intensive
world. State and national politics began to reflect the ground
swells of grass-roots support for rural electrification and economic
parity for agriculture. Cooke's report offered a practical alternative
to financing rural extensions by investor-owned utility companies.
The Giant Power Survey provided reformers with a model of regional planning under government sponsorship. In 1928, leadership on the state level passed to the new governor of New York,
Franklin D. Roosevelt. An ardent advocate of country life, Roosevelt would appoint Cooke three years later to an innovative
agency, the Power Authority of the State of New York. On a national level, the campaign for public power and regional planning
still centered on the fate of the federal power dam and fertilizer
plant at Muscle Shoals, Alabama. In Washington this fight was led
by a group of Republican senators from the Midwest and the West,
including George W. Norris of Nebraska, Burton K. Wheeler of
Montana, and Robert M. La Follette, Jr., of Wisconsin.[61]

The emergence of an urban, energy-intensive society over the
postwar decade played an integral part in creating a culture of consumption in the United States. Falling utility rates and cheap,

Table 34   Average per Capita Consumption of Light and Power, 1926
(Kilowatt-Hours)

District[a]	Population	Residential Light and Power	Commercial Light and Power	Total Light[b]	Total Power[c]	Total Electricity
Chicago	3,105,000	116.2		259.5	402.2	661.7
Residential district C (North Shore)	139,100	123.1	45.8	179.6	166.4	346.0
Industrial district F (Chicago Heights)	111,060	62.2	38.0	110.4	697.3	807.7
Satellite city district J (Joliet)	112,200	51.9	30.2	80.8	404.7	485.4
Rural district L (Lacon)	21,550	38.0	23.9	72.3	110.8	183.1
Rural district R (Pontiac)	33,620	29.9	20.9	58.5	43.6	102.1

Sources: Public Service Company of Northern Illinois, Statistical Department, "Average Annual Kilowatt Hours per Customer by Towns," (photocopy in my possession); H. A. Seymour, "History of Commonwealth Edison," 3 vols., manuscript, 1935, Appendix, 176, in CEC-HA, box 9001; Commonwealth Edison Company, Statistical Department, "Kilowatt Hours Sold per Capita," memo, 10 February 1927, in CEC-LF, F23-E3; U.S. Census, Population, 1920, 1930.
[a]See map 8 for the area encompassed by each statistical district.
[b]Includes municipal street lighting and miscellaneous municipal lighting.
[c]Excludes railroad power.

mass-produced appliances spurred the mechanization of home, shop, and factory. Electrical technology also encouraged suburbanization by furthering the regional deconcentration of the city's population and industry. By the end of the decade, this ubiquitous world of energy made the exclusion of the slum, the ghetto, and the countryside into a glaring failure of the American dream. Increasingly, liberal reformers called for regional planning under government supervision to bring the urban poor and rural farmers into the machine age. With the crash of the stock market in October 1929, these voices of protest would finally find a sympathetic national audience.

10

A Ubiquitous World of Energy

In 1928, the demolition of the old Loop headquarters of the Chicago Arc Light and Power Company marked the end of an era, a fifty-year period when electric lines spread throughout the region. Built a year after the Great Fire of 1871, the building at 28 Market Street (now Wacker Drive) on the Chicago River between Madison and Washington streets had been converted from a hotel into the "central manufacturing block" by the end of the decade. Its steam engine had supplied power to a variety of small factories and inventors' workshops, including the one occupied by Milan C. Bullock, who in 1880 had installed one of the city's first electric dynamos, a two arc lamp machine that became the centerpiece of a fast-growing enterprise to supply the central business district with better lighting. By the time Samuel Insull's Chicago Edison Company bought out the arc light company in 1892, the entire building was being used as a generating station. Its peculiar rope drive—linking engines and dynamos—produced a deafening shriek, and "the upper floors swayed as on shipboard, the vibration being particularly severe when, as periodically happened, most of the engines happened to chug in the same direction at the same time." Insull was having the historic building torn down to use the site for a grand skyscraper, a new home for Chicago's Civic Opera Company.[1]

Whereas the destruction of the old generating station symbolized the close of one period, Insull's inauguration of a series of major policy changes in 1928 announced a new era of planning for an energy—intensive society. The sixty-nine-year-old businessman's reform package centered on a new structure of residential rates. To promote the use of household appliances, Insull instituted an alternative method of estimating maximum demand. Scrapping his original formula for the demand system of rates, the utility czar now offered to calculate the primary charge based on the number of rooms in a dwelling rather than on the number of electrical sockets. The old method tended to inhibit home builders from installing an adequate number of convenience outlets, making the use of appliances tiresome and unsightly. With few dwellings left to wire, new business would have to come from rising demand among the existing pool of energy consumers.[2]

This first significant departure from Insull's marketing strategy of 1898 reflected an acknowledgment that the search for better lighting was for the most part over. The following year, in May 1929, Commonwealth Edison installed its one millionth meter during the celebration of the opening of the Daily News Building, directly across the river from the Civic Opera House.[3] The occasion was a fitting tribute to the master salesman and his gospel of consumption. Although the long campaign to sell more light fixtures and brighter lights was not abandoned, it settled down to a low-level program of gradually raising the standards of illumination in all types of interior space, commercial and industrial as well as residential. In the future, Insull believed, the most promising approach to stimulating household demand for electrical energy would be supplying customers with attractive "load building" products like the phenomenally successful radio.

In conjunction with the utility executive's shift in marketing strategy from lighting to appliances, the General Electric Company and other manufacturers began to create products that incorporated the machine-age aesthetics of industrial design. In 1925, for example, GE introduced the monitor-top refrigerator, so called because its round condenser resembled the gun turret of the famous Civil War ship. In the second half of the twenties, sophisticated advertising was not considered enough; the products themselves had to express high technology. Within two years, GE went into mass production of the all-steel, streamlined household appliance, which embodied the "image of the machine with its attributes of speed, efficiency, precision and reliability," according to a recent assessment.[4] Six years later, the company presented to Henry Ford the one millionth model to roll off the assembly line. In addition to refrigerators and radios for the residential consumer, improved air-conditioning equipment promised to build a similar demand for more energy in the commercial sector.[5]

Insull's planning initiatives were a response to the virtual completion of electrification in the Chicago region, except for the rural areas. In 1928 he opened a working demonstration farm in Lake County to help capture this last untapped market. In the first year and a half, the showcase country estate in Mundelein presented a wide array of rural activities and lectures, attracting over 50,000 visitors. More important, Insull launched an innovative five-year program to finance rural electrification. For the first time, the Public Service Company of Northern Illinois (PSCNI) offered to absorb the cost of line extensions if rural customers agreed to pay a minimum energy bill for a fifty-month period. The minimum-guarantee plan finally lifted an unbearable financial burden off rural families, spurring a fivefold increase in the number of electrified farms by 1935, to 11,000, and doubling the size of the rural distribution network.[6]

The new phase of Insull's planning for an energy-intensive society extended from boosting the consumption of electric light and power to promoting gas as a source of industrial and residential heat. In the postwar decade, the suburbanization of heavy industry furnished the economic foundations for the inexorable spread of gas lines throughout the territory served by the PSCNI. Although gas cost home customers more than coal or oil for heating, Insull was encouraged by the rapid growth of new installations, from 196 households in 1924 to 1,420 only three years later. To boost this trend, he restructured residential gas rates in 1927 to eliminate separate bills for heating and cooking gas. "The variety of types of homes now being served," the utility company reported the following year, "suggests the possibility that the cost of the service even now, does not place it beyond the reach of the average home-owner."[7] During that year, Insull coordinated the regional integration of the gas supply by interconnecting the lines of the city's Peoples Gas Company with those of the suburban PSCNI. The unification of the distribution grid was a preliminary step in an ambitious scheme to lay a 1,000-mile pipeline from the natural gas fields of Texas to the Midwest.[8]

The new directions in energy planning that Insull forged in 1928 were cut short less than two years later by the stock market crash. Suddenly the great financier was in serious trouble. For the past five years, in fact, political and economic problems had been mounting steadily for the electric utility industry's leading figure. On a local level, he was deeply embroiled in the endless struggle over the fate of the city's elevated and surface railway lines. Since the prewar period, Insull had favored a consolidation of the two services, but political wrangling over municipal ownership thwarted the acceptance of any reform package. In 1927 the utility czar tried again to gain acceptance of a transit consolidation plan by appealing directly to the state legislature. For the next two years, however, this effort was stymied by a coalition of civic reformers organized under the banner of the People's Traction League. Insull was thrown on the defensive by its able group of public advocates, including Paul Douglas, Charles Merriam, Donald Richberg, and Harold Ickes.[9]

The utility financier's maneuvering on a local and state level soon spilled over into national politics. In 1926 Insull helped elect Frank L. Smith, a former chairman of the Illinois public utility commission, to the United States Senate by contributing the princely sum of $125,000 to his campaign. In Washington, the revelation of an election scandal in Pennsylvania stirred the Senate to hold hearings on the influence of campaign funding on the democratic process. Led by Senator James A. Reed, the investigators forced Insull to testify and exposed the tremendous political power at the command of this leader of a vast utility empire of interlocking holding companies. Although Insull's campaign contributions were not illegal, the Senate refused to seat Smith on the grounds that taking such a huge sum of money from a single contributor was immoral.[10]

The public responded positively to the Senate hearings, encouraging George Norris, Thomas Walsh, and other champions of government-sponsored regional planning to press their attack on the "Power Trust." In 1927 they sponsored a new study of the pervasive corruption of the American economy and political process by the utility companies and their financial managers on Wall Street. The investigation by the Federal Trade Commission resulted in a seventy-volume report. More damaging, perhaps, for the utility industry was the ceaseless barrage over a seven-year period of shocking disclosures about the Power Trust. In 1928, for example, Franklin D. Roosevelt echoed the disturbing news from Washington to hammer home his own denunciation of the utility industry in a successful bid to become governor of New York. By the time of the great crash on Wall Street, then, the seemingly unlimited and hence dangerous influence of the utility industry had already matured into a potent political issue at all levels of the federal system.[11] Yet as long as Insull's far-flung financial empire of interstate holding companies continued to pay regular dividends, he could weather these political storm clouds.

For a while Insull seemed to hold up the pyramid of utility enterprises under his control by the sheer force of his personality and his reputation as a financial wizard. On the eve of the crash, this interlocking structure of local operating and interstate holding companies allowed him single-handedly to maintain an iron grip on the management of energy and transit services worth $3 billion. He affected the lives of over a million stock and bond holders as well as 41 million customers. For instance, he directed utility companies across the country that together generated one-eighth of the total output of electricity in the United States. In the year following the collapse of Wall Street, the continuing strength of consumer demand for central station service buoyed Insull's confidence in the future. As the depression deepened in 1931, however, the utility magnate's efforts to shore up the faltering pyramid of securities resting on this solid foundation of operating utility companies became increasingly desperate. According to his sympathetic biographer, Forrest McDonald, he became "possessed by delusions of grandeur . . . [taking] on greater and greater burdens until it appeared as if he were attempting to carry the entire American economy on his shoulders."[12]

The end came in June 1932, when other financiers forced Insull to surrender control of over sixty corporations, including his original springboard to the top of the utility industry, the Commonwealth Edison Company. Soon after resigning power, he became the target of both political persecution and criminal prosecution. Insull's long reign as one of the nation's leading businessmen made him the perfect scapegoat for the depression. On 23 September, for example, Franklin Roosevelt cast the fallen executive as society's enemy. Speaking before the Commonwealth Club of San Francisco, the presidential contender condemned "the lone wolf, the unethical competitor, the reckless promoter, the Ishmael or Insull whose hand is against every man's, [who] declines to join in achieving an end recognized as being for the public welfare, and [who] threatens to drag industry back to a state of anarchy."[13]

Less than two weeks later, a Cook County grand jury returned the first of several state and federal indictments against Insull for securities fraud. In Paris at the time, he sought to avoid prosecution in the highly charged atmosphere of the political campaign by taking refuge in Greece, which did not have an extradition agreement with the United States. But after Roosevelt's election, intense pressure from the State Department finally led to the old man's arrest as he fled to Turkey. In May 1934, Insull returned to his adopted country under police guard to stand trial in federal court. Five months later he testified on his own behalf to persuade the jury that though he might have made some big mistakes, he was not a criminal. He lost almost all of his personal wealth in a last-ditch effort to save the financial pyramid from collapse. After less than two hours' deliberation, the jury agreed he was innocent of any criminal intent. Insull returned to Paris to live out his final days. He died from a heart attack on 16 July 1938, while waiting for a subway train.[14]

After a half-century in the electric utility industry, Insull became a legendary figure who made an enduring impression on American public policy and culture. Of course, the most ephemeral aspect of the Chicagoan's legacy was his reputation as an infallible business leader and financier. Even today, the refrain "I lost a lot of money with Sam Insull" is not an uncommon response when Chicago's senior citizens are asked about their memories of the past. After 1929 he came to personify the New Era's broken dreams of perpetual prosperity. The smashed hopes of tens of thousands of local investors in his holding companies make his self-imposed exile abroad easy to understand. Like the great engineer Herbert Hoover, Insull had become such an outspoken champion of the machine age during the twenties that his historical standing cannot be divorced from the causes of the Great Depression of the thirties. Insull's financial schemes helped swell a speculative bubble on Wall Street that eventually burst, sending the economy into a tailspin.

In part his reputation as one of the country's most infamous swindlers stems from the New Deal reformers' focus on curbing the financial abuses of the utility industry. Policymakers reacted to the twin evils of the Money Power and the Power Trust by enacting regulatory legislation, including the Securities Exchange Act and the Public Utility Holding Company Act. Insull's old nemesis, Harold Ickes, headed the committee that drafted the original version of the holding company bill, with its "death sentence" proviso. It imposed a deadline for the breakup of these pyramids of power that created huge interstate monopolies of essential public service. Although loopholes in the final legislation allowed companies like Commonwealth Edison to maintain regional monopolies, they were finally made accountable to the public authority for their financial practices.[15]

In comparison with the general success of the New Dealers in imposing regulatory checks on the securities markets, the reformers had much more limited success in their attempts to restructure the technological and organizational foundations of the electric utility industry. To be sure, the creation of the Tennessee Valley Authority (TVA) and other public power facilities as "yardsticks" of fair energy rates was a major accomplishment. The inauguration of the Rural Electrification Administration (REA) by executive order in 1935 also represented a breakthrough in building public institutions that offered viable alternatives to the delivery of utility services by investor-owned companies. Frank Norris, Morris Cooke, and other veterans of the long struggle for public power deserve credit for translating the seminal ideas of the regional planners of the twenties into model public programs.[16] Yet these parallel institutions did little to change the monopolistic structures of the private utility companies.

Given the historic climax of popular antipathy toward the Power Trust, the antitrusters missed a monetary opportunity to restore market mechanisms in the delivery of central station service. The development of large-scale superpower and giant power systems provided working models that separated power generation, long-distance transmission, and local distribution into three distinct functional units. Structural reforms that provided incentives to continue this type of vertical fission might have encouraged the growth of competition in the supply and delivery of energy, ultimately resulting in lower consumer rates. In the mid-twenties, for example, Insull set up a new enterprise, the Superpower Company, to build a generator station 175 miles south of Chicago at Powerton, in the midst of the coal mines near Pekin, Illinois. Interchange energy agreements among several utilities in Illinois and Indiana provided for sharing the costs of transmitting Powerton's electrical energy across their service territories. In the Chicago region, the power was delivered to a convenient point on a suburban ring of high-voltage lines around the city. At this "Electric Junction," several distribution grids were linked by a sophisticated, computer-driven system of load dispatching. The success of the generator company concept soon led Insull to promote a second venture, appropriately called the State Line Generator Company. The new station was built in the midst of the steelyards on the border between Illinois and Indiana and contained the largest turbogenerator ever constructed, 200,000 kw. By the time it opened in 1929, Insull was planning for a future based on the interregional integration of the nation's electrical power supply.[17]

In a similar way, important aspects of the regional planning sponsored by the Roosevelt administration pointed toward a functional separation of the electrical utility industry into its three technological parts. In the case of the REA, the problem the planners faced was narrowed primarily to financing farmers and organizing them into cooperatives that could manage local distribution grids.

Having to focus on these specific tasks helped highlight differences between service functions to consumers and access to the supply system of transmission networks and generating stations. In the case of the TVA, the utility industry lobbied successfully against virtually every proposal to bring the government into direct competition with private companies in the transmission and distribution of electricity. Public power advocates were restricted largely to building hydroelectric dams to supply the region with cheap and abundant energy.[18]

Although neither public nor private planners sponsored the idea of separate agencies for the long-distance transmission of electricity, the natural gas pipeline companies provided a close analogy. They acted as common carriers between producers and distributors. This arrangement created a competitive market for gas supplies, since local utilities could seek out the lowest-priced fuel among a large number of producers. In 1938 Congress formalized this competitive structure in the Natural Gas Act, which gave the Federal Power Commission jurisdiction over the interstate transportation of the energy fuel. Despite the close parallel between the three technological components of gas and electric utilities, the New Deal did not pursue a similar public policy for the structural reform of central station service. On the one hand, popular attention was riveted on the sensational financial abuses of Wall Street. In the battle against the Power Trust, reformers had to maintain political pressure on this single issue in the face of formidable opposition from the utility industry lobby. On the other hand, a general consensus prevailed among all sides in the policy debate on a regional concept of electric utilities. Both the advocates of public power and their counterparts in the utility industry defined the structure of central station service in terms of geography rather than function. After Insull's fall, few reformers gave serious consideration to his last set of plans for an energy-intensive society. His structural innovations to build the organizational and technological foundations for the interregional movement of energy have been all but forgotten.

Insull's most enduring influence on public policy has been his first major contribution, the gospel of consumption. Public utility rates remain based on various demand systems that reward greater energy use at the expense of the small consumer. In the Chicago area, for example, residential customers now pay a set service fee of about $10 per month, followed by the highest unit prices for the first block of electricity and lower unit prices for subsequent blocks beyond average use. Industrial users and other large consumers, of course, pay much lower unit prices for equivalent blocks of energy. During the early stages of electrification, this type of rate discrimination was good public policy, because it benefited the entire community of energy consumers. The demand system of rates helped companies lower the cost of delivering central station service to all groups of customers. As we have seen, Insull used a sophisticated system of differential rates to diversify and build the electrical load to a point of efficient operation of the utility's very expensive generating equipment. The gospel of consumption set in motion a self-perpetuating cycle of rising use and declining rates that eventually enveloped an urban society in a ubiquitous world of energy.

But once a highly diversified community of energy consumers has been created, rate discrimination against small users has dubious value as public policy. The demand system forces those who can least afford it to pay the most for basic necessities of modern life: light, heat, and power. Insull's argument that this structure of charges was justified because it fairly apportioned the expenses of supplying central station service no longer pertains after a reasonable period of paying off the capital costs of the basic infrastructure. The completion of electrification in Chicago by 1928 might be taken as a date when traditional American ideals of equity and equality should have been given preference in rate making over the goal of maximizing energy consumption. At the same time, discriminatory rates should not be discarded entirely as a policy tool to advance social welfare and to manage the electrical load efficiently and economically. The Insull legacy has taught us that differential rate structures can have powerful effects on group patterns of energy consumption.

Until recently, however, there was little popular pressure to make policymakers question the gospel of consumption or its demand system of rates. As long as its two underlying conditions of abundant supplies of cheap fuel and technological improvement remained constant, the price of electricity continued to fall both in absolute terms and in relation to the cost of living. In the forty years between 1929 and 1969, for instance, the average unit price of electricity for residential customers in Chicago fell steadily without significant interruption from 4.8¢ to 2.5¢ per kwh. Since then, the cost of fossil fuels, environmental protection, and new generating equipment, especially nuclear reactors, has been rising dramatically. The energy transition of the 1970s represents a major reversal of the economic and technological conditions that prevailed throughout most of the country's history. By 1982 a new era of high-cost energy had driven the average unit price up to 8¢ per kwh. By the end of the decade, residential consumers were again paying charges exceeding the 12¢ per kwh that had prevailed at the turn of the century when Insull first formulated the demand system of rates.[19]

Before the oil embargo of 1973–74, Insull's gospel of consumption was as much an unquestioned part of modern American culture as the "invisible" world of energy this doctrine helped create. Even the Great Depression could not stop consumer demand for more electricity in the home. Once a certain level of comfort and convenience had been reached, it was not easy to return to darkness and drudgery. Most Chicagoans were unwilling to unplug their light fixtures, refrigerators, and radios, no matter what other sacrifices they might have to make in their family budgets. To be sure, some city dwellers lost everything, including their homes. Between the peak year of 1929 and 1932, about 56,500 households, or 7 percent of the city's 805,000 residential customers, had to discontinue service. Some families left Chicago, while others had to return to gaslights or kerosene lamps. As early as 1936, however, new customers and old ones able to restore service combined to push the total number of residential consumers above the previous record.

A more telling indication of society's dependence on energy to maintain the standards of daily life was the ceaseless rise in the average household use of electricity (see chart 18). In 1929 the average residential customer used 629 kwh per year, a figure that continued to increase over the next three years by almost 20 percent, to 749 kwh annually. In 1933 household use declined slightly by 1.2 percent, falling back to the averages of 1931 but staying substantially above predepression patterns. The following year residential consumption resumed its upward climb, reaching an average of 911 kwh by the end of the decade. Over the course of the depression, the average household recorded a 45 percent increase in its consumption of electricity.[20]

For the most part, this growth in consumption represented greater use of household appliances, just as Insull had predicted. A historic moment was reached in 1933, when appliances consumed over half the electricity in the home, surpassing energy demands for lighting for the first time. In that year, for example, Chicagoans purchased 50,000 refrigerators, and the number

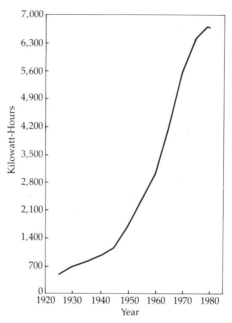

CHART 18.   Average annual consumption of electricity by residential customers, 1925–80 (kilowatt-hours). *Source:* Commonwealth Edison Company, *Annual Report,* 1925–80, in CEC-HA, boxes 113–16.

reached 130,000 new units only seven years later. In the home, there was no turning back; city dwellers had become accustomed to a level of energy use that they could no longer forgo without destroying their sense of personal dignity and collective hope. Chicago's World's Fair of 1933 brought reassurance to an anxious people. Futuristic exhibits mounted by the utilities industry and model homes of tomorrow designed by leading architects promised even greater comfort and convenience at the touch of a button.[21]

In contrast to the rising residential demand for energy, its use by commerce and industry closely paralleled changes in the national economy. Modern industrial society had become so dependent upon energy that the sudden plunge of the stock market in 1929 was soon reflected in a drop-off of store and factory customers. Two years later the decline in demand for central station service hit bottom, a 22.3 percent slide from the 1929 peak. Yet this drop represented a return to a level of consumption attained only a few years back, in 1928. And by 1935 Chicago's total demand for electricity was again setting new records. The muted fall and quick recovery of demand for central station service suggests that the electrical industry remained one of the economy's engines of growth, helping to pull the nation out of the depression. A decade after the crash on Wall Street, energy use by industrial and commercial consumers in Chicago had surged ahead approximately 45 percent, about equal to the gain recorded for residential consumption.[22]

On 21 October 1929, the United States celebrated "Light's Golden Jubilee" in honor of Thomas Edison's invention of the incandescent lamp fifty years earlier. Across the country, Americans heard President Hoover pay homage to the inventor in a radio broadcast from Dearborn, Michigan. In this suburban enclave outside Detroit, Henry Ford was building a bucolic fantasy, Greenfield Village, complete with an idealized reconstruction of the wizard's Menlo Park laboratories. The timing of the "light jubilee," only three days before "Black Thursday" on Wall Street, was filled with irony, especially for Insull and his "old chief." For Edison the mass-media event revealed how thoroughly the machine age had eclipsed the bygone days when the inventor found an answer to better lighting. A half-century later, his place in American mythology was firmly fixed. As historian David Nye points out, the actual presence of the eighty-two-year-old man was less important to the celebration than his homespun image as a self-made success story and a master of technology. Edison's wife was one of the few to notice that "the light jubilee is one grand advertisement for General Electric and the light companies, that dearie has just been made the excuse."[23] If Edison the man was no longer important, his symbolism as an American hero had become an indispensable part of our popular culture.

For Insull, Edison's natural heir to the leadership of the electrical industry, the grand ceremony marked the last few pleasant days of a long and triumphant career. He would soon have to suffer through the painful unfolding of the Great Depression, which would rewrite the final chapters of his legend. His reputation as an American hero was smashed along with the nation's dream of endless prosperity. Although Insull's image in the popular imagination has faded far more than Edison's, the businessman stands equal to the inventor in making an enduring impression on American society.

Ultimately, Insull had less to do with the creation of our world of energy than did the extraordinary ability of electricity to improve the quality of everyday life. The master salesman may have pulled Chicagoans into an early lead in gaining access to the benefits of energy at a reasonable price, but consumers in other urban places did not lag behind for long. The most basic reason for the success of this energy form has been its unparalleled flexibility in meeting city dwellers' demands for better sources of light, heat, and power as well as for faster modes of transportation and communication. The rise of the industrial city brought not only new levels of population density and economic activity but a related series of catastrophes that threatened the public health, safety, and welfare. The construction of an infrastructure of utility services, paved streets, and other public works went a long way toward making the city a good place to live. Beginning with the telegraph in the 1840s, electrical technology became increasingly central to this process of city building. To a large extent, then, electricity exerted a powerful influence in the urbanization of the United States simply because it became one of the most useful tools at the disposal of modern society in achieving its vision of the American dream.

But the rise of the "networked city" had implications that went beyond the technical arena of practical problem solving into the social realm of comfort, convenience, and conspicuous consumption.[24] In the early nineteenth century, for instance, laying a grid of water lines under the city streets provided protection for the entire community against terrible losses from fires in buildings packed tightly together. It also guarded the public health against the frightening spread of epidemic disease. At the same time, however, middle-class homeowners transformed the city's water supply into a private luxury. In the 1850s they began installing indoor plumbing and baths that rivaled the comforts of the emperors of ancient Rome. As the networked city continued to grow, the introduction of electrical technology tended to reinforce existing cultural trends in American society, which placed an increasing value on the material possessions each individual could amass. In this context, electricity supplied the technical means to construct not only a better physical environment for the entire community but also a particular cultural vision of the ideal city, one that embodied deep-seated values of privatism and social segregation.[25]

To an extent, then, electrical technology became more than simply a tool; it became a cultural icon of the machine age. The unique ability to satisfy such a wide variety of the city's public and private needs for essential utility services gradually made such technology a pervasive part of everyday life. In giving substance to a certain technological vision of the city, electricity became an active agent of cultural change that helped shape the physical form of the modern metropolis and the structure of the society living within it. The potential of energy to ameliorate so many of the city's functional problems influenced decision makers and consumers in one direction while reducing the chances that alternative, nontechnological paths to the future would be taken. By the 1920s, an intensive use of energy had became a crucial prerequisite for maintaining a highly deconcentrated, suburban style of life. As one of the twentieth century's most important symbols of progress, electricity played an integral part in the ascendancy of this culture of consumption.

To be sure, the dual roles of electricity as a better tool and as an active agent of change are closely intertwined and difficult to isolate, since one has tended to reinforce the other. Yet an effort to sort them out can improve our understanding of the impact of technology on the city and on contemporary life. Let us look first at electricity as a mechanical instrument designed to help solve the problems of the city. In this role, technology helped realize an industrializing society's pent-up demand, current wants, and future aspirations. The Chicago experience leaves no doubt that electrical technology brought about a significant improvement in the material conditions of the urban environment and the quality of daily life. It is safe to assume that very few critics of our culture of consumption have disconnected their utility meters and returned to the "good old days" of kerosene lamps, coal-burning stoves, and sadirons. City dwellers of the Progressive Era considered electricity an unqualified blessing in meeting long-standing needs for a superior source of illumination, a mechanical substitute for the horsecar, and a convenient means of running a virtually endless variety of machines and heating devices.

More than any other quality, this flexible versatility accounts for the commercial success of electrical technology in the rapidly growing industrial centers of the late nineteenth century. In the classic *Mechanization Takes Control*, for example, Siegfried Giedion shows that after the Civil War, inventors began flooding the United States Patent Office with thousands of ideas for tools designed to ease the drudgery of housework, from irons and washing machines to vacuum cleaners and dishwashers. The mechanical principles of these laborsaving devices were quickly mastered, but they lacked a practical source of power until the appearance thirty years later of the small electric motor. "It meant to the mechanization of the household," Giedion reasons, "what the invention of the wheel meant to moving loads. It set everything rolling. Without it, mechanical comfort in the house could have advanced little beyond its condition in the 'sixties."[26] In a similar way, the accumulating needs of post-Appomattox cities for modes of mass transit that were faster, cheaper, and more reliable than animals could provide spurred a wide-ranging search for mechanical methods of traction. Although the cable car proved a viable answer, the electric trolley furnished an even better way to move the masses.[27] In these and many other cases, more and more city dwellers adopted electrical technology because it satisfied a common striving to upgrade the material conditions of daily life.

As a multipurpose tool, the new technology gave substance to a particular version of the urban ideal drawn chiefly by the elite groups, which held the most economic and political power in society. Well before the energy revolution in the 1880s, they had set in motion the cultural trends that would result in the physical and social construction of the "private city," where success would be defined by material possessions and conspicuous consumption. The versatility of electricity was tailor-made to help realize contemporary society's prevailing values and goals. On the one hand, the Victorians desired technology that could enhance the welfare of the community at large. But on the other hand, they wanted ways to control access to its benefits. Electricity met this requirement, since its distribution could be regulated by individual meters that measured its use in discrete amounts. In an emerging culture of consumption, central station service of the kind Insull provided marked a major step toward spreading the benefits of the new technology to the city's people at prices they could afford to pay. Yet ability to pay remained the bottom line in securing access to the new standard of living.

The transformation of the United States from a land of agrarian "island communities" to a nation of densely packed cities also produced an unfortunate, if understandable, drive toward social segregation.[28] Even before they had the technical means to impose segregation in spatial terms, the city builders of the Victorian age carved out a set of psychological worlds or "separate spheres." These defined social space in the city based on gender, class, race, religion, and ethnicity. And in some cases, such as the red-light districts of commercialized sex, social segregation was given physical expression. By the time of the Columbian Exposition of 1893—the symbolic coming-of-age of electricity—the urge for a segregated city led to the formal separation of the Court of Honor from the other official buildings as well as to the banishing of activities deemed "low class" to a third area known as the Midway. In the "dream city," crueler psychological forms of segregation were played out in the exclusion of African-American contributions from the fairgrounds and in the live anthropological exhibits that treated native people like creatures in a zoo.

In the real cities of the late nineteenth century, the same cultural orientations defined the suburbs as insular bastions of social exclusion. At the time of the fair, Evanston's and Oak Park's rejection of annexation referenda signaled the beginning of this phase of suburbanization in Chicago. It was marked by higher and higher barriers of restricted entry, giving rise to political fragmentation and social segregation on a metropolitan scale. The social forces leading to this version of the suburban ideal predated the commercial use of electricity for light, heat, and power. For the most part, the revolt against the city was simply accelerated by the application of more and more energy. Electricity and gasoline-powered automobiles acted chiefly as tools, albeit extremely powerful ones, to further existing cultural orientations. Technology helped realize an explosive social impulse to sort the city's people into their constituent groups by physically spreading them out over a metropolitan expanse of separate, homogeneous enclaves.

At the same time, electricity shaped the architectural and geographical expression of the emerging metropolis of the twentieth century. In this role technology acted both as a tool and as an active agent of change. Over half a century ago, Lewis Mumford was among the first to propose that electricity and the automobile had the potential to speed up urban deconcentration. In *Technics and Civilization* and several succeeding books, Mumford investigated the effects of the highway and the internal combustion engine on American culture. Many others have added detail to his original hypothesis, confirming that the automobile has had profound effects on urban life and form in the United States.[29] In contrast, Mumford's suggestions about electricity and the city attracted little attention until the oil crisis of the mid-1970s, which gave dramatic punctuation to a historic transition from relatively low-cost to high-cost energy. Since then, studies like this one have been accumulating evidence that Mumford was also on target in assigning a cultural role to energy as an agent of social change that closely parallels the influence of the automobile.

If Chicago is a representative test case of the American experience, then Mumford's proposal that electrical technology would accelerate and transform urban deconcentration seems correct. Of course, the new forms of energy use neither created a suburban ideal nor initiated dispersed settlement over a metropolitan territory. Yet they reinforced this particular social orientation and helped shape the resulting physical environment. Like the automobile, electrical technology promoted a sprawling pattern of land use by sharply reducing the cost of bringing the public utilities and modern conveniences of city life to areas of low density. The overhead wires of the electrical grid tended to burst the geographical limits on suburbanization that had been imposed by steam technology and that confined most commuters and businesses to an area close to the railroad tracks. In addition to providing energy to any location at a lower cost than previous methods, electricity furnished economical power to run a host of other public utilities that contemporaries considered essential to modern life, including the telephone, street lighting, gas, water, sewerage, and rapid transit services. In short, energy made the suburbs more attractive and affordable to a enlarging segment of the population.

By significantly expanding the geographical boundaries of the city, electricity promoted the growth of a certain kind of twentieth-century metropolis while discouraging alternative models. In this way, technology encouraged the evolution of urban society into a fragmented pattern of suburban sprawl. To achieve this social vision of the good life, Americans became increasingly dependent on the automobile and on energy. Social choice accounts for this result, not the inevitable workings of a technological determinism. To find an example of a different route to the creation of a modern metropolis, one need look only as far as the garden city idea of Mumford's mentors, Patrick Geddes and Ebenezer Howard. England, Sweden, and Holland have followed their call to preserve open spaces or green belts around the cities. Rather than allowing housing subdivisions, paved roads, and parking lots to sprawl everywhere over the landscape, garden city plans of metropolitan growth encouraged a series of more compact communities at some distance from the built-up areas. Urban planners promoted higher-density settlement by first building convenient public transportation links to the city center, so that housing in the new towns tended to cluster around the rapid transit stations. Present-day advocates of the garden city idea are moving further in the direction of energy efficiency. Around the stations, they are redesigning the street to return it to the pedestrian as the quintessential experience of urban life.[30]

In the United States, technology promoted not only urban deconcentration and social segregation but cultural and economic centralization. Again, the underlying socioeconomic trends fueling the formation of large-scale organizations and the regional integration of the nation were set into motion well before electric light, heat, and power had a significant impact on American life. The same forces of industrialization that propelled city dwellers out to the suburbs produced a curious countercurrent of economic and political power gravitating toward centralized institutions of administrative control. In Chicago, electrical technology became a powerful means of integrating the surrounding hinterland. Telegraph, telephone, rapid transit, and power lines shrank time and distance between formerly semiautonomous localities. In the decade following the purchase of Insull's first suburban utility company in 1902, stringing a regional network of power lines was a graphic metaphor for weaving these places into a single, if unwitting, community.

To explain the role of technology in this seemingly paradoxical process of deconcentration and centralization, the perspectives offered by Marshall McLuhan seem especially pertinent. In his book *Understanding Media,* the communications guru of the 1960s rejected the fashionable view that technology was an alien force opposed to the humanistic values and traditions of Western civilization. On the contrary, McLuhan argued that technology represented "the extensions of man." He explored this novel theme by examining the ways various devices augmented the natural powers of our bodies, from the Roman road as an early extension of the foot to television as a far-seeing mechanical eyeball. Most important here, he drew a sharp distinction between mechanical and electrical technology. The former had long expanded the reach of society, whereas the latter was now reversing this historical trend by causing a radical implosion of society and culture. By externalizing the central nervous system in the form of electronic media, Americans wrapped themselves inside a new world of instant communications and information processing. McLuhan's insights on electrical technology help explain why the physical deconcentration of the city went hand in hand with the economic and cultural integration of the region and the nation. Within the "global village," localism and isolation give way to internationalism and interdependence.[31]

McLuhan's pathbreaking book also sheds light on the electronic media as agents of cultural change. Despite the cryptic obscurity of his famous phrase "the medium is the message," it opens up new ways of looking at technology and culture. In effect, McLuhan downplayed the functional purposes of technology in order to highlight its more subtle and direct influences on our perceptual and cultural orientations. In a classic case, for example, he argued that the program content of television is irrelevant compared with the profound impact the medium itself is having on everyday life. To McLuhan, television and other electronic media are creating a "totally new human environment" of involvement, participation, and commitment.[32]

From this point of view, electric lighting can be regarded as a "medium" of city building in the hands of architects and developers. In the 1880s, the new technology had the general effect of raising the standards for acceptable illumination. The universal demand for more light changed sensibilities and shifted priorities among architects and their clients to stress this aspect of building design. Changing sensitivities meant that both natural and artificial light became much more important elements of design in all structures, whether residential, commercial, or industrial. By the early twentieth century, the subtle impact of electric lighting was becoming evident in such architectural expressions as the ribbons of casement windows wrapping around Frank Lloyd Wright houses, the brighter lights in offices, and the sawtooth roofs of the new factories. By the 1920s, architects were working with electric lighting as a basic part of the skyscraper. Banks of floodlights shining on facades and hidden high in the towers helped sustain the skyscrapers' image as cultural icons of modernity, especially in a conservative age of neoclassical revival. In this way, technology influenced cultural change because architecture is the public sculpture of the city. Electric lighting became an integral design element in this art form, reshaping the urban environment and the aesthetics of the people who lived within it.

The influence of electricity as a medium of communication illustrates how technology affected culture more directly. The radio, for example, was active in creating an American popular culture. Spreading rapidly into the home during the late twenties, the radio provided listeners across the region (and the nation) with a new shared experience. For the first time, midwestern farmers, Italian immigrants, the suburban elite, small children, and myriad others were all spending leisure time in the same pursuit. Recent research cautions against attributing too much importance to the homogenizing effects of the mass media during the first half of the twentieth century. Historian Liz Cohen, for example, has found that in Chicago, different ethnic and racial groups among the working class generally listened to different programs. Our fragmented and segregated society sought out programming that reinforced separate group loyalties and identities.[33] Actually, this finding supports McLuhan's contention that the essence of the radio experience is a private, aural interplay between the individual listener and the source of the sound. The content of successful radio therefore tends to be intimate and conversations, such as the call-in talk shows and programs aimed at specific segments of the population.

But on a collective level, the radio caused a reformation of "tribal" communities throughout the society. In shrinking time and distance, the electronic media created new communications communities by bringing widely diverse people together in original ways. Between the wars, the radio changed America's perceptions of itself. The new self-image found expression in several forms. For instance, theories of communications and community came under new scrutiny among intellectuals, who realized these subjects badly needed redefinition. Their concerns grew out of an awareness of the acute social tensions that characterized the postwar decade. The period's historical labels as the era of a "clash of cultures" and the "tribal twenties" also reflect an appreciation for the transformation of American society that was occurring under the impact of modern technology.[34] In this context, the diversity of the radio's program content had far less influence in defining modern culture than the shared experience of listening to the new form of communication. Simply that more and more individuals who had previously engaged in different patterns of group activity were now doing the same thing helped to homogenize society and to promote the spread of a common popular culture. Although many shows may have appealed narrowly to special target groups, the medium evoked the same message of "tribal" feelings in all of them.

The radio's instant popularity is just one example of the endless ways electrical technology and intensive energy use became pervasive parts of American life by the time of the Great Depression. During the decade of hardship that followed, a consensus emerged that technology offered the best hope for pulling the country into the future. As we have seen, this consensus ranged from the intellectual circles of Mumford and Giedion to the workaday worlds of farmers and factory hands. Serving as the most important icons of progress in a culture of consumption, electricity and the automobile were assigned major tasks in the economic reconstruction of society. A wide spectrum of progressive reformers, including the champions of the New Deal, turned to electrical technology as a powerful instrument of change that could get the nation moving again toward a higher standard of living. Perhaps the two world's fairs of the thirties embodied the fullest cultural expression of these aspirations. Chicago's Century of Progress (1933) and New York City's World of Tomorrow (1939) both offered an anxious people reassuring visions of a suburban nation in which abundant cheap energy would support a life of comfort and convenience for everyone.

In this respect the contrast between the two fairs of the future and the Columbian Exposition of 1893 could not have been more complete. During the crisis of the nineties, a nation's doubts about the future were reflected in the Court of Honor's beaux-arts style of architecture, which harked back to a pristine, classical era of civilization. At the crossroads between an agrarian past and an urban-industrial future, traditional values soothed anxieties by wrapping the new products of the machine in a familiar setting. Forty years later, the depression evoked a very different and understandable urge to leave the past behind and get on with the future. In turning over the fairs of the 1930s to the industrial designer, the sponsors of these cultural events adopted the aesthetics of the machine age as the central motif of the American way of life. The processes of a technological society, not its products, were now given first importance. Significantly, the sponsors of the 1939 New York exhibition turned to a famous industrial designer, Henry Dryfuss, to draw up the master plan for a scale model of the community of the future, Democracity. Summing up the cultural meaning of the two "people's fairs" of the depression era, Eugene Santomasso concludes that "while the 1933 Fair had shown the future as science fiction, the New York planners conceived of it as an attainable goal, presented on a grand scale in the tangible form of elements from commerce and industry, expanded via architecture and art. Design and daily life were conceived of as being one. The Fair visitor was to be thrust into the full-blown Age of Consumerism and the Age of the Machine."[35] The decision of New York's Consolidated Edison Company to design the largest diorama ever built as an exact replica of the city's skyline reflected the tremendous change that had taken place since the world's fair of 1893, when electricity held its debutante's party. Less than fifty years later, the electric "City of Light" portrayed by the huge model was a fitting testimonial to the power of technology as a tool of modern society as well as a potent force for change in its own right. The creation of a ubiquitous world of energy played an integral part in the new culture of consumption, turning a technological vision of the American dream into everyday reality.

Notes

Bibliographical Note

The historical records of the Commonwealth Edison Company are kept in three distinct places: a central archive, an inactive collection of library files, and a working library. Each will be described below. In the notes, the citations for the three groups of records are abbreviated as follows:

1. Historical archives, cited as CEC-HA, box [no.].
   The historical archives are under the jurisdiction of the Office Systems Department and are kept at the company's record center, North Division Headquarters, 2800 North California, Chicago, Illinois.
2. Company Library [Historical] Files, cited as CEC-LF, F[ile] [no.]-E[nvelop] [no.].
   This collection of history files is under the care of the library, a branch of the industrial relations department. It is kept at a remote location. The library provides access to the history files.
3. Company library, cited as CEC-Library
   The company maintains an active library at its corporate headquarters, One First National Plaza, Chicago, Illinois.

The historical archives of the Commonwealth Edison Company (CEC) are the largest of the three collections and consist of four thousand linear feet (two thousand file-sized boxes) of records of the electric (gas and water) utilities of northern Illinois. This region currently encompasses 10,000 square miles. The collection includes the corporate records of the CEC, its direct predecessors and subsidiaries, and its affiliated companies, which have gradually been merged with the parent firm. Several of these affiliated companies themselves had predecessors whose records are also included in the collection. In addition, the archives contain large collections of photographs, company publications, and trade journals.

The archives span the period from 1849 to the present, containing the records of some water-power companies, post-Civil War gas companies outside Chicago, and almost all existing records of early (1878–1900s) local electric companies throughout northern Illinois. Until 1953 the system Insull built was operated as a series of interlocking corporations, including the CEC, the Public Service Company of Northern Illinois, the Illinois Northern Utilities Company, and the Western United Gas and Electric Company. All but the CEC conducted gas, water, and electric railway operations. All the affiliated companies and the CEC have also been engaged as owners, partners, or both, in the mining and transportation of coal. These coal operations took place mostly in central Illinois and Indiana.

In 1953 these companies were consolidated into the CEC. At the same time, the gas and water operations were divested (and their records were removed from the archives). During the 1950s and 1960s the CEC became affiliated with the Central Illinois Electric and Gas Company, and the two were consolidated in 1967. The records of this company and many of its local predecessors have become part of the CEC historical archives.

Approximately half of this collection consists of the records of the CEC, and the other half contains the records of the affiliated companies before they merged with the parent firm. Of the one thousand boxes of CEC materials, about half refer to the pre-1953 period and half to the era of consolidated operations. But the bulk of the engineering and construction records (contracts, work orders, purchase orders, line orders, etc.) have not been accessioned into the historical archives. These record groups are on permanent retention schedules and at present form part of the company's records management system. These records are kept at the same location as the historical archives. Both active and historical collections have been cataloged on sophisticated, computer data bases.

The library historical files originated in the work of previous librarians who acted as company historians from the 1910s to the 1950s, when the files were closed. The files are arranged by topics (files) and by subtopics (envelopes within each file). An extensive card index of the files is maintained in the library and provides the best means of access to this important collection of historical materials.

The company library contains a collection of books and technical journals stretching back to the 1880s.

Preface

1. See Joel Tarr and Gabriel Dupuy, eds., *Technology and the Rise of the Networked City in Europe and America* (Philadelphia: Temple University Press, 1988).

2. In this regard I disagree sharply with the best-known authority in the field, Thomas P. Hughes. See Thomas P. Hughes, *Networks of Power: Electrification in Western Society, 1880–1930* (Baltimore: Johns Hopkins University Press, 1983). Although my case rests on the study that follows, see also the insights on the fallacies of Hughes's brand of technological determinism in David Joravask, "Machine Dreams," *New York Review of Books,* 7 (December 1989): 11–15.

Chapter One

1. *Chicago Tribune,* 26 April 1878. A complete clipping file of electricity-related stories during this pioneer period can be found in CEC-LF, F5-E2.

Also see the letter by Charles Brush describing the involvement of the inventor of the device Barrett used, in CEC-LF, F6-E1.

2. *Chicago Tribune,* 27 April 1878.

3. *Chicago Times,* 3, 5 May 1878.

4. *Chicago Times,* 5, 15 May 1878, for the first quotation; *Chicago Tribune,* 5 May 1878, for the second. For the reported sighting in Englewood, see *Chicago Tribune,* 16 May 1878.

5. *Chicago Tribune,* May-July, October 1878, 1879; passim. See above, note 1.

6. A brief list of essential reading on the history of artifical illumination would include Arthur A. Bright, Jr., *The Electric-Lamp Industry: Technological Change and Economic Development from 1800 to 1947* (New York: Macmillan, 1949); Harold I. Sharlin, *The Making of the Electrical Age* (New York: Abelard-Schuman, 1963), 141–50; Harold C. Passar, *The Electrical Manufacturers, 1875–1900* (Cambridge: Harvard University Press, 1953), pt. 1; Thomas P. Hughes, *Networks of Power: Electrification in Western Society, 1880–1930* (Baltimore: Johns Hopkins University Press, 1983), chaps. 1–3; the critical assessment of Edison by Wyn Wachhorst, *Thomas Alva Edison: An American Myth* (Cambridge: MIT Press, 1981); and the careful study of Edison's light bulb by Robert Friedel and Jerry Israel, with Bernard S. Finn, *Edison's Electric Light: Biography of an Invention* (New Brunswick, N.J.: Rutgers University Press, 1986).

7. *Chicago Times,* 3 May 1878.

8. W. G. C. Rice, *Seventy-five Years of Gas Service in Chicago* (Chicago: Peoples Gas Light and Coke Company, 1925); Illinois Bureau of Labor Statistics, *Ninth Biennial Report, 1896, pt. 3, Franchises and Taxation: The Gas Companies of Chicago* (Springfield: Illinois Bureau of Labor Statistics, 1896), 239–320.

9. William J. Cronon, "To Be the Central City: Chicago, 1848–1858," *Chicago History* 10(Fall 1981): 130–40; Glen Holt, "The Birth of Chicago: An Examination of Economic Parentage," *Journal of the Illinois State Historical Society* 76(Summer 1983): 83–94; Jones Piety, "The Illinois and Michigan Canal and the Early Historical Geography of Chicago," *Geographical Perspectives* 47(Spring 1981): 30–37; Bessie Louise Pierce, *A History of Chicago,* 3 vols. (New York: Knopf, 1957), vol. 1; and the appropriate sources listed in Frank Jewell, comp., *Annotated Bibliography of Chicago History* (Chicago: Chicago Historical Society, 1979).

10. Alfred Chandler, Jr., "Anthracite Coal and the Beginnings of the Industrial Revolution in the United States," *Business History Review* 46(Summer 1972): 141–81; Dolores Greenberg, "Reassessing the Power Patterns of the Industrial Revolution: An Anglo-American Comparison," *American Historical Review* 87(December 1972): 1237–61.

11. Cronon, "To Be the Central City," 130–40; Wyatt W. Beecher, *The Economic Rivalry between St. Louis and Chicago, 1850–1880* (New York: Columbia University Press, 1947); and the sources on railroad history listed in Jewell, *Bibliography,* 50–54.

12. For the Southwest, see Harold L. Platt, "Energy and Urban Growth: A Comparison of Houston and Chicago," *Southwestern Historical Quarterly* 91(July 1987): 1–18.

13. Chandler, "Anthracite Coal," 141–81; Alfred Chandler, Jr., *The Visible Hand: The Managerial Revolution in American Business* (Cambridge: Harvard University Press, 1977), chap. 3.

14. Chicago Board of Trade, *Annual Report*, 1858, 37. Also see Rice, *Seventy-five Years*, 4.

15. J. W. Foster, *Report upon the Mineral Resources of the Illinois Central Railroad* (New York: Roe, 1856), 27, 9–30.

16. A. Alexander Bischoff, *Coal Trade at Chicago* (Chicago: Donohue, Henneberry, 1885), 16; A. T. Andreas, *History of Chicago*, 3 vols. (New York: Arno, 1975 [1884–86]), 2:673–74; Illinois Department of Mines and Minerals, *A Compilation of the Reports of the Mining Industries of Illinois from the Earliest Records to the Year 1930* (Springfield: Illinois Department of Mines and Minerals, 1930).

17. Chandler, *Visible Hand*, 245, pt. 3.

18. Andreas, *History of Chicago*, vol. 2, passim; *Industrial Chicago*, 6 vols. (Chicago: Goodspeed, 1891–96); Pierce, *Chicago*, vol. 3, passim. Meat packing was a major exception involving intense use of labor for butchering and for refrigeration in the form of ice cutting, storage, and transportation. See Louise Carroll Wade, *Chicago's Pride: The Stockyards, Packingtown, and Environs in the Nineteenth Century* (Urbana: University of Illinois Press, 1987). In the twentieth century, the application of electrical technology to ice making would have profound effects on daily life. Nutritional standards, health, and domestic routines would all undergo major transformations as a result of the widespread diffusion of iceboxes and refrigerators in the home. These developments will be considered in chapter 8.

19. David A. Hounshell, *From the American System to Mass Production* (Baltimore: Johns Hopkins University Press, 1984); Siegfried Giedion, *Mechanization Takes Command* (New York: Norton, 1948), pts. 4–5. In Chicago, German immigrants with craftsmen's skills emerged in the postwar era as an important source of entrepreneurial talent in the rapidly growing field of manufacturing. See John B. Jentz, "Class and Politics in Gilded-Age Chicago, 1869 to 1875," paper presented to the annual meeting of the Social Science History Association, Chicago, 1988. In addition to the growth of small manufacturing firms staffed by German immigrants, Jentz also notes that all groups of Chicago's industrial workers suffered an early and rapid decline of skilled jobs. At the same time, a related rise in unskilled jobs at relatively large factories reflected the trend toward mechanization and the substitution of machine power for craft skills.

20. Chandler, *Visible Hand*, chaps. 4–5; see above, note 10. Coal dealers also organized to uphold the price of coal and to regulate the practices of its retailers. See *The Constitution and By-Laws of the Chicago Coal Exchange* (Chicago: Bliss, Barnes and Gritzner, 1874, 1889), which lists the names

and locations of the membership, including a map. The exchange had 15 "first class" large wholesalers, 74 members in the "second class," and about 750 in the "third class" of small retailers. The map clearly shows that small retailers were scattered throughout the city along the major railroad tracks. Photo surveys of railroad rights-of-way confirm this pattern of local coal distribution. See the photographic collection at the Chicago Historical Society for these surveys.

21. See above, note 6; and consider the astute analysis of the underlying social sources of the demand for more light in Mark J. Bouman, "Luxury and Control: The Urbanity of Street Lighting in Nineteenth-Century Cities," *Journal of Urban History* 14(November 1987): 7–37.

22. *Chicago Tribune*, 7 September 1850, as quoted in Rice, *Seventy-five Years*, 7, 3–8. On urban rivalry, see Bayrd Still, "Patterns of Mid-Nineteenth Century Urbanization in the Middle West," *Mississippi Valley Historical Review* 28 (September 1941): 187–206; Harry N. Scheiber, "Urban Rivalry and Internal Improvements in the Old Northwest, 1820–1860," *Ohio History* 71(October 1962): 227–39, 290–92.

23. *Chicago Evening Journal*, 5 September 1850, as quoted in Rice, *Seventy-five Years*, 7. For a fascinating contemporary description of the gas-making process, see J. W. Watson, "Gas and Gas-Making," *Harper's New Monthly Magazine* 26(December 1862): 12–28.

24. Rice, *Seventy-five Years*, passim; Bouman, "Luxury and Control," 7–37.

25. Bouman, "Luxury and Control," 4–24; Andreas, *History of Chicago*, 3:126–28.

26. Rice, *Seventy-five Years*, 9–30; Andreas, *History of Chicago*, 3:126–28; Chicago Department of Public Works, *Annual Report*, 1876–80, passim, for statistics on the number of gas and oil streetlights in each of the city's three division, north, south, and west.

27. Illinois Bureau of Labor Statistics, *Ninth Biennial Report*, 277–83.

28. The best source of information about the gaslight industry is the proceedings of the annual meetings of the industry's trade association, the American Gas-Light Association. Formed in 1876 in response to political pressure during the depression, the association was not only the first national urban utility group, but also a center of its most innovative ideas about public services, government-business relations, and the effect of artificial lighting on the daily lives of city dwellers.

29. Giedion, *Mechanization Takes Command*, 364–88.

30. Charles E. Gregory, "Comments on the Growth of Arc Lighting in Chicago," paper read before the Chicago Electric Club, Chicago, 1889, in CEC-LF, F6-E1 (reprinted in *Western Electrician*, 21 December 1889, 320–21).

31. Hughes, *Networks of Power*, chaps. 2–3; Bright, *Electric Lamp Industry*, 21–70; W. Paul Strassmann, *Risk and Technological Innovation: American Manufacturing Methods during the Nineteenth Century* (Ithaca: Cornell University Press, 1956), 158–83; and Joel A. Tarr, with Thomas Finholt and David

Goodman, "The City and the Telegraph: Urban Telecommunications in the Pre-Telephone Era," *Journal of Urban History* 14(November 1987): 38–80. Consider also Friedel and Israel, *Edison's Electric Light.*

32. W. James King, *The Development of Electrical Technology in the Nineteenth Century: 3. The Early Arc Light and Generator,* United States National Museum Bulletin 228 (Washington, D.C.: Smithsonian Institution, 1962).

33. See Wachhorst, *Edison,* and Tarr, "City and the Telegraph," 38–80.

34. Tarr, "City and the Telegraph"; *The Biographical Dictionary and Portrait Gallery of Representative Men of Chicago. . . . ,* 3 vols. (Chicago: American Biographical, 1892), 1:39–40; Albert Nelson Marquis, ed., *The Book of Chicagoans* (Chicago: Marquis, 1911), 39. For a description of the fire alarm system, see Andreas, *History of Chicago,* 3:123–25.

35. Andreas, *History of Chicago,* 2:123–25, 3:593–95.

36. On Gray, see Andreas, *History of Chicago,* 3:593–94. On the Western Electric Company, see Pierce, *History of Chicago,* 3:230, and *A History of the City of Chicago: Its Men and Institutions* (Chicago: Inter Ocean, 1900), 245–46. On the notion of a community of talent, see Sam Bass Warner, *The Province of Reason* (Cambridge: Harvard University Press, 1984). The pool of talent is described in Jane Jacobs, *Cities and the Wealth of Nations* (New York: Random House, 1984). She argues that innovation and economic growth stem from the close proximity of talented entrepreneurs who are able to replace goods imported into the urban economy with "import substituting activity."

37. Pierce, *History of Chicago,* 3:228–30.

38. Andreas, *History of Chicago,* 2:125–26, 3:597; William E. Keily, "General Anson Stager," typescript, 1918, in CEC-LF, F6A-E3.

39. *Chicago Times,* 10 February 1880, for the quotation; see CEC-LF, F5-E2A, for the demonstration of Edison's light by George Bliss on 12 December 1878. As I will describe in the following chapter, Bliss would become the first salesman of Edison lighting equipment in the Midwest, an enterprise that eventually became the Commonwealth Edison Company.

Chapter Two

1. *Chicago Tribune,* 15 January 1880; William E. Keily, Memorandum on the Palmer House Installation, ca. 1916, in CEC-LF, F6-E1. The Keily memo is based on a letter from Charles H. Wilson, who installed the two ten-arc dynamos made by Weston. Also see the article by the son of the Brush's patent attorney, William M. Porter, *"In Its Infancy": Pioneers Days of Practical Electric Lighting* (Detroit: Porter, 1917); E. A. Edkins, "History of Early Lighting in Chicago," manuscript, ca. 1923, in CEC-LF, F6-E1.

2. Charles E. Gregory, "Comments on the Growth of Arc Lighting in Chicago," paper presented to the Chicago Electric Club, Chicago, 1889, in CEC-LF, F6-E1; W. C. Jenkins, *Chicago's Marvelous Electric Development* (Boston: Chapple, 1911), 6.

3. Stephen Davis, " 'Of the Class Denominated Princely': The Tremont

House Hotel," *Chicago History* 11(Spring 1982): 26–36; Perry Duis, "Whose City? Public and Private Places in Nineteenth-Century Chicago," *Chicago History* 12(Spring 1983): 2–27, 12(Summer 1983: 2–23; Gunther Barth, *City People* (New York: Oxford University Press, 1980).

4. William R. Leach, "Transformations in a Culture of Consumption: Women and Department Stores," *Journal of American History* 71(September 1984): 319–42; Ann Douglas, *The Feminization of American Culture* (New York: Knopf, 1977); Stuart Ewen, *Captains of Consciousness: Advertising and the Social Roots of the Consumer Society* (New York: McGraw-Hill, 1976). By the 1920s, contemporary social critics were drawing links between consumption and leisure; cf. Thorstein Veblen, *The Theory of the Leisure Class* (New York: Macmillan, 1899), and Harlan Paul Douglass, *The Suburban Trend* (New York: Century, 1925).

5. Harold M. Mayer, *The Railroad Pattern of Metropolitan Chicago* (Chicago: University of Chicago Department of Geography, 1943). On architecture, see Carl W. Condit, *The Chicago School of Architecture* (Chicago: University of Chicago Press, 1964), 51–57, 80–88. The true progenitor of the steel-frame skyscraper, the Home Insurance Building, followed three years later, in 1884.

6. Frank Lloyd Wright, *An Autobiography* (New York: Duell, Sloan and Pearce, 1943), 70, for the quotation. For an introduction to Wright and his architecture, see Robert C. Twombly, "Saving the Family: The Middle Class Attraction to Wright's Prairie House, 1901–1909," *American Quarterly* 27(March 1975): 57–72.

7. The history of Chicago's environment and environmental reform remains to be written. One important source is the Citizens Association *Annual Report*, 1877, which records the three-year-old group's initial concerns about protecting the environment. In this case the civic group was concerned about obnoxious smells from fertilizer plants and sewage. In 1881 it also became concerned about smoke pollution and sponsored an abatement ordinance. See ibid., 1882, 1884, 1888. Another important source of information about air pollution in the Loop is the large collection of photographs at the Chicago Historical Society. I have made an extensive study of these images of the Loop, which often show a visibility of only three to four blocks, especially during winter. Recent research confirms that air pollution from coal-fired furnaces was a severe problem. See Christine M. Rosen, "Chicago's Society for the Prevention of Smoke: Education, Technology, and the Law in the Fight against Air Pollution in the 1890s," unpublished manuscript, July 1989, in my possession.

8. Gregory, "Arc Lighting," 1–4. The Globe Hotel was at State and Harrison. The Buckingham Theater was a block away on Third Avenue.

9. A. T. Andreas, *History of Chicago*, 3 vols. (New York: Arno, 1975 [1884–86]), 3:598; *Edison Round Table*, 31 January 1922, 4, in CEC-Library.

10. W. L. Abbott Letter to William E. Keily, 29 June 1917, in CEC-LF, F6-E1, for the quotation; Gregory, "Arc Lighting."

11. Gregory, "Arc Lighting"; Edgar A. Edkins, "The Beginning of Elec-

tric Lighting in Chicago," manuscript, ca. 1914, in CEC-LF, F5-E3. According to Edkins, the installation at the Academy of Music was the first theater in Chicago to use incandescent lights, and it may have been the first in the world. The strike was in reaction to the installation of 150 incandescent lights in the auditorium of the theater. Since there were as yet no dimmer switches, the lights were not used to illuminate the stage.

12. Thomas P. Hughes, *Networks of Power: Electrification in Western Society, 1880–1930* (Baltimore: Johns Hopkins University Press, 1983), chap. 1, for a model of technological innovation that employs the concept of reverse salients; W. James King, *The Development of Electrical Technology in the Nineteenth Century: 3. The Early Arc Light and Generator,* United States National Museum Bulletin 228 (Washington, D.C.: Smithsonian Institution, 1962); Harold C. Passer, *The Electrical Manufacturers, 1875–1900* (Cambridge: Harvard University Press, 1953), pt. 1.

13. *Chicago Tribune,* 14 May 1881, for the quotation; *Chicago Tribune,* 17, 19 December 1880.

14. Chicago, *Report of the Fire Marshall,* 1884, 104–5, for statistics; William E. Keily Memorandum on Early Lighting Systems, in CEC-LF, F6—E1; *Edison Round Table,* January 1914, 384–85, in CEC-Library, on Van Depoele's station, including a photograph of the site at 64 North Canal Street. Also see Andreas, *History of Chicago,* 3:597–98.

15. Keily Memorandum, in CEC-LF, F6-E1. In February 1883 Van Depoele first demonstrated his electric railway on a four-hundred-foot length of track in front of his factory at 15–21 Clinton Street. In September he made a second demonstration in conjunction with the city's Interstate Fair and Exposition. For assessments of Thomson-Houston's improvements of the arc lamp and Van Depoele's work on electric traction, see Passer, *Electrical Manufacturers,* 21–31, 216–36. The sale of Van Depoele's company is noted in *Western Electrican* 5(13 July 1889): 16.

16. *Chicago Tribune,* 21 May 1880 for the first quotation, 19 December 1880 for the second.

17. Robert Friedel and Paul Israel, with Bernard S. Finn, *Edison's Electric Light: Biography of an Invention* (New Brunswick, N.J.: Rutgers University Press, 1986); Hughes, *Networks of Power,* chaps. 2, 4; Passer, *Electrical Manufacturers,* pt. 2; Payson Jones, *A Power History of the Consolidated Edison System, 1878–1900* (New York: Consolidated Edison Company, 1940), 151–56; Arthur A. Bright, Jr., *The Electric Lamp Industry* (New York: Macmillan, 1949), 32–89.

18. *Chicago Tribune,* 18 November 1881. Earlier, kerosene lamps had been used in the machine room, but work had to stop at nightfall in the erecting room because its paints, varnishes, and lumber posed too great a risk of fire.

19. *Chicago Tribune,* 11 November 1882, for the story on the party. For the cost of Doane's plant, see J. M. Clark, "List of Contracts Taken by the Western Electric Company," manuscript n.d., in CEC-LF, F6A-E1.

20. Cf. Clark, "List of Contracts," and the slightly different figures in

J. H. Goehst, "List of Isolated Plants and Wiring Jobs Installed by the Western Edison Company from 1882 to 1887 in Cook County, Illinois," manuscript, n.d., in CEC-LF, F6A-E1. On the origins and the founders of the Western Edison Company, see Edkins, "Electric Lighting"; Andreas, *History of Chicago*, 3:598; "The Story of the Commonwealth Edison Company," manuscript, 1922, 1–8, in CEC-HA, box 9001.

21. Louis Sullivan, *The Autobiography of an Idea* (New York, AIA Press, 1924), 258. See Robert Twombly, *Louis Sullivan: His Life and Work* (New York: Viking, 1986), 137–60, for an insightful analysis of the architect's intellectual evolution. Much of the material below on the theaters comes from the Twombly study. For an introduction to modern architecture, see Condit, *Chicago School*.

22. *Chicago Tribune*, 7 December 1884 for the first quotation, 26 August 1884 for the second quotation, as cited in Twombly, *Sullivan*, 146–47.

23. *Chicago Tribune*, 2 July 1885, for the first quotation; *Real Estate and Building Journal* 27(18 July 1885): 348, for the second quotation, as cited in Twombly, *Sullivan*, 146, 137–48.

24. *American Architect and Building News* 26(28 December 1889): 299–300, as quoted in Twombly, *Sullivan*, 192. On the Auditorium Building, see Twombly, *Sullivan*, 160–95; and Condit, *Chicago School*, chaps. 2, 7.

25. Twombly, *Sullivan*, 173, on Wright; for the Monadnock Building, see Condit, *Chicago School*, 65–69; "A History of Electric Service in Chicago," manuscript, Commonwealth Edison Company, 1966, in CEC-HA, box 9001.

26. Gregory, "Arc Lights," 5, for the first quotation; H. Jampolis Letter to E. W. Lloyd, 31 January 1914, in CEC-LF, F6-E1, for the second quotation. For statistics, see John Barrett's compilation in Chicago, *Report of the Fire Marshall*, 1884, 104–5.

Chapter Three

1. A. A. Griffith, "Lighting by Electricity," *Inland Architect* 1(March 1883): 21.

2. Robert H. Wiebe, *The Search for Order, 1877–1920* (New York: Hill and Wang, 1967). For a discussion of Wiebe's contribution in the context of the complicated historiography of progressivism, see Daniel T. Rodgers, "In Search of Progressivism," *Reviews in American History* 10(December 1982): 113–32.

3. Unfortunately, there is little modern scholarship on urban politics in Chicago during the Gilded Age. For the best of the new political history, see Richard Schneirov, "Class Conflict, Municipal Politics, and Governmental Reform in Gilded Age Chicago, 1871–1875," in *German Workers in Industrial Chicago, 1850–1910: A Comparative Perspective*, ed. Hartmut Keil and John B. Jentz (De Kalb: Northern Illinois University Press, 1983), 183–205, and John B. Jentz, "Class and Politics in Gilded Age Chicago, 1869–1875," paper presented at the annual meeting of the Social Science History

Association, Chicago, 1988. For the 1880s, the best approach is to study the career of Carter Harrison, who served five out of eight terms as mayor between 1879 and 1893. See Willis J. Abbot, *Carter Henry Harrison: A Memoir* (New York: Dodd, Mead, 1895); and Claudius O. Johnson, *Carter H. Harrison I: Political Leader* (Chicago: University of Chicago Press, 1928). For general perspectives, see Bessie Louise Pierce, *A History of Chicago*, 3 vols. (New York: Knopf, 1957), vol. 3, passim, and Ray Ginger, *Altgeld's America* (New York: Funk and Wagnalls, 1958). For a closer look at Chicago's municipal institutions, see Samuel Edwin Sparling, *Municipal History and Present Organization of the City of Chicago* (Madison: University of Wisconsin, 1908).

4. William E. Keily, Memorandum, ca. 1916, in CEC-LF, F6-E1. Bullock took the prudent step of enlisting the backing of the business community by selling shares in the Brush Electric Light Company to influential members of the social elite such as Marshall Field, Robert Law, and Joseph Sears. However, this offer to share in the profits of the new technology was only a token, because Brush continued to hold 75 percent of the stock. For more details on Bullock's early activities in the electric utility business, see Commonwealth Edison Company, *Yearbook*, 1928, 23, in CEC-HA, box 113. His pioneering efforts would lay the foundations of the Chicago Arc Light and Power Company.

5. *Chicago Tribune*, 21 September, 17 October 1881.

6. *Chicago Tribune*, 3 December 1881, 25–29 April 1882. For confirmation of gas company influence in the Brush bid for a franchise, see A. T. Andreas, *History of Chicago*, 3 vols. (New York: Arno, 1975 [1884–86]), 3:597, where the chronicler states that "the entrance of the electric-light companies was opposed by the old gas light companies, and it was charged that electric light wires were dangerous to human life, and were, besides, a fruitful source of fires." For a similar assessment, see W. C. Jenkins, *Chicago's Marvelous Electrical Development* (Boston: Chapple, 1911), 6.

7. Chicago, *Report of the Fire Marshall*, 1884, 105–6, for a list of improper installation problems. Also see the complaints of Chicago architects, who deplored their lack of power to set the standards for electrical installations in comparison with their European counterparts, in *Inland Architect* 3(March 1884): 15–16.

8. Essential introductions to this topic include Morton Keller, *Affairs of State: Public Life in Late Nineteenth Century America* (Cambridge: Harvard University Press, 1977), and Charles W. McCurdy, "Justice Field and the Jurisprudence of Government-Business Relations," *Journal of American History* 61(March 1975): 970–1005. Two important and useful guides to contemporary law are John F. Dillon, *Commentaries on the Law of Municipal Corporations*, 5th ed. (Boston: Little, Brown, 1911); and Thomas M. Cooley, *Treatise on the Constitutional Limitations Which Rest upon the . . . States . . . ,* 5th ed. (Boston: Little Brown, 1883). In the mid-eighties, the Supreme Court worked out a public utilities concept. See *Spring Valley Water Works vs. Schottler*, 110 U.S. 347 (1884); *Butchers' Union Company vs. Crescent City*

Company, 111 U.S. 746 (1884); and *New Orleans Gas Company vs. Louisiana Light Company,* 115 U.S. 650 (1885). Beginning in 1884, the annual reports of the fire marshall make it clear that the municipal government was capable of performing its inspection duties. Yet safety problems continued, as might be expected from any complex technology like electrical utility service. By 1891, for example, Chicago architects were willing to admit they needed expert advice in designing the electrical wiring requirements of large office buildings. See L. K. Comstock, "The Poor Man's Lamp," *Inland Architect* 18(September 1891); 19–20.

9. Chicago, *Report of the Fire Marshall,* 1888, 80; Chicago, *Proceedings of the City Council,* 1883–84, 244, 268, 402, 538. The ordinance of 9 April 1884, which passed 28 to 0, is reported in full on the last page listed. On 31 July 1882 the Sectional Underground Company received a twenty-five-year franchise from the city council.

10. Chicago, *Proceedings of the City Council,* 1881–82, 347, 519–20, 549–54; *Chicago Tribune,* 30, 25–30 April 1882; *Chicago Daily News,* 25, 29–30 April 1882.

11. Thomas Turner, "Presidential Address," American Gas-Light Association, *Proceedings* 8(1888): 201–8, and note 1 above. It is outside the scope of this book to document further changes in perceptions about lighting during the 1880s, but the annual meetings of the gas men during this period provide a very fruitful source of information on the effect of electricity on city dwellers. In the process, the gas industry began to change its own self-image from a lighting service to an energy business. Also see Mark J. Bouman, "Luxury and Control: The Urbanity of Street Lighting in Nineteenth Century Cities," *Journal of Urban History* 14(November 1987): 7–37.

12. See the full-page article "How Gas Is Made," *Chicago Tribune,* 21 June 1896, for a detailed description of production methods in Chicago. For a brief history of the "Lowe process" of making water gas, see American Gas-Light Association, *Proceedings* 3(1879): 116–26.

13. *Chicago Tribune,* 23 June 1886, for the quotation; see W. G. C. Rice, *Seventy-five Years of Gas Service in Chicago,* (Chicago: Peoples Gas Light and Coke Company, 1925), 33–36; Illinois Bureau of Labor Statistics, "The Gas Companies of Chicago," in *Ninth Biennial Report, 1896* (Springfield: Illinois Bureau of Labor Statistics, 1896), 277–80, for details on the gas war by one of the era's foremost experts, Edward W. Bemis. Perhaps the best insight on Chicago's utility speculators is provided by Joel Tarr, *A Study of Boss Politics: Williams A. Lorimer of Chicago* (Urbana: Univeristy of Illinois Press, 1971), 75–88, and Joel Tarr,"John R. Walsh of Chicago: A Case Study in Banking and Politics, 1881–1905," *Business History Review* 40(Winter 1966): 451–66. For a fascinating look into the craft of fitting gas service pipes in the home, see *Industrial Chicago,* 6 vols. (Chicago: Goodspeed, 1891), vol. 2, *The Building Interests.*

14. *Chicago Tribune,* 21 May 1886 for the quotation, 2 May 1886 for first notice of the "invasion." For recent interpretations of the labor clash, cf.

Paul Avrich, *The Haymarket Tragedy* (Princeton: Princeton University Press, 1984); and Bruce C. Nelson, "Anarchism: The Movement behind the Martyrs," *Chicago History* 15(September 1986): 4–19.

15. *Chicago Tribune,* 2 October 1886. The rival firm was the newly reorganized Consumers company. Its financial revival will be considered in the following section.

16. *Chicago Tribune,* 2, 21 May, 9 July, 3 October 1886.

17. *Chicago Daily News,* 11 August 1885; *Chicago Tribune,* 11, 12 August 1886.

18. *Chicago Tribune,* 2 May 1886 for the quotation, 22, 23 June, 15–16 July 1886 for the Consumers gas company receivership, 9 July 1886 for an editorial complaining about the inconvenience of having a third set of gas pipes laid in the streets of the South Side.

19. *Chicago Tribune,* 8 July 1886 for the quotation, July-August 1886, passim. Full documentation on the links between the newspapers and the rise of progressivism in Chicago lies outside the scope of this book, but this important topic is the subject of Royal J. Schmidt, "The *Chicago Daily News* and Traction Politics, 1876–1920," *Journal of the Illinois State Historical Society* 64(Autumn 1971): 312–26, and an excellent book by David Nord, *Newspapers and New Politics* (Ann Arbor, Mich.: UMI Research, 1981), 66–69, 127–30, passim. Nord concludes that "by far the most compelling issue in most cities was public utility regulation, especially street railways" (127). Also see below, notes 23–24.

20. *Chicago Tribune,* 23 May, 15–16 July 1886; *Chicago Daily News,* 9 August 1886, on the receiver's sale; *Chicago Tribune,* 3 October 1886, on the court injunction case.

21. See note 2 above and the perspectives offered on this concept of progressivism by James Weinstein, *The Corporate Ideal in the Liberal State: 1900–1918* (Boston: Beacon, 1968).

22. Norman C. Fay, "Plain Tales from Chicago," *Outlook,* March 1909, 547–49. Fay resigned in 1893 after a self-confessed career as a "monopolist and extortioner, engaged in robbing the people of Chicago."

23. Fay, "Plain Tales"; *Chicago Tribune,* 27 April, 1 May 1887; Illinois Bureau of Labor Statistics, "Gas Companies," 269–84; Sidney I. Roberts, "Portrait of a Robber Baron: Charles T. Yerkes," *Business History Review* 35(Fall 1961): 344–71.

24. Fay, "Plain Tales," 547; *Chicago Tribune,* 27 April 1887, for the newspaper's statement.

25. For a explanation of the civic group's involvement, see Citizens Association, *Annual Report,* 1888. The Citizens Association had resorted to the courts before, suing for relief against individual business, such as soap factories, that were causing obnoxious water and air pollution. In this instance the reformers were acting in a semipublic capacity to help the city enforce health and license ordinances that had originally been sponsored by the association. The suit to destroy the Gas Trust was different because it represented a moral position against a type of business practice rather

than a simple effort to force individuals to conform to community standards. See ibid., 1876–80, passim; Louise Wade, *Chicago's Pride: The Stockyards, Packingtown, and Environs in the Nineteenth Century* (Urbana: University of Illinois Press, 1987), 130–43, 313–21; and Harold L. Platt, "Creative Necessity: Municipal Reform in Gilded Age Chicago," paper presented at a Conference to Honor Harold M. Hyman, New York, 1989. Cf. Fay, "Plain Tales," 547–49, because the new president of the Gas Trust was also a member of the association's executive committee. Fay recalled his disagreement with the head of the civic group about the likely results of the suit. For the court battle, see *The People ex rel Francis B. Peabody vs. The Chicago Gas Trust Company,* 130 Ill. 268 (1889). The role of businessmen in reform has been a favorite topic of dissertations, including Donald B. Marks, "Polishing the Gem of the Prairies: The Evolution of Civic Reform Consciousness in Chicago, 1874–1900" (Ph.D. diss., University of Wisconsin, 1974); and Michael P. McCarthy, "Businessmen and Professionals in Municipal Reform: The Chicago Experience, 1877–1920" (Ph.D. diss., Northwestern University, 1970), esp. 30–43, where McCarthy discusses the generational change in reform outlooks.

26. *Chicago Tribune,* 27 April 1887; Citizens Association, *Annual Report,* 1890; 1893. Also consider Sidney I. Roberts, "The Municipal Voters League and Chicago's Boodlers," *Journal of the Illinois State Historical Society* 53(Summer 1960): 117–48; John Buenker, "Dynamics of Chicago Ethnic Politics, 1900–1930," *Journal of the Illinois State Historical Society* 67(April 1974): 175–99; and Lloyd Wendt and Herman Kogan, *Bosses in Lusty Chicago* (Bloomington: Indiana University Press, 1967 [1943]). In 1885 Massachusetts set up the first state commission to regulate electric and gas companies. Significantly, the timing coincides closely with the new competition in the industry. The source of the initiative is also important, because the gas companies themselves sponsored the measure to escape from the uncertain forces of the maket. The companies' expectations that the commission could be used to stifle competition were fulfilled during the ensuing years. See Joseph B. Eastman, "The Public Utilities Commissions of Massachusetts," in *The Regulation of Municipal Utilities,* ed. Clyde Lyndon King (New York: Appleton, 1912), 284–85.

27. *Chicago Tribune,* 7 May 1887; *Western Electrician* 1(3 September 1887): 16, 3(September 22 1888): 155; *Economist* 3(May 1890): 535, 5(7 March 1891): 372, 7(14 May 1892): 726, 8(22 October 1892): 584. Also see the records of the Chicago Arc Light company and other early utility companies in the corporate secretary's alphabetical file of "Important Papers," in CEC-HA, boxes 1–2.

28. The technological and business history of the Edison utility is well documented in contemporary journals and successive generations of company histories. Perhaps the best contemporary source is the *Western Electrician,* which started publication in 1887 in Chicago. This timing reflects the mature state of the city's community of electricians. For the rise and expansion of the station at the northeast corner of Adams and La Salle

streets, see "The Chicago Edison Station," *Western Electrician* 3(1 September 1888): 103–4, and "Dynamo Room of the Chicago Edison Company's Station," *Western Electrician* 6(1 February 1889): 49. A second fruitful source is the annual reports to the stockholders, which contain statistics on finances and output besides commentary on the problems of trying to keep up with the demand for more energy. The annual reports of the Chicago Edison Company are contained in CEC-HA, boxes 1–2. After conferring with the parent company, Chicago Edison decided to draw upon local talent to design the station, including architect S. S. Beman (the builder of Pullman) and engineer Fred Sargent. Planned for a capacity of 40,000 lights, the plant started with four 200 HP steam engines and eight dynamos that could each power 1,200 lights. This story can be found in two company-sponsored histories: H. A. Seymour, "History of Commonwealth Edison," manuscript, 1935, and "History of Commwealth Edison Company, manuscript, 1952, both in CEC-HA, box 9001; they are also available at the company headquarters' downtown library.

29. Chicago Edison Company, *Annual Report,* 1887–93, in CEC-HA, box 1; Seymour, "History of Commonwealth Edison," 192; Chicago Arc Light and Power Company Letter to Samuel Insull, 31 January 1893, in Chicago Arc Light and Power Company file, CEC-HA, box 1.

30. Chicago, *Report of the Fire Marshall,* 1888, 81–83, 1890, 107, 1892, 129; John P. Barrett, "Electric Lighting in Chicago," *Western Electrician* 6(15 February 1890): 77–78. The Chicago facility became an important national model for advocates of the municipal ownership of essential urban services and utilities. See, for example, William J. Meyers, "Municipal Electric Lighting in Chicago," *Political Science Quarterly* 10(March 1895): 87–94; John R. Commons, "The Social and Economic Factors in Chicago's Municipal Lighting," *Municipal Affairs* 6(Spring 1902): 109–18; and Charles W. Haskins and Elijah W. Sells, "Municipal Electric Lighting in Chicago: Accountants' Report," *Municipal Affairs* 6(Spring 1902): 88–108.

31. See "Chicago Central Station Development Chart," in CEC-LF, F6C-E4, for a complete list of these companies. Also cf. *A History of the City of Chicago: Its Men and Institutions* (Chicago: Inter-Ocean, 1900), 263–64, for a second list. Some primary source material on these companies is contained in CEC-HA, boxes 1–3.

32. "Chicago Electric Franchises," *Western Electrician* 6(21 June 1890): 339 for the quotation, 339–40. Also see *History of the City of Chicago,* 263–64. The annexations also had a major effect on the movement for municipal reform by bringing white middle-class suburbanites into the polity. See Michael P. McCarthy, "Chicago, the Annexation Movement and Progressive Reform," in *The Age of Urban Reform,* ed. Michael H. Ebner and Eugene M. Tobin (Port Washington, N.Y.: Kennikat, 1977), 43–54. The boundaries of the area where underground distributors were required were Thirty-ninth Street on the south, Ashland Avenue (1600 west) on the west, North Avenue (1600 north) on the north, and the lakefront on the east.

33. Chicago, *Report of the Fire Marshall*, 1890, 101, 1888, 80, where 53 motors are reported to be in service. In 1890 the city inspected 275 new motors.

34. Chicago Arc Light and Power Company, *Annual Report*, 1889, 2; "History of Commonwealth Edison" (1952), 8–9.

Chapter Four

1. R. Commerford Martin and Luther Stieringer, "City of Light," National Electric Light Association, *Proceedings of the Seventeenth Annual Convention*, 1894, 189, 189–209 (hereafter cited as NELA, *Proceedings*).

2. The historical literature on the crisis of the nineties and the World's Fair is extensive. For a general synthesis of the period, one can start with Robert H. Wiebe, *The Search for Order, 1877–1920* (New York: Hill and Wang, 1967). On the crisis of the nineties, Henry Steele Commager, *The American Mind: An Interpretation of American Thought and Character since the 1880s* (New Haven: Yale University Press, 1950); John Higham, "The Reorientation of American Culture in the 1890s," in *Writing American History: Essays on Modern Scholarship*, ed. John Higham (Bloomington: Indiana University Press, 1970), 73–102; David Noble, *The Progressive Mind, 1890–1917*, rev. ed. (Minneapolis: Burgess, 1981), 1–22; Arnold M. Paul, *Conservative Crisis and the Rule of Law: Attitudes of Bar and Bench, 1887–1895* (New York: Harper and Row, 1960); and T. J. Jackson Lears, *No Place of Grace: Antimodernism and the Transformation of American Culture, 1880–1920* (New York: Pantheon, 1981). On the World's Fair, see David F. Burg, *Chicago's White City of 1893* (Lexington: University of Kentucky Press, 1976), and R. Reid Badger, *The Great American Fair: The World's Columbian Exposition and American Culture* (Chicago: Nelson Hall, 1980).

3. NELA, *Proceedings*, 1891, 85. Also see NELA, *Proceedings*, 1890, 46–55, where the experience of the Paris exposition of 1889 is used not as an ideal model but only as a starting point for the 1893 fair. Also see John P. Barrett, *Electricity at the Columbian Exposition* (Chicago: Donnelley, 1894). On the Paris exposition, see Richard D. Mandell, *Paris 1900* (Toronto: University of Toronto Press, 1967). I wish to thank Paul Barrett of Illinois Institute of Technology, who provided me with insights on the goals of the planners for surroundings that harked back to an idealized vision of classical antiquity.

4. Henry Adams, *The Education of Henry Adams*, Sentry ed. (Boston: Houghton Mifflin, 1946 [1918]), 343. For an insightful analysis of Adams and his essay, cf. Lears, *No Place of Grace*, 262–97, and Lynn White, Jr., *Machina ex Deo* (Cambridge: MIT Press, 1968), 57–73.

5. See John F. Kasson, *Civilizing the Machine: Technology and Republican Values in America, 1776–1900* (New York: Penguin, 1977 [1976]), chap. 5, and Noble, *Progressive Mind*, 37–52, on the linkage between technology, social harmony, and political democracy. See Asa Biggs, *The Power of Steam: An Illustrated History of the World's Steam Age* (Chicago: University of

Chicago Press, 1982), 72–92; Wyn Wachhorst, *Thomas Alva Edison: An American Myth* (Cambridge: MIT Press, 1981), 3–46; Cecelia Tichi, *Shifting Gears: Technology, Literature, Culture in Modernist America* (Chapel Hill: University of North Carolina Press, 1987); and John R. Stilgoe, *Metropolitan Corridor: Railroads and the American Scene* (New Haven: Yale University Press, 1983), 105–32, on the transition from steam to electricity as a cultural symbol. As Stilgoe points out, steam continued to fascinate Americans for another forty years as a symbol of power. Yet by the 1890s, steam technology had lost its mystical, almost divine imagery because it had become so familiar and commonplace. A similar process of demystification would occur for electricity as it too came to be taken for granted in the cities by the time of the Great Depression. Only in rural areas did electricity retain its initial hold on the imagination. See Richrd A. Pence, ed., *The Next Greatest Thing* (Washington, D.C.: National Rural Electric Cooperative Association, 1984).

6. Mark J. Bouman, "Luxury and Control: The Urbanity of Street Lighting in Nineteenth-Century Cities," *Journal of Urban History* 14(November 1987): 7–37; Howard P. Segal, *Technological Utopianism in American Culture* (Chicago: University of Chicago Press, 1985).

7. NELA, *Proceedings*, 1891, 46–66.

8. Murat Halstead, "Electricity at the Fair," *Cosmopolitan* 15(September 1893): 577, 577–83, for the first quotation; *Electrical Engineering* 1(1893): 25, for the second quotation.

9. Halstead, "Electricity at the Fair," 578, for the long quotation; *Harper's Bazaar*, 9 September 1893, for the phrase used at the beginning of the paragraph.

10. Barrett, *Electricity*, passim; *Electrical Engineering* 1(1893), passim.

11. Adams, *Education*, 380 for the first quotation, 379 for the second quotation, 379–90. Although Adams wrote the essay after attending another exposition in 1900, it was the Chicago experience that set off his train of thought on the relation between modern technology and American culture. Also see above, note 4.

12. Lears, *No Place of Grace*, 32, for the quotation; Ray Ginger, *Altgeld's America: The Lincoln Ideal versus Changing Realities* (New York: Funk and Wagnalls, 1958), for an overview of Chicago history during this period; Richard Sennett, *Families against the City: Middle Class Homes of Industrial Chicago, 1872–1890* (Cambridge: Harvard University Press, 1970), for middle-class responses to industrialization; and the sources listed above, note 2.

13. Gwendolyn Wright, *Moralism and the Model Home: Domestic Architecture and Cultural Conflict in Chicago, 1873–1913* (Chicago: University of Chicago Press, 1980), 232, 199–227; Jeanne Madeline Weimann, *The Fair Women* (Chicago: Academy, 1981), 338–41, 458–63, 539–43; Barrett, *World's Fair*, 402–3; *Electrical Engineering* 2(1893): 1–6; Dolores Hayden, *The Grand Domestic Revolution* (Cambridge: MIT Press, 1981), chap. 6. Hayden, a latter-day cooperationalist, is a harsh critic of the reformers who opted for

technological solutions to the problems of liberating housewives and domestic servants from the drudgery of the kitchen. Cf. Ruth Schwartz Cowan, *More Work for Mother: The Ironies of Household Technology from the Open Hearth to the Microwave* (New York: Basic, 1983), for a more balanced analysis. Contemporary sources of feminist thought about household technology will be explored in the following section. For perspectives on the links between the scientific management and home economics movement, see Bettina Berch, "Scientific Management in the Home: The Empress' New Clothes," *Journal of American Culture* 3(Fall 1980): 440–45, and Samuel Haber, *Efficiency and Uplift: Scientific Management in the Progressive Era, 1890–1920* (Chicago: University of Chicago Press, 1964).

14. William R. Leach, "Transformations in a Culture of Consumption: Women and Department Stores, 1890–1925," *Journal of American History* 71(September 1984): 324–25, 319–42; Russell Lewis, "Everything under One Roof: World's Fairs and Department Stores in Paris and Chicago," *Chicago History* 12(Fall 1983): 28–47.

15. Insull Letter to Chester E. Tucker, 3 November 1926, Samuel Insull Papers, Loyola University of Chicago, Chicago, Illinois, box 2, for the quotation. The letter to the president of Cornell University also contains an account of Insull's employment by Gouraud and an Edison engineer, E. H. Johnson. Most historical information on Insull's early life comes from the "Memoirs of Samuel Insull," 1–55, in Samuel Insull Papers, box 17, which was written in 1934 during Insull's extradition back to the United States to face federal charges of mail fraud. Also see several chapters in his two collections of lectures in which he recounts his experiences. See William E. Keily, ed., *Central-Station Electrical Service: Its Commercial Development and Economic Significance as Set Forth in the Public Addresses (1897–1914) of Samuel Insull* (Chicago: privately printed, 1915), passim; and William E. Keily, ed., *Public Utilities in Modern Life: Selected Speeches (1914–1923) by Samuel Insull* (Chicago: privately printed, 1924), 184–98, 332–33, 342–43. For a sympathetic biography that stresses Insull's business and financial accomplishments, see Forrest McDonald, *Insull* (Chicago: University of Chicago Press, 1962), 1–24.

16. Letter to J. E. Kingsbury, [1881], reprinted in Keily, *Central-Station Service*, xxxvii, for the quotation; Samuel Insull, "Possibilities of the Central Station Business," from an article in *Western Electric* 20(28 September 1907), in ibid, 48–53. For Insull's account of the Pearl Street Station, see Samuel Insull, "The Development of the Central Station," address to the Electrical Engineering Department, Purdue University, 17 May 1898, in Keily, *Central-Station Service*, 17–20.

17. McDonald, *Insull*, 25–39.

18. Insull, "Memoirs," 57, for the first quotation; "Organization of the General Electric Company," *Electric World* 19(14 May 1982): 331, for the second quotation. Also see McDonald, *Insull*, 39–54. On the merger of the manufacturing firms, see Harold C. Passer, *The Electrical Manufacturers, 1875–1900: A Study in Competition, Entrepreneurship, Technical Change and*

Economic Growth (Cambridge: Harvard University Press, 1953), 41–43, 78–104; Arthur A. Bright, Jr., *The Electric-Lamp Industry: Technological Change and Economic Development from 1800 to 1947* (New York: Macmillan, 1949), 89–104. As Bright points out, the urge to escape from direct competition provided an impetus for the mergers of 1892 as well as for a secret patent-pooling agreement between GE and Westinghouse four years later.

19. Insull, "Memoirs," 56, for the quotation, 56–62. Insull accepted a salary of $12,000 a year for the first three years, approximately one-third to one-half his annual pay at GE.

20. Chicago *Report of the Fire Marshall*, 1892, 121, for statistics on the city's electrical services. The report includes a description of the municipally owned facilities for street lighting. The report lists two municipal plants that powered 1,102 arc lamps and 1,100 incandescent lights. The incandescent lights were used to illuminate the waterworks and the city hall.

21. "Prospectus of the Northwestern Electric Light and Power Company," 1892, in Important Papers Files, CEC-HA, box 2. The promoter's promises of forthcoming right-of-way franchises in the Loop were fulfilled. Before the company was sold to Insull in January 1894, it obtained four franchise extensions, on 6 June, 20 June, 3 October 1892, and 30 January 1893. See ibid. A complete list of utility franchises was compiled by the city clerk; see James R. B. Van Cleave, comp., *List of Franchises* (Chicago: Higgins, 1896).

22. The boundaries of the underground district (and the Chicago Edison Company's right-of-way franchise) were North Avenue (1600 north), Ashland Avenue (1600 west), Thirty-ninth Street on the south, and the lake on the east.

23. Frank Lloyd Wright, *An Autobiography* (New York: Duell, Sloan, and Pearce, 1943), 70; Vivian Palmer, "Documents: History of Chicago," 6 vols., Chicago Historical Society, 1925–30, vol. 2 (Uptown). This set of documents contains scrapbooks of communities on the North Side as well as a series or oral histories collected by Palmer, a University of Chicago graduate student, in 1925–27. For insight on the process of house building with more and more infrastructure improvements included, see Ann Durkin Keating, "The Role of Real Estate Subdividers and Developers in Nineteenth Century Service Provision: The Case of Chicago," paper presented at the Social Science History Association, October 1983, and Roger D. Simon, *The City Building Process: Housing and Services in New Milwaukee Neighborhoods, 1880–1910*, Transactions of the American Philosophical Society 68, pt. 5 (Philadelphia: American Philosophical Society, 1978).

24. *Chicago Journal*, Summer 1887, for the first quotation; *Chicago Journal*, 23 April 1894, for the second quotation, included in Vivian Palmer, "Documents: History of Chicago," vol. 2, doc. 12; ibid., doc. 14, for the third quotation. The interviewee remained unidentified. Also see the description of the community and its electric services in "Edgewater Electric Light Station," *Western Electrician* 12(14 December 1889): 305; and the materials

contained in "Edgewater Light Company," in Important Papers Files, CEC-HA, box 2.

25. E. C. Ward, "Report to the Stockholders," 12 September 1891, in "The Hyde Park Thomson-Houston Light Company," in Important Papers Files, CEC-HA, box 2. The service area of the company was bounded by Forty-third Street on the north, Fifty-ninth Street on the south, Drexel Boulevard on the west, and the lake on the east.

26. The electrification of the suburbs and satellite cities of the Chicago area will form the subject of the following chapters. Source material will be detailed there for the most part. For an introduction to the origins of electric service outside the city, see Imogene E. Whetstone, "Historical Factors in the Development of Northern Illinois and Its Utilities," Public Service Company of Northern Illinois, 1928, in CEC-HA, box 9003. Primary sources include the extensive collection of early franchises printed as "The Important Papers of . . . ," in CEC-HA, boxes 4045–46, and the corporate records of the early companies, in CEC-HA, boxes 4206–15, and in CEC-LF, F35.

27. "Report upon the Conditions of the Underground System and the Station Building of the Chicago Edison Company to the Stockholders," 3 March 1888, in CEC-HA, boxes 1–2; ibid., 8 March 1889, for cost statistics; "Kwh Output Data, 1888–1902" (photocopy in my possession) for transmission-loss statistics. The best source of technical data on the history of the company's generating stations is a set of files known as "The History of the Substation Department of the Commonwealth Edison Company," CEC-HA, boxes 177–78.

28. See "History of the Substation" for the substation; and CEC-LF, F6B-E3, for samples of the petitions from Prairie Avenue residents.

29. For an introduction to AC technology, see Passer, *Electrical Manufacturers*, 128–50; Thomas P. Hughes, *Networks of Power: Electrification in Western Society, 1880–1930* (Baltimore: Johns Hopkins University Press, 1983), 79–139.

30. "Englewood Electric Light Company," in Important Papers Files, CEC-HA, box 2; "Evanston Electric Illuminating Co.," in Important Papers Files of the Public Service Company of Northern Illinois, CEC-HA, box 4208.

31. For a history of the street railway, see Robert David Weber, "Rationalizers and Reformers: Chicago Local Transportation in the Nineteenth Century" (Ph.D diss., University of Wisconsin, 1971), 101–12. On 15 July 1895, the first trolley entered the Loop. A year later the city council passed an ordinance allowing all transit companies entering the CBD to use electric motive power. For an introduction to the technology of the electric motor, see Passer, *Electrical Manufacturers*, 211–76. The electrical industry was among the first to institutionalize the process of research and development. See George Wise, "A New Role for Professional Scientists in Industry: Industrial Research at General Electric, 1900–1916," *Technology and Culture* 21(July 1980): 408–29; and Hughes, *Networks of Power*, 140–74. On

the public policy of rapid transit in Chicago, see Paul Barrett, *The Automobile and Urban Transit: The Formation of Public Policy in Chicago, 1900–1930* (Philadelphia: Temple University Press, 1983).

32. As a starting point, see Forrest McDonald, "Samuel Insull and the Movement for State Utility Regulatory Commissions," *Business History Review* 32 (Autumn 1958): 241–54. However, McDonald gives Insull far too much credit as the original author of the "natural" monopoly theory. Actually, the idea had its roots in the post-Civil War era as the problems of railroad economics and regulation pressed for solution. For early expressions of the economic concept in a context of urban public utilities, the annual meetings of the gaslight fraternity must be considered. See, for example, Theobald Forstall, "Some Reasons Why Gas Consumers Should Not Favor Competiton in Its Supply," *Proceedings of the Fifth Annual Meeting of the American Gas-Light Association,* 1877, 24–30; F. C. Sherman, "The Relations of Municipalities to Gas Companies," *Proceedings of the Seventh Annual Meeting of the American Gas-Light Association,* 1879, 60–63; and J. C. Pratt, "The Future of the Gas Interests," *Proceedings of the Tenth Annual Meeting of the American Gas-Light Association,* 1882, 156–65. These views gained legitimacy when professional economists added their support for "natural" monopoly theories. See Richard T. Ely, "Report of the Organization of the American Economic Association," *Publications of the American Economic Association* 1, no. 1(1886): 5–40; Edmund T. James, "The Relation of the Modern Municipality to the Gas Supply," *Publications of the American Economic Association* 1, nos. 2, 5 (1886): 53–122; and Henry Carter Adams, "The Relation of the State to Industrial Action," *Publications of the American Economic Association* 1, no. 6(1886): 471–549. By 1890 the idea had gained wide currency. See, for example, E. Benjamin Andrew, "The Economic Law of Monopoly," *Journal of Social Science* 26(1890): 1–12, and Arthur T. Hadley, "Private Monopolies and Private Rights," *Quarterly Journal of Economics* 1(1887): 28–44. In the utilities industry, the leaders of the gas business were among the first to advocate state regulatory commissions as a way to escape local politics and franchise competition. See M. S. Greenough, "Presidential Address," *Proceedings of the Fourteenth Annual Meeting of the American Gas-Light Association* 7(1886): 225. In the legal profession, similar currents of thought were gaining the ascendency at the same time. See *Sinking Fund Cases,* 99 U.S. 700 (1878); *New Orleans vs. Clark,* 95 U.S. 644 (1877); *New Orleans Water-Works Company vs. Rivers,* 115 U.S. 674 (1885). For historical perspectives on the relation between monopoly and regulatory ideas, see Charles W. McCurdy, "Justice Field and the Jurisprudence of Government-Business Relations," *Journal of American History* 61(March 1975): 970–1005; and Morton Keller, *Affairs of State: Public Life in Late Nineteenth Century America* (Cambridge: Harvard University Press, 1977).

33. Ernest Edkins, as quoted in John Hogan, *A Spirit Capable: The Story of Commonwealth Edison* (Chicago: Mobium, 1986), 24. Edkins served as company historian into the 1910s. The Adams Street Station followed the

model of Pearl Street with engines on one floor and dynamos on another. There were, in fact, several fires at the Adams Street Station. See Hogan, *Spirit Capable*, 24–26. The company's rivals suffered the same problems. See, for example, "Fire in the Station of the Chicago Arc Light and Power Company," *Western Electrician* 3(22 September 1888): 155, and the description of an even worse fire three years later, "Chicago Arc Light and Power Company," in Important Papers Files, CEC-HA, boxes 1–2; additional materials on the company are found in CEC-LF, F6-E2, including a photograph of the company's station. Also see descriptions of the CALPC's central station at the corner of Market and Washington streets, which was remarkably similar to its Edison counterpart. One account revealed that "the station contained what was probably the most amazing conglomeration of equipment that was ever assembled under one roof. It was overloaded practically all of the time, and when darkness came on and all the engines were running, the rope drives and counter shafts thundering and the series machines shrieking like fiends, speech was drowned out, and all communication in the plant had to be in sign language," as quoted in [J. F. Rice], *History of the Commonwealth Edison Company* (Chicago: Commonwealth Edison Company, 1952), 106. Also see below, chapter 10, for the demolition of the building at the end of the 1920s.

34. Frank Gorton, Letter to Samuel Insull, 19 December 1890, in "Chicago Edison Company," Important Papers Files, in CEC-HA, boxes 1–2, for the quotation. Significantly, Gorton drew a close parallel between the telephone and electric business on the need to plan for expansion. However, he prudently cautioned that he also obtained enough contracts to ensure a profitable return on the investment before commencing construction. As we shall see in the following chapters, Insull would adhere strictly to these guidelines in building the Fisk Street Station and subsequent plants. For a brief biography of Gorton, see *A History of the City of Chicago: Its Men and Institutions* (Chicago: Inter-Ocean, 1900), 262–65. According to McDonald, Insull and Villard became close friends during Insull's New York period. See McDonald, "Samuel Insull and Regulatory Commissions," 243. For a description of the Harrison Street Station, see *The Sargent and Lundy Story* (Chicago: Sargent and Lundy, 1961), chap. 1. A fuller discussion of the innovative station will follow below. Statistics for coal consumption come from [Rice], *History of Commonwealth Edison*, 59. In 1888 the Adams Street Station used about twelve tons of coal daily, a figure that rose rapidly, to fifty tons per day in 1894.

35. *Sargent and Lundy Story*, 2–12; "The Lighting System of the Chicago Edison Company," *Electrical Review* 40(January 1902): 42–51; [Rice], *History of Commonwealth Edison*, 7–8, 42–43. On fuel consumption statistics, see E. J. Fowler Letter to Insull, 5 February 1937, in Insull Papers, box 3, for the earlier figures for Adams Street Station in 1893. For Harrison Street, see Insull's remarks in NELA, *Proceedings*, 1898, 68.

36. Insull Letter to Fort Wayne Electric Company, ca. August 1892, in "Central Electric Light and Power Company," Important Papers Files, in

CEC-HA, box 1, for the quotation; Insull Letter to C. A. Coffin, 22 August 1892, in "Chicago Edison Company," in CEC-HA, box 1. The agreement with Coffin covered a broad range of topics, including royalties to GE and restrictions on the sale of light bulbs by GE within Cook County.

37. Coffin Letter to Insull, 25 August, 1892, Insull correspondence with McDonald, 1–5 August 1892, and "Chicago Arc Light and Power Company," all in Important Papers Files, in CEC-HA, box 1; "Arc Light and Power," *Economist* 8 (22 October 1892): 584, for the report of a merger.

38. *Economist* 9(21 January 1893): 78, for Beale's quotation; William E. Keily, "How the Small Electric Light Companies of Chicago Were Merged into One Great Electric Service Corporation," manuscript, 22 January 1916, in CEC-LF, F6C-E4, for a complete list of companies Insull purchased. The calculation of the tangible assets and "goodwill" of these firms was made for the 1907 rate hearing before the city council by the city electrician, the city comptroller's office, and the accounting firm of H. M. Byllesby and Company.

39. [Rice], *History of Commonwealth Edison,* 43, for Harrison Street Station statistics; "Kwh Output Data, 1888–1902," for generator output statistics. Also see the statistics and other useful information in Chicago Edison Company, *Annual Report,* 1888–1907, in CEC-HA, boxes 1–2.

40. Insull, "Memoirs," 70, for the quotation. For a similar recollection, see Samuel Insull, "Stepping Stones of Central-Station Development through Three Decades," address before the Brooklyn Edison Company Section of the National Electric Light Association, 26 June 1912, in Keily, *Central-Station Service,* 351.

41. On complaints of high rates during the 1890s, see John F. Gilchrist, "Cheap Power: A Great Community Asset," paper presented to the Junior Association of Commerce, Chicago, 1937, in CEC-LF, F5B-E1.

42. Insull, "Memoirs," 71, for the quotation; H. A. Seymour, "History of Commonwealth Edison," manuscript, 1934, 214–16, in CEC-HA, box 9001, for early rate schedules; C. B. Crother, "Construction Department," manuscript, 1913, 4, in CEC-LF, F3-E1, for details on the Great Northern Hotel contract. In 1893 the company announced a new schedule of discounts based on this model. The discounts increased with the size of the customer's annual bills. In exchange, customers agreed to purchase all their artificial illumination from Chicago Edison for a three- to five-year term. Known as the "guaranteed discount schedule," it began to give breaks to customers with a minimum annual bill of $2,000. For these large consumers, a rate of 12¢ per kwh was charged, with rates falling to 5¢ per kwh for those with $12,000 in yearly billings. See Seymour, "History of Commonwealth Edison," 218. For perspectives on the adoption of meters by the electrical industry in imitation of the gas utility practice, consider the discussion in NELA, *Proceedings,* 1894, 336–40.

43. *Electrical Engineering* 1(1893): 15, 9–26; Barrett, *World's Fair,* 6–9.

44. Hughes, *Networks of Power,* 120–22, 209–11. The rotary convertor was invented in 1888 by Charles S. Bradley. By the time of the fair, both GE and Westinghouse were beginning to manufacture it. Hughes argues

forcefully that the fair demonstrated the possibilities of the rotary convertor to the electrical fraternity. But neither the reports of *Electrical Engineering* nor Chief Electrician John Barrett in his book, *Electricity at the Columbian Exposition,* pay much attention to the device. In fact, most electrical needs were supplied in the traditional manner of creating completely separate systems (circuits) for the different lighting and power needs, as well as the two currencies. If Insull recognized at the fair that the rotary convertor could answer the problems of a universal electrical supply system, it represents another instance of his ability to be among the first to exploit new innovations to best advantage. The debate over AC versus DC continued for several years. For a good debate by strong advocates of the two systems, compare Herbert A. Wagner, "General Distribution from Central Stations by A.C.," in NELA, *Proceedings,* 1898, 135–64, and Louis Ferguson, "General Distribution from Central Stations by D.C.," NELA, *Proceedings,* 1898, 165–75. Significantly, Ferguson did not disagree about the need to use AC in the outlying residential districts of the city. He ably defended the DC system in the CBD, where Edison companies had already sunk a huge investment in distributor conduits.

45. [Rice], *History of Commonwealth Edison,* 69–72; Insull, "A Quarter Century Central-Station Anniversary Celebration in Chicago, 1887–1912," address to Commonwealth Edison Company, 20 April 1912, in Keily, *Central-Station Service,* 320–21. The equipment used consisted of a 250 kw converter wound in reverse for use at the Harrison Street Station to turn DC into AC. At the substation, two 100 kw converters were installed to reverse the current from AC to DC. See Crother, "Construction Department." Apparently Boston Edison was the first to use a rotary converter. See Hughes, *Networks of Power,* 209.

46. For example, see Sidney Roberts, "Portrait of a Robber Baron: Charles T. Yerkes," *Business History Review* 35 (Winter 1961): 344–71; Ginger, *Altgeld's America,* 89–113; Weber, "Rationalizers and Reformers"; Lloyd Wendt and Herman Kogan, *Bosses in Lusty Chicago: The Story of Bathhouse John and Hinky Dink* (Bloomington: Indiana University Press, 1967 [1943]), 184–99; and Edward R. Kantowicz, "Carter H. Harrison II: The Politics of Balance," in *The Mayors: The Chicago Political Tradition,* ed. Paul M. Green and Melvin G. Holli (Carbondale: Southern Illinois University Press, 1987), 16–32.

47. McDonald, *Insull,* 81–89. McDonald accepts the account of Samuel Insull, Jr., who reported that his father paid $50,000 for the company, but the city report of 1907 put the price at $170,000. See Keily, "Small Electric Light Companies," 4. For all practical purposes, the two companies operated as one. For the organization of the merger, see "Operating Contract between Chicago Edison Company and Commonwealth Electric Company," 11 July 1898, in CEC-HA, boxes 1–2 (copy in my possession).

48. Samuel Insull, "Dinner [Address] in Honor of Messrs. S. Z. De Ferranti, C. Ch. Merz, and Arthur Wright, of London," New York, 28 September 1911, in Keily, *Central-Station Service,* 217, for the quotation.

49. See McDonald, *Insull,* 67–69, for the story of the utility executive's

trip to England; Seymour, "History of Commonwealth Edison," 278–80, for the company's adoption of the meter. The demand meter worked like a recording thermometer. The peak demand for electricity created the greatest amount of heat in the measuring device and was recorded on a paper disk. See NELA, *Proceedings, 1898,* 76–77, for a nontechnical description of the meter.

50. Insull, "Central Station Development," 25. By 1907 it was firmly settled that storage batteries were uneconomical. See Insull, "Elucidation of Electric-Service Rates for Businessmen," address presented to the City Club of Chicago, 1907, in Keily, *Central-Station Service,* 58.

51. In the simplest case, for example, the regular watt meter might read 7,000 kwh, and the demand reading might read 3,000 kwh. The customer would be charged the primary rate of 20¢ per kwh for the first 3,000 kwh and the secondary rate of 10¢ per kwh for the remaining 4,000 kwh. In 1898 Chicago Edison began to use this type of ratestructure for residential customers, with a few modifications. Most important were two mechanisms to guarantee a minimum return to the company. A minimum bill of $1 per month was imposed regardless of consumption. In addition, the primary rate was charged for a minimum of forty-five hours' use of the maximum demand in the winter and fifteen hours' use in the summer. For instance, a residential consumer in the winter has a maximum demand of 2.5 kwh (or fifty fifty-watt bulbs burning at the same time) and a total consumption of 100 kwh during the month. The customer would be charged the prime rate of 20¢ per kwh times the maximum demand of 2.5 kwh, times forty-five hours, or $22.50. For the remaining 55 kwh the customer would be charged the secondary rate of 10¢ per kwh, or $5.50. The net bill would total $28. A different schedule of rates applied to power applications, with a sliding scale of primary charge was 10¢ per kwh for the first thirty hours of use, 8¢ per kwh for the next thirty hours of maximum demand, and so on. See Seymour, "History of Commonwealth Edison," 279–81.

52. Arthur Wright, "Profitable Extension of Electricity Supply Stations," NELA, *Proceedings, 1897,* 162 for the first and second quotations, 174 for the third quotation, 173 for the fourth quotation, and 166 for the final quotation. Also see 213–14 for a description of how the rate system worked in England, where over forty cities were using it. Perhaps reflecting the rapid adoption of the demand meter, in 1897 Wright was elected president of the British Municipal Electric Light Association.

53. NELA, *Proceedings, 1897,* 200 for the quotation, 191–99 for the discussion.

54. NELA, *Proceedings, 1897,* 75 for the Insull quotation, 83 for the Ferguson quotation, 67–91 for the general debate. Also see Insull, "Presidential Address," paper presented to the twenty-first annual meeting of NELA, 1898, in Keily, *Central-Station Service,* 34–42.

55. Insull, "Presidential Address," 45 for the quotation, 34–47 for the larger discussion by Insull of the direct links between the new economics

and political approaches to government-business relations. Also see McDonald, "Insull and Regulation," and above, note 32.

56. NELA, *Proceedings*, 1898, 73 for the Ferguson quotation, 73–91 for Insull's and Ferguson's participation in the discussion of ratemaking. In calling for a "wholesale discount" for large power customers, Ferguson was careful to point out that this rate break applied only to the secondary charge for actual consumption and never to the primary charge for readiness to serve. "The intention," Ferguson emphasized, "was to give the low rate to the long-hour." See NELA, *Proceedings*, 1898, 91 for the quotation, 90–91.

57. Insull, "Development of the Central Station," 31 for the first quotation, 33 for the second quotation.

58. On new the new rates and free wiring offer, see Seymour, "History of Commonwealth Edison," 149–61, 279–80; on the creation of the statistics department, see E. J. Fowler, "History of the Statistical Department," manuscript, 1926, in CEC-LF, F23-E2, and [J. F. Rice], "Materials collected by J. F. Rice in Preparing the Company History," manuscript, 1952, 198, in CEC-HA, box 9001.

59. For a general history, see M. Luckiesh, *Artificial Light: Its Influence upon Civilization* (New York: Century, 1920), 99–102. For contemporary perspectives on the mantle and its influence in changing perceptions of gas from a lighting fuel to an energy fuel, see Arthur E. Boardman, "Presidential Address," *Proceedings of the Twenty First Annual Meeting of the American Gas-Light Association*, 1893, 191–99; George B. Ramsdell, "Presidential Address," *Proceedings of the Twenty Eighth Annual Meeting of the American Gas-Light Association*, 1900, 149–57; Edward G. Pratt, "Presidential Address," *Proceedings of the Twenty Ninth Annual Meeting of the American Gas-Light Association*, 1901, 170–88; and Alton D. Adams, "Gas or Electricity for Heating, Light, Power?" *Gassier's Magazine* 17(1900): 513–17.

60. W. G. C. Rice, *Seventy-five Years of Gas Service in Chicago* (Chicago: Peoples Gas Light and Coke Company, 1925), 33 for the quotation, 32–33.

61. Rice, *Seventy-five Years*, 35–41, for a brief overview; Illinois Bureau of Labor Statistics, *Ninth Biennial Report, 1896*, pt. 3, *Franchises and Taxation: The Gas Companies of Chicago* (Springfield: Illinois Bureau of Labor Statistics, 1896), 239–320, for the most detailed narrative of the Gas Trust. The report was written by Edward W. Bemis, one of the Progressive Era's most noted authorities on urban public utilities. For the predictions of the Gas Trust president, see Norman C. Fay, "Plain Tales from Chicago," *Outlook*, March 1919, 549, and above, chapter 3. Both the state and the national courts upheld the "Consolidation Act" of 1897. See *Deenan vs. Peoples Gas Light and Coke Company*, 205 Ill. 482 (1903); and *Peoples Gas Light and Coke Company vs City of Chicago*, 194 U.S. 1 (1904). In the latter case, the high court also upheld the city's power to regulate the rates of the new company and restricted its special privileges to the territory served in its original franchise. On Altgeld's antimonopoly views, see Ginger, *Altgeld's America*, 168–88.

62. *History of the City of Chicago* (1900), 261 for the first quotation, 260 for the second quotation.

63. Stilgoe, *Metropolitan Corridor*, 109 for the quotation, 105–32 for a fuller analysis of the cultural symbolism of electrical technology and the engineers who harnessed its power.

Chapter Five

1. H. A. Seymour, "History of Commonwealth Edison Company," manuscript, 1934, Appendix D, in CEC-HA, box 9001, for the wiring scheme; Commonwealth Edison Company, Departmental Correspondence, E. Fowler to J. Gilchrist, 4 December 1916, in CEC-LF, F23-E4. On the emergence of a consumer economy, see Thorstein Veblen's classic *The Theory of the Leisure Class* (New York: Macmillan, 1899); Stuart Ewen, *Captains of Consciousness: Advertising and the Social Roots of the Consumer Society* (New York: McGraw-Hill, 1976); T. J. Jackson Lears, *No Place of Grace: Antimodernism and the Transformation of American Culture, 1880–1920* (New York: Pantheon, 1981); Willian R. Leach, "Transformations in a Culture of Consumption: Women and Department Stores, 1890–1925," *Journal of American History* 71 (September 1984): 319–42.

2. Seymour, "History of Commonwealth Edison," 287, Appendix, 81, 174, 176.

3. Samuel Insull, "The Larger Aspects of Making and Selling Electrical Energy," address presented to the Association of Edison Illuminating Companies, 1909, in William E. Keily, ed., *Central-Station Electric Service: Its Commercial Development and Economic Significance as Set Forth in the Public Addresses (1897–1914) of Samuel Insull* (Chicago: privately printed, 1915), 79.

4. Seymour, "History of Commonwealth Edison," Appendix, 176.

5. Unfortunately, historians have dealt piecemeal with this important period of Chicago history. Several scholars have illuminated special topics of municipal reform, but with a myopic focus that ignores all the related aspects of the Progressive Era. The best of these monographs examines public transportation policy. See Paul Barrett, *The Automobile and Urban Transit: The Formation of Public Policy in Chicago, 1900–1930* (Philadelphia: Temple University Press, 1983). On the effort to rewrite the municipal charter and give Chicago "home rule," see the careful political study by Maureen A. Flanagan, *Charter Reform in Chicago* (Carbondale: Southern Illinois University Press, 1987). For a brief survey by different authors of the mayors during this period (Carter Harrison II [1897–1905], Edward F. Dunne [1905–7], and Fred Busse [1907–11]), see Paul M. Green and Melvin G. Holli, ed., *The Mayors: The Chicago Political Tradition* (Carbondale: University of Southern Illinois Press, 1987), 16–60. A synthetic treatement remains to be written that links charter reform, municipal ownership, utility regulation, and ethnocultural politics. On the debate among progressive reformers on the issue of urban utilities, see David Nord, "The Experts versus the Experts: Conflicting Philosophies of Municipal Utility

Regulation in the Progressive Era," *Wisconsin Magazine of History* 58(Spring 1975): 219–36.

6. Insull, "Larger Aspects," 81, for the quotation.

7. The number of customers is calculated by combining the patrons of the Commonwealth Edison Company within the city and those of the Public Service Company of Northern Illinois (PSCNI) in the surrounding territory. See Seymour, "History of Commonwealth Edison," Appendix, 174, and Public Service Company of Northern Illinois, *Yearbook, 1924*, 26 (hereafter cited as PSCNI *Yearbook*), in CEC-HA, box 4094.

8. U.S. Census, Population, 1890, 1, pt. 1: cxcii–cxciii; Population, 1900, 1, pt. 1:116; Population, 1910, 2:512; Manufactures, 1900 (1905), 232–38; Manufactures, 1910 (1914), 319–20, 339–61. Urban historians are making important strides in linking migration movements to jobs in the city. Two of the best studies of this kind are John Bodnar, Roger Simon, and Michael P. Weber, *Lives of Their Own: Blacks, Italians, and Poles in Pittsburg, 1900–1960* (Urbana: University of Illinois Press, 1982); and Theodore Hershberg, ed., *Philadelphia: Work, Space, Family, and Group Experience in the Nineteenth Century* (New York: Oxford University Press, 1981).

9. Some participants in the National Electric Light Association (NELA) debates of 1898 recognized the psychological incentive inherent in the two-tier system of rates. In particular, see the remarks of Detroit's Alexander Dow in *Proceedings of the National Electric Light Association*, 1898, 76–82 (hereafter cited as NELA, *Proceedings*).

10. John Gilchrist, "Campaigning for Business," paper presented to the annual meeting of the Association of Edison Illuminating Cmpanies, 1902, reprinted in John Gilchrist, ed., *Public Utility Subjects*, 2 vols. (Chicago: privately printed, 1940), 1:1–20. Gilchrist, like most of Insull's other top aides, started on the ground floor of the Chicago Edison Company in the late 1880s and rose through the ranks to become a department head/vice president. In this case, Gilchrist grew up in a middle-class family on the South Side of the city. In the early 1880s, when Gilchrist was a teenager, he became fascinated with electricity and started selling battery-operated doorbells to neighbors in Hyde Park. Upon graduating from high school, he joined the Chicago Edison Company and continued his role as a salesman of electricity. See his brief autobiographical sketch in Gilchrist, *Public Utility Subjects*, 1:iii–v. For a similar promotional effort to link electrical service with a suburban ideal, cf. Mark H. Rose, " 'There Is Less Smoke in the District': J. C. Nichols, Urban Change, and Technological Systems," *Journal of the West* 25(January–February 1986): 44–54.

11. Gilchrist, *Public utility Subjects*, 1:11 for the quotation, 10–11 for an explanation of the apartment building gambit. As my friend and colleague Paul Barrett reminded me, there were still a "fair number" of apartment buildings with gaslights in the hallways during the early 1950s when he was growing up. His observation underscores the important point that the diffusion of technology is rarely absolute or complete as long as substitutes are available.

12. In that pivotal year, 1898, the city council institutionalized its in-

volvement in the inspection, regulation, and provision of electricity by creating a department of electricity separate from the fire marshall's office. See Chicago Department of Electricity, *Annual Report*, 1898; William J. Meyers, "Municipal Electric Lighting in Chicago," *Political Science Quarterly* 10(March 1895): 87–94; Charles W. Haskins and Elijah W. Sells, "Municipal Electric Lighting in Chicago: Accountants' Report," *Municipal Affairs* 6(Spring 1902): 88–92; John Commons, "Social and Economic Factors in Chicago Municipal Lighting," *Municipal Affairs* 6(Spring 1902): 109–18.

13. Seymour, "History of Commonwealth Edison," 217, 284–85; C. B. Crothers, "[History of the] Construction Department," manuscript, 1913, 7–8, in CEC-LF, F3-E1. After the purchase of the Chicago Arc Light and Power Company, arc light rates were set at 65¢ per lamp per night from dusk to midnight, with a 15¢ discount for prompt payment of bills. The new rate offered to the businessmen's association was $1.82 per lamp per week or 26¢ per night.

14. This area was essentially the central business district. It was bounded by the north branch of the Chicago River on the north, the south branch of the river on the west, Twelfth Street on the south, and Lake Michigan on the east.

15. Gilchrist, "Campaigning for Business," 11–17; Seymour, "History of Commonwealth Edison," 281–82.

16. Insights on the economics of the self-contained electric plant are provided by the central station men who were trying to undercut it. See, for example, Insull, "Larger Aspects," 73–96, and John Gilchrist, "The Competition of Isolated Plants," paper presented to the annual meeting of the Association of Edison Illuminating Companies, 1901, in Gilchrist, *Public Utility Subjects*, 1:21–35.

17. [J. F. Rice], *History of the Commonwealth Edison Company* (Chicago: Commonwealth Edison Company, 1952), 110. The company's historical archives contain an extensive collection of the records of the Illinois Maintenance Company. See CEC-HA, boxes 3050–56.

18. Seymour, "History of Commonwealth Edison," 215–19, 280–81. In 1898 the power rate consisted of primary rate of 10¢ per kwh for the first 30 hours of maximum demand and secondary rates of 8¢ per kwh for the next 30 hours, 6¢ for the next 30 hours, 4¢ for the next 450 hours, and 3¢ for the remaining 180 hours. In addition, discounts were given on the secondary rate based on its total amount. The discounts ranged from 15 percent discount for bills over $40 to 40 percent for bills over $200. In 1900, further cuts reduced the secondary rate cut to 5¢ for the second 30 hours of use and 3¢ for the last 660 hours, while the discounts were increased to as much as 50 percent for bills over $1,000.

19. Seymour, "History of Commonwealth Edison," 226, 325; "Kilowatt Hour Output Data, 1888–1902" in CEC-HA, box 205.

20. B. G. Lamme, as quoted in Seymour, "History of Commonwealth Edison," 324. Lamme was the chief engineer of the Westinghouse Electric and Manufacturing Company.

21. John Gilchrist, "Electric Elevators and Conveyors," *Proceedings of the Association of Edison Illuminating Companies,* 1900, 3–9. On the search for a mechanical substitute for the horse-drawn streetcar, see William D. Middleton, *The Time of the Trolley* (Milwaukee: Kalmbach, 1967), 23–51. Important background on the Chicago experience is supplied by Robert David Weber, "Rationalizers and Reformers: Chicago Local Transportation in the Nineteenth Century" (Ph.D. diss., University of Wisconsin, 1971).

22. Louis Ferguson, "Central Station Advancement," 1898, as quoted in Seymour, "History of Commonwealth Edison," 322; Gilchrist, "Elevators," 3–9. Also see Chicago, *Report of the Fire Marshall,* 1890, 101, where John P. Barrett noted that "the variety of small industries to which this species of power is applicable is constantly and rapidly changing. A few years ago, the only use for which a motor seemed fit was its employment as a fan, and only the smaller models were in demand. Today, the electric motor is crowding out the steam plant and taking its place for driving small machinery, printing presses, elevators, restaurant fans, embossing and stamping machines, polishing, etc."

23. Gilchrist, "Elevators," 9, 3 for the first and second quotations, 3–26 for the full report.

24. E. W. Lloyd, "Notes on the Sale of Electric Power" Chicago 1906, a pamphlet in CEC-LF, F4-E2. This sophisticated analysis of 110 industrial customers provides statistics on individual industries, including their use of group or individual drive, their load factors, their consumption patterns, and the horsepower rating of over 600 motors they used. Lloyd was one of the top "general contract agents" of the contract department of the Chicago Edison Company and the Commonwealth Electric Company. On the inefficiency of steam power systems, see Louis C. Hunter's assessment that "as industrial establishments grew in scale and in complexity of operations, such gains [in the efficiency of prime movers] were offset by difficulties inherent in a system of power distribution which became more cumbersome and less efficient with greater extent and which imposed a straitjacket upon the organization of production. The requirements of power transmmisson before the coming of electricity imposed a heavy burden on engineers in mill design and construction," in Louis C. Hunter, *A History of Industrial Power in the United States, 1780–1930,* vol. 1, *Waterpower in the Century of the Steam Engine* (Charlottesville: University Press of Virginia, 1979), 478. On the use of industrial power in the late nineteenth century and the transition from steam and water power to electricity, see Warren D. Devine, "From Shafts to Wires: Historical Perspectives on Electrification," *Journal of Economic History* 43(June 1983): 347–72; Arthur G. Woolf, "Energy and Technology in American Manufacturing" (Ph.D. diss., University of Wisconsin, 1980); Richard DuBoff, "The Introduction of Electric Power in American Manufacturing," *Economic History Review* 20(December 1967): 509–18. During this period, inventors created a wide variety of machines that remained hand powered until the advent of the small electric motor. See Sigfried Giedion, *Mechanization Takes Command*

(New York: Oxford University Press, 1948), 512–96. On the influence of fire insurance underwriters on the change in factory design, see John Stilgoe, *Metropolitan Corridor: Railroads and the American Scene* (New Haven: Yale University Press, 1983), 82–85.

25. Lloyd, "Notes"; Seymour, "History of Commonwealth Edison," 325.

26. Seymour, "History of Commonwealth Edison"; W. I. Miskoe, "The Centenary of Modern [Electrical] Welding, 1885–1985: A Commemoration," *Welding Journal* 65(April 1986): 19–26.

27. E. W. Lloyd, "Memo on the Contract Department," ca. 1920, in CEC-LF, F4-E1; Lloyd, "Notes"; John F. Gilchrist, "The Isolated Plant," in *Electric Service in Chicago: A Series of Advertisements by the Commonwealth Edison Company*, Chicago, 1916, no. 17 (in my possession), 21–35. C. B. Crothers, "Construction Department," 1913, in CEC-LF, F3-E1, lists some of the conversions to electric power. In 1903, for example, Crothers notes that his department removed a gasoline engine at the Haffner Furniture Store and replaced it with electric motors. The following year a similar conversion took place at the Illinois Seed Company. In 1905 he noted a major victory in the battle against the self-contained system at the Fair department store. The combined connected load of the store's light and power equipment was equivalent to 20,000 light bulbs. Similar gains, large and small, are listed for the subsequent years.

28. "Kilowatt Hour Output, 1888–1902."

29. "Net Increase of Contracted Business from 1889 to 1925 Inclusive," in CEC-LF, F4-E5. The standard used was a sixteen-candle power incandescent bulb, which used about 50 kw per hour.

30. For perspectives on modern advertising, see Roland Marchand, *Advertising the American Dream: Making Way for Modernity, 1920–1940* (Berkeley, University of California Press, 1986).

31. Gilchrist, "Campaigning for Business," 8, 19–20; Seymour, "History of Commonwealth Edison," Appendix D. A complete set of *Electric City* magazine is found in CEC-LF. Also see the narrative history of the advertising department in CEC-LF, F4-E1. By 1910 the utility had obtained space in about seventy-five drugstores on the South Side and in the CBD alone. A complete list of these locations is contained in the Commonwealth Edison Company "Advisory Reports," CEC-HA, box 34.

32. Commonwealth Electric Company, Board of Directors, Meeting of 16 January 1900, in CEC-HA, box 300. In 1900 the company had to build a small central station on the South Side to accomodate the growth of AC service. For a brief sketch of the Fifty-sixth Street Station, see [Rice], *History of Commonwealth Edison*, 44, and "Substation History," CEC-HA, box 177.

33. See Bion Joseph Arnold, *Report on the Engineering and Operating Features of the Chicago Transportation Problem . . . 1902* (New York: McGraw, 1905), 66, 97, for the engineer's projections of the growth of Chicago's population and ridership over the next fifty years. Significantly, Arnold

predicted that ridership would increase faster than population. With a population estimate of 5.2 million people in 1952, he calculated that minimum gross annual reciepts on the surface lines alone would increase from $13.3 million to $70 million (based on a 5¢ fare). These projections must have encouraged Insull to assume whatever investment risks were involved in obtaining this class of business.

34. "The Lighting System of the Chicago Edison Company," *Electrical Review* 40(January 1902): 48–50.

35. Frank Lloyd Wright, "The Art and Craft of the Machine," in *Frank Lloyd Wright: Writings and Buildings,* ed. Edgar Kaufman and Ben Raeburn (New York: Horizon, 1960), 72, for the quotation; [Rice], *History of Commonwealth Edison,* 43, for the description of the Harrison Street Station. The Wright essay will be discussed in chapter 7.

36. Edwin T. Layton, Jr., "Scientific Technology, 1845–1900: The Hydraulic Turbine and the Origins of American Industrial Research," *Technology and Culture* 20 (January 1979): 64–89; Thomas P. Hughes, *Networks of Power: Electrification in Western Society, 1880–1930* (Baltimore: Johns Hopkins University Press, 1983), 195–97, 210–12.

37. Commonwealth Electric Company, Board of Directors, Meetings of 17 December 1901, 15 January 1902, in CEC-HA, box 300. The story of the historic Fisk Street Station has been well told several times. See Samuel Insull, "Massing of Energy an Economic Necessity," paper presented to the General Electric Company, Boston, 25 February 1910, reprinted in Keily, *Central-Station Service,* 137–38; Samuel Insull, "Memoirs of Samuel Insull," 74–78, in Samuel Insull Papers, box 17, Loyola University of Chicago; Seymour, "History of Commonwealth Edison," 340–44; [Rice], *History of Commonwealth Edison,* 45–46; *The Sargent and Lundy Story* (Chicago: Sargent and Lundy, [1961]), chap. 2; Hughes, *Networks of Power,* 209–14.

38. On the decision to purchase the second-generation equipment, see Commonwealth Electric Company, Board of Directors, Meetings of 15 January 1902, 20 December 1904, 9 May 1904, 17 October 1905, in CEC-HA, box 300. On the technological progress of the steam turbine, see "The Commonwealth Edison Company," *Electrical Review* 52 (11 January 1908): 65–69; "The Commonwealth Edison Company," *Electrical Review* 52 (16 May 1908): 756–62; "New Features of Central-Station Service in Chicago," *Electrical Review* 54 (29 May 1909): 968–76.

39. See note 37 above; Ernest F. Smith, "The Development, Equipment, and Operation of the Sub-stations and Distributing Systems of the Chicago Edison Co. and Commonwealth Electric Co.," *Journal of the Western Society of Engineers* 8 (March-April 1903): 209–22; Louis A. Ferguson, "Distribution of Electrical Energy in Large Cities," *Transactions of the American Institute of Electrical Engineers* 18(November 1901): 813–20. For a list of the locations of the substations in 1908, see Commonwealth Edison Company, *Annual Report,* 1908, in CEC-HA, box 113. For a detailed technical history

of the distribution system, see "Substation History," in CEC-HA, boxes 177–78. In 1904 the load dispatcher's position was created. See "The Evolution of the Load Dispatcher," manuscript, 1918 (in my possession). The narrative history was written by one of the department's managers. The dispatcher idea owes its origins to the railroads, which put managers in charge of directing traffic within their major terminal areas.

40. Copies of the original illustrations are found in Keily, *Central-Station Service*, 437, 469. For perspectives on recent developments in the electrical industry, see Mark Hertsgaard, *Nuclear Inc.: The Men and Money behind Nuclear Power* (New York: Pantheon, 1983); and Richard Hirsh, *Technology and Transformation in the American Electric Utility Industry* (New York: Cambridge University Press, 1989). Hirsh provides convincing proof that the turbogenerator in the mid-sixties finally reached the limits of technical improvement in terms of scale economies and thermal efficiency. Despite this "technological statis," however, utility operators blindly continued to build bigger and bigger electrical plants, producing a dinosaur generation of faulty equipment.

41. Insull, "Larger Aspects," 81, for the quotation. A copy of the original diagram is found in Keily, *Central-Station Service* 436.

42. Commonwealth Electric Company, Board of Directors, Meeting of 19 December 1905, in CEC-HA, box 300; Insull, "Memoirs," 63–66; [Rice], *History of Commonwealth Edison*, 59–62; Forrest McDonald, *Insull* (Chicago: University of Chicago Press, 1962), 108–10. A copy of the contract is found in Commonwealth Edison Company, *Important Papers of the Company Compiled to July 1, 1911* (Chicago: Commonwealth Edison Company, 1911), 376–80. The contract specified that Peabody would provide the utility with coal on a cost-plus basis. Half of the mines' output would be earmarked for the utility and the other half for the coal dealer. Insull also purchased the Midland Railroad to transport the coal from downstate to the Illinois River, where it was then shipped through the waterway to the company's generating stations. The coal contract and railroad venture are classic examples of the business practice of vertical integration. Also see Hirsh, *Technology and Transformation,* pt. 2.

43. *Sargent and Lundy Story,* 14,

44. See U.S. Census, Special Reports, Electric and Street Railways, 1902, 332–33, for statistics on the equipment and energy consumption of the fifteen companies listed as urban or suburban railways entering Chicago. Several interurban lines are also listed but not included in the figures listed here. See Arnold, *Report on the Chicago Transportation Problem,* 32, 100–101, for data on the Loop's traffic congestion and the operating expenses of the companies. Also see above, note 33, which outlines Arnold's long-term projections that ridership will increase faster than population. I draw heavily here and in the following section upon Paul Barrett's perceptive analysis of the city's transportation problems. See Barrett, *Automobile and Urban Transit,* 3–66.

45. On the first street railway service, see Insull, "Memoirs," 75; Sey-

mour, "History of Commonwealth Edison," 77–80; "Substation History," in CEC-HA, box 177. On the Hopkinson rate system, see the running debate over its merits versus the Wright demand system in NELA, *Proceedings*, 1900, 291–335, 1902, 1–9, 400–426.

46. Insull, "Larger Aspects," 85, for the quotation. For the construction of the Quarry Street Station, see "Commonwealth Edison Company" (16 May), 756–62; "New Features," 970–72; [Rice], *History of Commonwealth Edison*, 47–49.

47. Statistics from figure 28 in Samuel Insull, "A Quarter-Century Central-Station Anniversary Celebration in Chicago, 1887–1912," in Keily, *Central-Station Service*, 325.

48. See Thomas P. Hughes, "The Electrification of America: The System Builders," *Technology and Culture* 20 (January 1979): 124–61.

49. For a general introduction to Chicago politics at the turn of the century, see note 5 above.

50. Flanagan, *Charter Reform*, 10–97.

51. Barrett, *Automobile and Urban Transit*, 9–45. For perspectives on the knotty issue of municipal ownership, see Nord, "Experts versus the Experts," 219–36.

52. James Weinstein, *The Corporate Ideal in the Liberal State, 1900–1918* (Boston: Beacon, 1968), 3–39; McDonald, *Insull*, 118–20.

53. Barrett, *Automobile and Urban Transit*, 31 for the quotation, 27–37 for a careful analysis of the decision for regulation. For additional political context, cf. Flanagan, *Charter Reform*, 1–63; McDonald, *Insull*, 74–124.

54. Arnold, *Report*, 231–32.

55. See for example, Commonwealth Edison Company, Board of Directors, Meeting of 4 February 1908, in CEC-HA, box 300. At this meeting Insull makes it clear that a guarantee of demand came before he would make the financial commitment to provide additional generating equipment. In the case at hand, he was awaiting a contract with the Chicago City Railway Company before placing orders for two turbogenerators that would be installed at the new Quarry Street Station.

56. On the settlement, see Barrett, *Automobile and Urban Transit*, 37–45; for Insull's contract negotiations, see Commonwealth Edison Company, Board of Directors, Meetings of 16 January 1906, 4 February 1908, in CEC-HA, box 300; for a sample of Insull's speechmaking, see his address to the presitigious City Club on 19 October 1908, in Keily, *Central-Station Service*, 65–72. An analysis of U.S. census reports confirms that Insull's contracts took on the burden of supplying more and more electrical energy to the transit system in Chicago. Although the companies generated far more electricity in 1907 than in 1902 (see table 12 for output statistics), they did it with the same equipment. In fact the capacity of their dynamos actually declined from 78,570 HP to 76,575 HP. In the short run, the contracts were highly beneficial to the transit companies becaue they were able to run their own equipment to the maximum of its capacity while the Edison company absorbed the "overload" during periods of peak demand. At the

same time, the contracts were beneficial to Insull's utility company because this "overload" was so large compared with other aspects of the light and power business. See U.S. Census, Special Reports, Electric and Street Railways, 1902, 332–33, 1907, 460–63, 534.

57. For a brief history of the gas company controversy, see the report of Major Dunne to the city council, Chicago, *Journal of the Proceedings of the City Council*, 1905, 1419–20. Also see *Peoples Gas-Light and Coke Company vs. Chicago*, 194 U.S. 1 (1904). The high bench upheld the circuit court's dismissal of the injunction against the enforcement of the city's 1900 rate ordinance. However, Chief Justice Melville Fuller limited the ruling to the franchise privileges of the original company, which did not extend to the consolidated 1897 version of the corporation. The Court avoided ruling on the rate-making powers of the city and directed the litigants to the state courts to settle this issue.

58. For a brief overview of the problems of the Peoples Gas Light and Coke Company in the first quarter of the twentieth century, see W. G. C. Rice, *Seventy-five Years of Gas Service in Chicago* (Chicago: Peoples Gas Light and Coke Company, 1925), 39–45. For a closer look, see the full-blown investigation in 1915–17 by the city council and its special counsel, Donald Richberg, in Chicago, *Proceedings of the City Council*, 1916–17, 1297–1305, 1916–17, 4325–32. For Insull's increasing role in the affairs of the gas company, see McDonald, *Insull*, 205–8. The impact of a national coal shortage during World War I on Chicago's electric and gas companies will be examined in chapter 8.

59. Chicago, *Proceedings of the City Council*, 1905–6, 1533–34, 2356–66.

60. Ibid., 2373 for the quotation, 2356–74. The rate figures used here differ slightly from those in the report. Cowderly and Humphreys added 10¢ MCF for a depreciation reserve, but Bemis did not include this category in his proposal. Since a prudent reserve of this type soon became an accepted practice in the utility industry, I have taken the liberty of adding a like amount to Bemis's figures to make the experts' figures comparable. The reports of the experts include some calculations of the immediate savings in fuel from the reform.

61. Chicago, *Proceedings of the City Council*, 1905–6, 2389–93, 2409–12, 2626, 2635–39; the text of the final ordinance is found at 2629–32.

62. Rice, *Seventy-five Years*, 38–45; Chicago, *Proceedings of the City Council*, 1916–17, 1279–1304, 4325–32A.

63. For example, see "Water Meters in Favor," *Chicago Tribune*, 5 June 1906; "Cables to Stop; Tunnels to Sink," *Chicago Tribune*, 12 June 1906; "Fisher Makes Draft for Car Settlement," *Chicago Record Herald*, 8 June 1906; "The Phone Grab," *Chicago Tribune*, 2 June 1907; "Phone Agitation to Bring Changes," *Chicago Tribune*, 6 June 1907. Also see the sources on the traction settlement and charter convention listed above, note 5.

64. See Commonwealth Electric Company, Board of Directors, Meetings of 5 May 1905, 6 July 1906, 11 June 1907, in CEC-HA, box 300, for insight into the links between the rate investigation and voluntary rate cuts by the utility company.

65. Chicago, *Proceedings of the City Council*, 1905–6, 3221 for the Beale quotation, 3226–31 for the report of the engineers.

66. Ibid., 3218 for the quotation, 3226–30 for the engineers' concurring opinions. Also see ibid., 1908, 4175–94, for a similar legal opinion of the first assistant corporation counsel, George Miller.

67. Ibid., 1905–6, 3219–26. The company asked for several minor concessions in addition to the major proposal for city permission to consolidate the two companies. As will be noted immediately below, Edison attorney Beale was the primary author of the resulting ordinance-contract. See "Talk Electricity at the City Club," *Chicago Tribune*, 8 March 1908.

68. "Commonwealth Edison Again," *Chicago Record Herald*, 13 February 1908. Also see "The Electric Ordinance," ibid., 7 June 1906; "They Have Been Warned," ibid., 18 June 1906; "The Edison Company," ibid., 26 January 1907; "Discrimination and Mr. Miller," ibid., 6 December 1907; "Flaw in Light Plan," ibid., 9 March 1908.

69. Chicago, *Proceedings of the City Council*, 1906–7, 854 for the Dunne quotation, 451–62, 813–16 for the passage of the ordinance, and 852–54 for the Dunne veto message and the vote of thirty-eight to thirty-one to override the veto. The ordinance provided for a rate that gradually fell from a net high of 14¢ per kwh and 10¢ per kwh for primary and secondary charges in the first year to a low of 12¢ and 7¢ in the fourth and fifth years. Also see Dunne's explanation of his veto in "Dunne Raps Light Law," *Chicago Record Herald*, 11 June 1906; "May Alter Light Law," *Chicago Record Herald*, 18 June 1906; and "Electric Is Dead; Trolleys Safe," *Chicago Tribune*, 19 June 1906.

70. "Talk Electricity at the City Club," *Chicago Tribune*, 8 March 1908, for the revelation, the Beale quotation, and the Insull quotation; "The Corrupt and the Blind," *Chicago Record Herald*, 15 June 1906, for the editorial quotation.

71. Chicago, *Proceedings of the City Council*, 1908, 4151–57, 4175–4201, 4662–75; "Discuss Electric Act," *Chicago Tribune*, 12 March 1908; "Light Bill Peppered: Club Report Stings," *Chicago Record Herald*, 12 March 1908; "To Change Light Bill," *Chicago Record Herald*, 18 March 1908; "Vote on Light Bill: Adoption Imminent," *Chicago Record Herald*, 24 March 1908.

72. *Chicago Record Herald*, 7 March 1908, for the quotation.

73. "The Commonwealth Edison Ordinance," *Chicago Record Herald*, 25 May 1908, for the quotation. For the concerns of the outlying districts, see "More Light Bill Raps," *Chicago Record Herald*, 16 February 1908; "Aldermen Adopt Light Ordinance," *Chicago Tribune*, 24 March 1908.

74. John Gilchrist, "Report of the Rate Research Committee," paper presented to the National Electric Light Association, 1911, in Gilchrist, *Public Utility Subjects*, 1:92–93.

75. National Civic Federation, *Report on Municipal and Private Operation of Public Utilities*, 3 vols. (New York: National Civic Federation, 1907); Weinstein, *Corporate Liberalism*, 3–39. Gilchrist, reflecting on the trend toward state regulation by commission in 1911, urged fellow members of the industry to lead the reform. "It is certainly much safer," he asserted,

"that such a movement should be started by the National Electric Light Association than to have the initiative taken by the city government or commissions, who at the present time have but little organization for acting in concert" (Gilchrist, *Public Utility Subjects*, 1:92, for the quotation). For general perspectives on the growth of utility regulation at the state level, see Clyde Lyndon King, ed., *The Regulation of Municipal Utilities* (New York: Appleton, 1912), and Leo Sharfman, "Commission Regulation of Public Utilities: A Survey of Legislation," American Academy of Political Science and Society, *Annals* 53 (May 1914): 1–18. The origins of state utility regulation in Illinois will be discussed in chapter 7.

76. Insull, "Selling of Electricity in London and Chicago Compared," paper presented to the H. M. Byllesby and Company, 1911, in Keily, *Central-Station Service*, 168.

77. George Wise, "A New Role for Professional Scientists in Industry: Industrial Research at General Electric, 1900–1916," *Technology and Culture* 21 (July 1980): 408–29. Also see *Sargent and Lundy Story*, chaps. 2–3, which details the work of the engineering firm in cities other than Chicago.

78. For example, see E. W. Lloyd, "Compilation of Load Factors," NELA, *Proceedings*, 1909, 586–600; George McKana, "Significance of Statistics," NELA, *Proceedings*, 1910, 291–305; H. B. Gear, "Diversity Factor in the Distribution of Electric Light and Power," *Journal of the Western Society of Engineers* 15 (September–October 1910): 572–86. Also see E. J. Fowler, "History of the Statistical Department," manuscript, 1926, in CEC-LF, F23-E2; and NELA, *Electrical Solicitors' Handbook* (New York: NELA, 1909), in CEC-HA, box 9046.

79. Insull, "Larger Aspects," 81, for the quotation. The phrase "engineering of selling," was coined in the early 1920s, but it seems appropriate here. See Samuel Insull, "Address before the Western Society of Engineers," 1923, as quoted in William E. Keily, ed., *Public Utilities in Modern Life: Selected Speeches (1914–1923)* (Chicago: privately printed, 1924), 383.

Chapter Six

1. "Retail Center at Night," *Chicago: The Electric City* 1 (April 1903): 5, in CEC-LF (hereafter cited as *Electric City*). Also see Murray Melbin, "Night as Frontier," *American Sociological Review* 43 (February 1978): 3–22, for a fascinating perspective on the meaning of urban activities after dark.

2. "Electric Light and Power in Building Operations," *Electric City* 1 (November 1903): 2, for the quotation and story. For a complete guide to the buildings of the Loop and the construction techniques used in their erection, see Frank A. Randall, *History of the Development of Building Construction in Chicago* (Urbana: University of Illinois Press, 1949). The best source of visitor observations is Bessie Pierce, ed., *As Others See Chicago: Impressions of Visitors* (Chicago: University of Chicago Press, 1933).

3. Brian Cudahy, " Chicago's Early Elevated Lines and the Construction of the Union Loop," *Chicago History* 7 (Winter 1979–80): 194–205.

4. "Formal Opening of Another Great Store," *Electric City* 1 (November 1903): 2, and John Vinci, "Carson Pirie Scott: 125 Years in Business," *Chicago History* 8 (Summer 1979): 92–97. For perspectives on Sullivan's architecture, see Robert Twombly, *Louis Sullivan: His Life and Work* (New York: Viking, 1986), 378–84 passim; and Carl W. Condit, *The Chicago School of Architecture: A History of Commercial and Public Buildings in the Chicago Area, 1875–1925* (Chicago: University of Chicago Press, 1964), 160–74. For broader historical perspectives, see Russell Lewis, "Everything under One Roof: World's Fairs and Department Stores in Paris and Chicago," *Chicago History* 12 (Fall 1983): 28–47, and William R. Leach, "Transformations in a Culture of Consumption: Women and Department Stores, 1890–1925," *Journal of American History* 71 (September 1984): 319–42.

5. See "New Marshall Field Store," *Electric City* 1 (April 1903): 1, for a description of the store's electrical services. Also consider Thomas S. Hines, *Burnham of Chicago: Architect and Planner* (Chicago: University of Chicago Press, 1979 [1974]), 302–7, and Lloyd Wendt and Herman Kogan, *Give the Lady What She Wants: The Story of Marshall Field and Company* (Chicago: Rand McNally, 1952).

6. "The Shanghai Restaurant," *Electric City* 1 (October 1903): 3, 6.

7. Chicago Department of Electricity, *Annual Report*, 1907, for the number of movie projectors. I have relied on an interview with Robert Brubaker, 13 October 1987, for the history of movie theaters in Chicago. Brubaker is a curator at the Chicago Historical Society and an expert on the city's cultural history. Also see Robert Sklar, *Movie-Made America: A Cultural History of American Movies* (New York: Vintage, 1975), 7–14, on early motion picture technology and the introduction of movies in vaudeville performances, and Kathleen D. McCarthy, "Nickel Vice and Nickel Virtue: Movie Censorship in Chicago, 1907–1915," *Journal of Popular Film and Television* 5 (1976): 37–55, on middle-class reactions to the new popular entertainment. On the emergence of more respectable movie theaters, see George D. Bushnell, "Chicago's Magnificent Movie Palaces," *Chicago History* 6 (Summer 1977): 99–106. For a broad perspective on the transformation of urban popular entertainment at the turn of the century, see Lewis A. Erenberg, "Ain't We Got Fun?" *Chicago History* 14 (Winter 1985–86): 4–21.

8. Imogene E. Whetstone, "Historical Factors in the Development of Northern Illinois and Its Utilities," typescript, Chicago, Public Service Company of Northern Illinois, 1928, 67–74, in CEC-HA, box 9003. A much closer examination of the electric utilities of the North Shore will follow in the next chapter.

9. "West Side Street Fair," *Electric City* 1 (September 1903): 7. A worker in the construction department of the utility, C. B. Crothers, noted that the first electric light festoons at a street carnival were used a year earlier at the South Chicago street fair. See C. B. Crothers, "Construction Department," manuscript, 22 April 1913, in CEC-LF, F3-E1. Also see Perry Duis, "Whose City? Public and Private Places in Nineteenth-Century Chi-

cago," *Chicago History* 7 (Spring 1983): 2–27, and 7 (Summer 1983): 2–23; Sklar, *Movie-Made America*, 16–32; and McCarthy, "Nickel Vice," on the nickelodeon and middle-class reactions against this working-class form of entertainment. June Sawyers, "The Way We Were," *Chicago Tribune*, 10 October 1987, sect. 10, p. 9, confirms that as early as 1907, the respectable press was condemning the nickelodeon. "The 5-cent theater," the *Tribune* complained, "[is] without a redeeming feature to warrant [its] existence." On the origins of the amusement park in Chicago, see Al Griffin, "The Ups and Downs of Riverview Park," *Chicago History* 4 (Spring 1975): 14–22, and Chuck Wlodarczyk, *Riverview: Gone but Not Forgotten, a Photo-History, 1904–1967* (Evanston, Ill.: Schori, 1977). Also consider John F. Kasson, *Amusing the Millions: Coney Island at the Turn of the Century* (New York: Hill and Wang, 1978).

10. U.S. Census, Population, 1900, 1, pt. 1:116, and 1, pt. 2:606. On the desire of the middle class to escape from the turmoil of Chicago during the late nineteenth century, see Richard Sennett, *Families against the City: Middle Class Homes of Industrial Chicago, 1872–1890* (Cambridge: Harvard University Press, 1970); Gwendolyn Wright, *Moralism and the Model Home: Domestic Architecture and Cultural Conflict in Chicago, 1873–1913* (Chicago: University of Chicago Press, 1980); Robert C. Twombly, "Saving the Family: Middle Class Attraction to Wright's Prairie House, 1901–1909," *American Quarterly* 27 (March 1975): 57–72; and Margaret Marsh, "From Separation to Togetherness: The Social Construction of Domestic Space in American Suburbs, 1840–1915," *Journal of American History* 76 (September 1989): 506–27. For a useful introduction to the process of suburbanization, see Kenneth T. Jackson, *The Crabgrass Frontier: The Suburbanization of the United States* (New York: Oxford University Press, 1985). For two early examples in Chicago, see Barbara Posadas, "A Home in the Country: Suburbanization in Jefferson Township, 1870–1889," *Chicago History* 7 (Fall 1978): 132–49; and Barbara Posadas, "Suburb into City: The Transformation of Urban Identity on Chicago's Periphery: Irving Park as a Case Study, 1870–1910," *Journal of the Illinois State Historical Society* 76 (Autumn 1983): 162–76. On the emergence of a "flat city" of middle-class residential areas, see Wim De Wit, "Apartment Houses and Bungalows: Building the Flat City," *Chicago History* 12 (Winter 1983–84): 18–29; and on subdivision planning in Chicago, see two brilliant contributions by Ann Durkin Keating, "From City to Metropolis: Infrastructure and Residential Growth in Chicago," *Essays in Public Works History* 14 (December 1985): 3–27, and Ann Durkin Keating, "Real Estate Developers and the Emergence of Suburban Government: The Case of Nineteenth Century Chicago," paper presented at the annual meeting of the American Historical Society, 1986.

11. Sam Bass Warner, Jr., *Streetcar Suburbs: The Progress of Growth in Boston, 1870–1900* (Cambridge: Harvard University Press, 1962), 154, for the quotation, and Michael H. Ebner, *Creating Chicago's North Shore: A Suburban History* (Chicago: University of Chicago Press, 1988).

12. Warner, *Streetcar Suburbs*, 155.

13. Herbert J. Gans, "Urbanism and Suburbanism as Ways of Life: A Reevaluation of Definitions," in *American Urban History*, 2d ed., ed. Alexander B. Callow, Jr. (New York: Oxford University Press, 1973), 507–21; Kenneth T. Jackson, "Urban Deconcentration in the Nineteenth Century: A Statistical Inquiry," in *The New Urban History*, ed. Leo F. Schnore (Princeton: Princeton University Press, 1975), 110–42. For perspectives on the case of the North Side of Chicago, compare Michael P. McCarthy, "Chicago, the Annexation Movement, and Progressive Reform," in *The Age of Urban Reform*, ed. Michael H. Ebner and Eugene M. Tobin (New York: Kennikat, 1977), 43–54, and Michael H. Ebner, "The Result of Honest Hard Work: Creating a Suburban Ethos for Evanston," *Chicago History* 13 (Summer 1984): 48–65.

14. For a useful taxonomy of urban space, see Duis, "Whose City?"

15. Investigations of housing conditions in Chicago were led by the social settlement movement. For an introduction, see Allen F. Davis, *Spearheads for Reform: The Social Settlements and the Progressive Movement, 1890–1914* (New York: Oxford University Press, 1967), and Thomas Lee Philpott, *The Slum and the Ghetto: Neighborhood Deterioration and Middle Class Reform, Chicago, 1880–1930* (New York: Oxford University Press, 1978). For the reformers' direct testimony, see Jane Addams, *Twenty Years at Hull-House* (New York: Macmillan, 1910); [Robert Hunter], *Tenement Conditions in Chicago: Report by the Investigating Committee of the City Homes Association* (Chicago, 1901); and Sophonisba P. Breckinridge and Edith Abbott, "The Housing Problem in Chicago," reprinted from *American Journal of Sociology* 26–27 [September 1910–September 1911]: passim [copy at the Chicago Historical Society].

16. For example, see Joel Tarr, "From City to Suburb: The 'Moral' Influence of Transportation Technology," in Callow, *American Urban History*, 202–12; Glen E. Holt, "The Changing Perception of Urban Pathology: An Essay on the Development of Mass Transit in the United States," in *Cities in American History*, ed. Kenneth T. Jackson and Stanley K. Schultz (New York: Knopf, 1972), 324–43. The classic study on the effects of mass transit on land-use patterns is Homer Hoyt, *One Hundred Years of Land Values in Chicago* (Chicago: University of Chicago Press, 1933), 128–200.

17. On the improvement of the light bulb, see the contemporary reports by Francis W. Willcox, "The Practical Side of the Incandescent Lamp," National Electric Light Association, *Proceedings of the National Electric Light Association*, 1901, 132–79 (hereafter cited as NELA, *Proceedings*); and Francis W. Willcox, "Higher Efficiency Incandescent Lamps," NELA, *Proceedings*, 1906, 597–633. Also see Paul W. Keating, *Lamps for a Brighter America: A History of the General Electric Lamp Business* (New York: McGraw-Hill, 1954), 61–79, and John W. Hammond, *Men and Volts: The Story of General Electric* (New York: Lippincott, 1941), 259–66, 333–36. On the introduction of the tungsten-filament bulb and cluster fixture, see H. A. Seymour, "History of Commonwealth Edison Company, manuscript, 1934, Appendix D, in CEC-HA, box 9001; Commonwealth Edison Company, Reports

of the Advisory Committee, 21 (5 October 1908, 30 (19 October 1908), 58 (1 March 1909), 72 (6 April 1909), 87 (2 July 1909), 253 (27 December 1910), in CEC-HA, box 34. On the number of fixture rentals, see untitled manuscript, ca. 1917, in CEC-LF, F5-E2 (photocopy in my possession). On the impact of the tungsten-filament bulb in Chicago, see George E. McKana, "The Significance of Statistics," NELA, *Proceedings,* 1910, 291–303. McKana discounts utility operators' fears that the new bulb would reduce consumption and cut into profits. On the contrary, he asserts a faith in the gospel of consumption, a belief that better service at a lower price will result in higher levels of consumption. McKana also argues that "the relatively large amount of tungsten cluster [fixture] business should be noted, as all of the business has been obtained in the last two years and has proven a very effective method of competing with the gas arcs, and has helped also the introduction of tungsten lamps generally among the small store customers" (see NELA, *Proceedings,* 1910, 303, for the quotation).

18. Erenberg, "Ain't We Got Fun?" 12 for the quotation, 4–21 for a excellent review of the emergence of new forms of popular entertainment in Chicago from the World's Fair to the Great Depression; see also the sources listed above, notes 7 and 9. On the White City substation, see "Substation History (1887–1910)," in CEC-HA, box 177.

19. Jane Addams, *The Spirit of Youth and the City Streets* (New York: Macmillan, 1909), 85–86 for the quotation, 12, 78–79 for similar observations of nightlife in the city. Lizabeth Cohen warnes against assuming that the new institutions of mass culture produced a homogenized mass society. She shows how different class, ethnic, and racial groups experienced these popular entertainments in different ways. See Lizabeth Cohen, "Encountering Mass Culture at the Grassroots: The Experience of Chicago Workers in the 1920s," *American Quarterly* 41 (March 1989): 6–33.

20. On Chicago as sign capital and on the Harrison Street Station sign, see "Chicago: An Electric City," *Electrical Review* 52 (11 January 1908): 68. On the total number of electric signs in Chicago, see Chicago Department of Electricity, *Annual Report,* 1902–12, passim.

21. Unfortunately there is no precise series of statistics on the number of residential customers before 1920, except for a few reports that cover a limited number of years between 1908 and 1916. Instead, the statistics lump residential and commercial lighting customers together. The figures used here were extrapolated from a comparison of these two sets of statistics. For statistics from 1898 to 1922, see the report of the Statistical Department, Commonwealth Edison Company, manuscript, 2 February 1922, in CEC-LF, F23-E3. For a statistical series from 1908 to 1933, see Seymour, "History of Commonwealth Edison," Appendix D. For statistics on the company's residential customers between 1908 and 1916, see Commonwealth Edison Company, Departmental Correspondence, E. Fowler to J. Gilchrist, 4 December 1916, in CEC-LF, F23-E4 (some of the data in this report are reproduced as table 15. The method of extrapolation used

here receives some confirmation from a report in May 1905 that lists 14,000 residential customers. The estimated figures for 1905 compare favorably with this number. The year begins with 12,650 residential customers and ends with 17,150. The number of dwellings in 1911 was extrapolated from U.S. Census, Population, 1910, 2:512, 1920, 3:274. For an introduction to the problems of defining an urban middle class, see Stephen Thernstrom, *The Other Bostonians: Poverty and Progress in the American Metropolis, 1880–1970* (Cambridge: Harvard University Press, 1973), 47–59, 289–302.

22. Ernest F. Smith, "The Development, Equipment, and Operation of the Sub-station and Distributing Systems of the Chicago Edison Co. and Commonwealth Electric Co.," *Journal of the Western Society of Engineers* 8 (March–April 1903): 209–22. Half of the utility's twenty-two substations served this area. On the distributor grid in 1908, see "The Commonwealth Edison Company," *Electrical Review* 52 (11 January 1908): 67.

23. See H. B. Gear, "Diversity Factor in the Distribution of Electric Light and Power," *Journal of the Western Society of Engineers* 15 (September–October 1910): 586. For an excellent description of land-use patterns on the North Side of Chicago, see James Leslie Davis, *The Elevated System and the Growth of Northern Chicago* (Evanston, Ill.: Northwestern University Press, 1965). For insight into the social and political composition of the North Side, see McCarthy, "Chicago, the Annexation Movement." Also see the sources listed above, note 16.

24. Davis, *Elevated System,* 11–19 passim.

25. See above, chapter 4, for a description of Edgewater. On the origins of the community builders, see Marc A. Weiss, "Planning Subdivisions: Community Builders and Urban Planners in the Early Twentieth Century," *Essays in Public Works History* 15 (September 1987): 21–46, and Marc A. Weiss, *The Rise of the Community Builders: The American Real Estate Industry and Urban Land Planning* (New York: Columbia University Press, 1987).

26. Keating, "From City to Metropolis," 3–27.

27. Ibid., for examples of planned subdivisions in Chicago. For a case study of a mixed neighborhood, see Posadas, "Home in the Country." On patterns of housing by rich and poor, see the fine study of Milwaukee by Roger Simon, *The City-Building Process: Housing and Services in New Milwaukee Neighborhoods, 1880–1910,* Transactions of the American Philosophical Society 68, pt. 5 (Philadelphia: American Philosophical Society, 1978). For the workingman's fixtures, see Commonwealth Edison Company, Reports of the Advisory Committee, September–October 1912, in CEC-HA, box 34.

28. On the links between the World's Fair and the Chicago plan of 1909, see Thomas S. Hine, *Burnham of Chicago: Architect and Planner* (Chicago: University of Chicago Press, 1974), 92–124, 312–45. The Victorians created a concept of "separate spheres" to divide gender roles. See Barbara Welter, "The Cult of True Womanhood: 1820–1860," in *The American Family in Social-Historical Perspective,* ed. Michael Gordon (New York: St. Martin's, 1973), 224–50.

29. See the sources listed above, note 10.

30. "Electrical and Mechanical Equipment of the Blackstone Hotel, Chicago," *Electrical World* 54 (November 1909): 1099–1101. Other Chicago hotels were making similar intensive use of electricity. For example, see the description of the Palmer House and the Plaza Hotel in *Electrical World* 54 (September 1909): 723. For a description of the Sherman House, see Commonwealth Edison Company, Board of Directors Meeting, 18 January 1910, in CEC-HA, box 300. On the role of the hotel as the innovator of household technology, see Stephen Davis, "'Of the Class Denominated Princely': The Tremont House Hotel," *Chicago History* 11 (Spring 1982): 26–36, and Gwendolyn Wright, *Building the Dream: A Social History of Housing in America* (New York: Pantheon, 1981), 135–51.

31. On the origins of the modern apartment building, see Carroll William Westfall, "Home at the Top: Domesticating Chicago's Tall Apartment Buildings," *Chicago History* 14 (Spring 1985): 20–39; and De Wit, "Apartment Houses and Bungalows." On the new homes of the elite, see Celia Hillard, "'Rent Reasonable to Right Parties': Gold Coast Apartment Buildings," *Chicago History,* 8 (Summer 1979): 66–77. Also see Wright, *Building the Dream,* 135–52.

32. Wright, *Building the Dream,* 160 for the quotation, 158–76 for a larger discussion of the bungalow. Also see Daniel J. Prosser, "Chicago and the Bungalow Boom of the 1920s," *Chicago History* 10 (Summer 1981): 86–95. For perspectives on the role of "masculine domesticity" in the new architecture of the single-family dwelling at the turn of the century, see Marsh, "From Separation to Togetherness," 506–27.

33. *The Electrical Solicitors' Handbook* (New York: NELA, 1909), 1:24. A copy of this book is found in CEC-HA, box 9046.

34. Commonwealth Edison Company, Department Correspondence, 4 December 1916, for household statistics. Also see McKana, "Significance of Statistics," 291–303; and E. W. Lloyd, "Compilation of Load Factors," NELA, *Proceedings,* 1909, 586–600. On the slow development of the modern electrical plug, see Fred E. H. Schroeder, "More 'Small Things Forgotten': Domestic Electrical Plugs and Receptacles, 1881–1931," *Technology and Culture* 27 (July 1986): 525–43.

35. Henry B. Fuller, *The Cliff Dwellers* (Ridgewood, N.J.: Gregg, 1968 [1893], 134, for the first quotation; Gear, "Diversity Factor," 582, for the second quotation. By the late 1890s, observers of household routines noticed that "electricity is making its way into the kitchen through the parlor and dining room." See Anna Leach, "Science in the Model Kitchen," *Cosmopolitan* 27 (May 1899): 96, for the quotation.

36. James Ayer, "Electric Heating and the Residence Customer," NELA, *Proceedings,* 1906, 192–93, for the quotation. See Charles A. Thrall, "The Conservative Use of Modern Household Technology," *Technology and Culture* 23 (April 1982): 175–94, for recent sociological insights on the effects of new technology in the home. Also useful is an earlier study, Joann Vanek, "Household Technology and Social Status: Rising Living Standards

and Status and Residential Differences in Housework," *Technology and Culture* 19 (July 1978): 361–75.

37. H. B. Gear, "A Survey of Electrical Appliances Used in Chicago Residences," manuscript, March 1923, in CEC-LF, F4A-E2. This report was part of a larger national survey of eight cities. See "Survey of Residence Electrical Installations: Part II," report presented at the annual meeting of the Association of Edison Illuminating Companies, September 1923, in *Minutes*, 609–37. A more careful analysis of these surveys will follow in later chapters. For insight on the primitive state of electrical appliances, see Leach, "Science in the Model Kitchen," 95–104.

38. David M. Katzman, *Seven Days a Week: Women and Domestic Service in Industrializing America* (New York: Oxford University Press, 1978), 125, for the quotation. Other scholars confirm that household servants considered laundry the worst job. See, for example, Faye E. Dudden, *Serving Women: Household Service in Nineteenth Century America* (Middletown, Conn.: Wesleyan University Press, 1983), 131–54, and the compelling description of washday in Robert A. Caro, *The Years of Lyndon Johnson: The Path to Power* (New York: Vintage, 1981), 502–15. The fan was the first appliance using a small motor to gain significant consumer sales. Obviously, fans were most popular in cities with hot, humid climates, such as New Orleans and Houston. See *Electrical Review* 38 (9 February 1901): 187–205, for several stories on the history and contemporary use of fans.

39. Ayer, "Residential Customer," 195, for the quotation. Cf. Gear, "Survey," 7, where he shows that as late as the 1920s "[there] is direct evidence that irons are used by many during the summer who use other heat during the cooler months."

40. William E. Keily, "History of the Electric Shop," manuscript transcript, 29 July 1920, in CEC-LF, F4B-E4. A photographic survey of the store is included in this file.

41. Ibid.; "Electric Shop—Chicago," sales booklet, 1909, in CEC-LF, F4B-E4. Also see the Christmastime copy, "A Christmas Gift for the Whole Family," *Chicago Tribune*, 8 December 1909, 17; and "Christmas Gifts Such as These Become Family Heirlooms," *Chicago Tribune*, 12 December 1909.

42. Samuel Insull, "Massing of Energy Production an Economic Necessity," paper presented to the General Electric Company, 25 February 1910, reprinted in William E. Keily, ed., *Central-Station Service . . . as Set Forth in the Public Addresses (1897–1914) of Samuel Insull* (Chicago: privately printed, 1915), 140, for the quotation. For similar statements, see ibid., 116, 168.

43. Ibid., 95–102. In the 1900s Insull saw advertising primarily as an educational vehicle, not as means to persuade the householder through envy and other psychological tactics. Insight on the educational role of utility companies is provided by a report of Insull's publicity consultant, William E. Keily, on the magazine, *Popular Electricity*. Keily first underscored the need for education among the "great middle classes . . . the great non-electrical public, a large portion of which feels a certain fascination and a rather fearful admiration for electricity." The new magazine

would help in overcoming this apprehension. "The great field of *Popular Electricity* is educating the masses of people—men, women and young people—in the manifold uses to which electricity may be put in the home, in the small shop, on the farm—everywhere it touches the daily lives of the average citizen and his family." Significantly, Keily focused on young boys, "an unfailing corp of readers . . . [because they are] the buyers of the future." See William E. Keily, "Report on *Popular Electricity-Magazine*," manuscript, 10 October 1908, in the Samuel Insull Papers, Loyola University of Chicago, box 84.

44. On the history of the advertising department, see the narrative history and materials in CEC-LF, F4. Between 1901 and 1908, Insull almost doubled the amount of money spent on getting more business. For each dollar of income, the amount spent on advertising increased from 0.375¢ to 1.97¢, while soliciting expenses declined from 2.2¢ to 1.55¢. Combining the two promotional expenses, the amount spent increased from 2.2¢ to 3.6¢. See Commonwealth Edison Company, Reports of the Advisory Committee, 50 (1 February 1909). For a brief description of the sales force, see "Chicago: An Electrical City," 69.

45. *Chicago Tribune*, 6 January 1902, 10 for the first quotation, 6 December 1909, 18.

46. Seymour, "History of Commonwealth Edison," Appendix D; F. H. Bernhard, "New Features of Central-Station Service in Chicago," *Electrical Review* 54 (29 May 1909): 976, for the 1908 iron campaign. The scheme worked well enough that the company repeated the offer four years later. See Commonwealth Edison Company, Reports of the Advisory Committee, 426 (15 April 1912). The irons were sold for $3 each, which covered the $2.25 cost of the appliance plus the promotional expenses. Also see "Thirty Days Free Trial," *Electric City* 10 (August 1912): 31; and "Marketing over Thirty Tons of G-E Electric Flatirons," *Electrical Merchandise* 11 (April 1912): 341. Significantly, this marked the end of the era of the giveaway. See Frank B. Rae, Jr., "Reforming the Appliance Policy," *Electrical Merchandise* 11 (April 1912): 195–96.

47. Louis Ferguson, in NELA, *Proceedings*, 1898, 331, for the first quotation; F. H. Golding, "How to Get the Old Buildings Wired," NELA, *Proceedings*, 1907, 71, for the second quotation. Although Golding came from Dayton, Ohio, his conclusions were similar to those of salesmen in Chicago. This approach received official sanction by NELA in its 1909 publication, *Electrical Solicitors' Handbook*.

48. Gear, "Diversity Factor," 576–81.

49. In 1898 Insull set up car-battery-charging stations and established a special low rate. See Seymour, "History of Commonwealth Edison," 280; the materials collected in the transportation file of the CEC-LF, F3; and "New Electric Truck Service Ready in Chicago," *Electrical Review* 71 (29 December 1917): 1091–94. Beginning in 1903, Insull became involved with a manufacturer of electric vehicles, the Walker Vehicle Company. See CEC-LF, F3A-E4. He was also involved in the "lead cab trust" to create a taxi

monopoly. See John Rae, *American Automobile Manufacturers: The First Forty Years* (Philadelphia: Chilton, 1959), 67–71. Efforts to revive the electric vehicle in Chicago during and after World War I will be examined more carefully in chapter 9.

50. *Chicago Tribune*, 2 December 1909 for the first quotation, 9 December 1909 for the second quotation. On the cost of operating gas and electric heating devices, see Ayer, "Residential Customer," 197–201; and NELA, *Electrical Solicitors' Handbook*, 2:39–57, 4:24–38. For a history of the electric stove and an explanation of its failure, see Thomas C. Martin and Stephen L. Coles, eds., *Story of Electricity*, 2 vols. (New York: Story of Electricity, 1919), 1:460–71.

51. See Commonwealth Edison Company, Department Correspondence, 4 December 1916, for average use and cost data. In 1908 Chicagoans living in flats paid an average of $1.59 per month for service, while homeowners paid $3.02.

52. Seymour, "History of Commonwealth Edison," 176, 325, and Appendix D; and Commonwealth Edison Company, Statistics Department, untitled manuscript, 10 March 1926, in CEC-LF, F23-E3, for electrical consumption statistics. See [J. F. Rice], *History of the Commonwealth Edison Company* (Chicago: Commonwealth Edison Company, 1952), 59, for coal statistics. U.S. Census, Population, 1910, 2:512, 1920, 3:274.

Chapter Seven

1. Compare Joel Tarr, "From City to Suburb: The 'Moral Influence' of Transportation Technology," in *American Urban History*, 2d ed., ed. Alexander B. Callow, Jr. (New York: Oxford University Press, 1973), 202–12, and Glen E. Holt, "The Changing Perception of Urban Pathology: An Essay on the Development of Mass Transit in the United States," in *Cities in American History*, ed. Kenneth T. Jackson and Stanley K. Schultz (New York: Knopf, 1972), 324–43. For the case of Chicago, see Paul Barrett, *The Automobile and Urban Transit: The Formation of Public Policy in Chicago, 1900–1930* (Philadelphia: Temple University Press, 1983). For the phrase "crabgrass frontier," see the best synthetic treatment of the history of the suburbs, Kenneth T. Jackson, *The Crabgrass Frontier: The Suburbanization of the United States* (New York: Oxford University Press, 1985). On antiurban thought in American history, see Morton White and Lucia White, *The Intellectual versus the City: From Thomas Jefferson to Frank Lloyd Wright* (Cambridge: Harvard University Press, 1962). On residential segregation as a major force in American suburbanization, see Frank J. Coppa, "Cities and Suburbs in Europe and the United States," in *Suburbia: The American Dream and Dilemma*, ed. Philip Doce (Garden City, N.Y.: Anchor, 1976), 167–92, and Sam Bass Warner, Jr., *The Urban Wilderness: A History of the American City* (New York: Harper and Row, 1972), 15–38. On the emergence of metropolitan forms of the city, see Edward K. Spann, *The New Metropolis: New York City, 1840–1857* (New York: Columbia University Press, 1981). Chi-

cago had many of the characteristics Spann noted for the late nineteenth century. Perhaps "metropolis of the twentieth century" is a more accurate description of the urban form that emerged after the turn of the century.

2. Gwendolyn Wright, *Moralism and the Model Home: Domestic Architecture and Cultural Conflict in Chicago, 1873–1913* (Chicago: University of Chicago Press, 1980), 235, for the quotation. For complementary perspectives, see Margaret Marsh, "From Separation to Togetherness: The Social Construction of Domestic Space in American Suburbs, 1840–1915," *Journal of American History* 76 (September 1989): 506–27. Also see the seminal essay by Robert C. Twombly, "Saving the Family: Middle Class Attraction to Wright's Prairie House, 1901–1909," *American Quarterly* 27 (March 1975): 57–72.

3. Public Service Company of Northern Illinois, "Report of the Public Service Company of Northern Illinois," 31 December 1912, in CEC-HA, box 4222 (hereafter cited as PSCNI).

4. For an introduction to the electrification of Chicago's suburbs, see Imogene Whetstone, "Historical Factors in the Development of Northern Illinois and Its Utilities" typescript, Chicago, PSCNI, 1928, in CEC-Library. For important insights on the growth of a modern infrastructure in the downtown districts of these communities, see Carole Rifkind, *Main Street: The Face of Urban America* (New York: Harper and Row, 1977).

5. A. B. Fitzgerald, quoted in Whetstone, "Historical Factors," 74.

6. See Jackson, *Crabgrass Frontier,* 3–11, for discussion of the difficulties involved in defining the suburb.

7. See Michael P. Conzens, "The Historical and Geographical Development of the Illinois and Michigan Canal–National Heritage Corridor," in *The Illinois and Michigan Canal–National Heritage Corridor: A Guide to Its History and Sources,* ed. Michael P. Conzens and K. J. Carr (De Kalb: Northern Illinois University Press, 1988), chap. 1, for current perspectives on the regional integration of the Chicago area. Also see two valuable contemporary accounts, Graham R. Taylor, *Satellite Cities: A Study of Industrial Suburbs* (New York: Arno, 1970 [1915]), and Helen Corbin Monchow, *Seventy Years of Real Estate Subdividing in the Region of Chicago* (Evanston, Ill.: Northwestern University Press, 1939), 88–118.

8. Monchow, *Seventy Years;* Michael H. Ebner, "The Result of Honest Hard Work: Creating a Suburban Ethos for Evanston," *Chicago History* 13 (Summer 1984): 49.

9. On the triumph of antiannexation sentiment at the turn of the century, see Ebner, "Result of Honest Hard Work"; Michael P. McCarthy, "Chicago, the Annexation Movement and Progressive Reform," in *The Age of Urban Reform: New Perspectives on the Progressive Era,* ed. Michael H. Ebner and Eugene M. Tobin (Port Washington, N.Y.: Kennikat, 1977), 43–54; Ann Durkin Keating, "Real Estate Developers and the Emergence of Suburban Government: The Case of Nineteenth Century Chicago," paper presented at the annual meeting of the American Historical Association, Chicago, 1986; and Jackson, *Crabgrass Frontier,* 138–56. On the fragmentation

of local government, see Jon C. Teaford, *City and Suburb: The Political Fragmentation of Metropolitan America, 1850–1970* (Baltimore: Johns Hopkins University Press, 1979); Milton Rakove, *Don't Make No Waves—Don't Back No Losers: An Insider's Analysis of the Daley Machine* (Bloomington: Indiana University Press, 1975), 198–284; and Anthony Downs, *Opening up the Suburbs: An Urban Strategy for America* (New Haven: Yale University Press, 1973).

10. On the middle-class appeal of Wright's homes, see Twombly, "Saving the Family," 59; the testimony of the architect's clients in Leonard K. Eaton, *Two Chicago Architects and Their Clients: Frank Lloyd Wright and Howard Van Doren Shaw* (Cambridge: MIT Press, 1969), 65–134; and the general perspectives on the suburban school provided by Harold Allen Brooks, *The Prairie School: Frank Lloyd Wright and His Midwest Contemporaries* (Toronto: University of Toronto Press, 1972). On the timing of housing reform, most scholars agree that the period from 1901 to 1909, when Wright left Oak Park, was the formative era of modern suburban architecture. See Robert C. Twombly, *Frank Lloyd Wright: An Interpretive Biography* (New York: Harper and Row, 1973), 31, 51–89, as well as Eaton, *Two Chicago Architects,* 5, who argues that the period 1890–1913 represents a third great "revolution" in the architecture of Western civilization. Hull House became a center of housing reformers. See Thomas Lee Philpott, *The Slum and the Ghetto: Neighborhood Deterioration and Middle Class Reform, Chicago, 1880–1930* (New York: Oxford University Press, 1978), 89–109.

11. See Frank Lloyd Wright, *An Autobiography* (New York: Duell, Sloan, and Pearce, 1943), 141–53; Robert Crunden, *Ministers of Reform: The Progressives' Achievement in American Civilization, 1889–1920* (New York: Basic, 1982), 151–52; Twombly, "Saving the Family," 57–72.

12. Frank Lloyd Wright, "The Art and Craft of the Machine," in *Frank Lloyd Wright: Writings and Buildings,* ed. Edgar Kaufman and Ben Raeburn (New York: Horizon, 1960), 53–73; Wright, *Autobiography,* 131–32; and Crunden, *Ministers of Reform,* 133–49, for a discussion of the concept "innovative nostalgia" as it applies to Wright; Robert Fishman, *Urban Utopias in the Twentieth Century: Ebenezer Howard, Frank Lloyd Wright, Le Corbusier* (New York: Basic Books, 1977), 91–110, for an evaluation of the Hull House paper and its larger context within the arts and crafts movement in Chicago. Also see T. J. Jackson Lears, *No Place of Grace: Antimodernism and the Transformation of American Culture, 1880–1920* (New York: Pantheon, 1981), 83–96, on antimodernism, the arts and crafts movement, and the role of technology in modern life. On middle-class anxieties at the turn of the century, see Richard Hofstadter, *The Age of Reform: From Bryan to F.D.R.* (New York: Vintage, 1955), 131–85. For perspectives on the antiurban manifestations of these anxieties, see White and White, *Intellectual versus the City,* 147–78, 189–220, and Peter J. Schmitt, *Back to Nature: The Arcadian Myth in Urban America, 1900–1930* (New York: Oxford University Press, 1969), 3–19.

13. Twombly, "Saving the Family," 59.

14. Wright, *Autobiography*, 79, for the quotation, which referred to Oak Park. For insight on Wright's Hull House paper, see Fishman, *Urban Utopias*, 91–114. For a complete list of Wright's houses, see William Allen Storrer, *The Architecture of Frank Lloyd Wright: A Compete Catalog* (Cambridge: MIT Press, 1974). Wright's revolt against the city helps account for his noticeable lack of success in attracting commissions to design commercial high-rise structures in downtown areas, yet several of his housing clients were manufacturers and inventors involved in complex technological processes. See Eaton, *Two Chicago Architects*, 41–52.

15. Henry Demarest Lloyd, as quoted in Wright, *Moralism and the Model Home*, 235. On Wright's defense of the machine as an agent of democratic values, see Fishman, *Urban Utopias*, 104–14, and Cruden, *Ministers of Reform*, 155–61. On Lloyd's role as a suburban reformer, see Michael Ebner, "Henry Demarest Lloyd's Winnetka," *Chicago History* 15 (Fall 1986): 20–29. Significantly, both Wright and Lloyd were members of Chicago's Art and Craft Society, which was centered at Hull House. For broader perspectives, see Tarr, "Moral Influence," 202–12; Jackson, *Crabgrass Frontier*, 157–71; and Wright, *Moralism and the Model Home*, 105–49. For insight on the women of Hull House, see Kathleen D. McCarthy, *Noblesse Oblige: Charity and Cultural Philanthropy in Chicago, 1849–1929* (Chicago: University of Chicago Press, 1982), 99–112, and Helen Lefkowitz Horowitz, "Hull-House as Women's Space," *Chicago History* 12 (Winter 1983–84): 40–55.

16. Wright, *Autobiography*, 141, where the architect describes how "the walls were beginning to go as an impediment to outside light and air." Also see Wright's reflections on his efforts to incorporate the most modern technology, materials, and construction techniques in his designs, in ibid., 139–49. On his clients' view of the casement window as an early "trademark" of his work, see Eaton, *Two Chicago Architects*, 104–5.

17. Wright, *Moralism and the Model Home*, 150–70; Dolores Hayden, *The Grand Domestic Revolution: A History of Feminist Designs for American Homes, Neighborhoods, and Cities* (Cambridge: MIT Press, 1981), 151–74. On the links between scientific management, domestic servants, and home economics, see Samuel Haber, *Efficiency and Uplift: Scientific Management in the Progressive Era, 1890–1920* (Chicago: University of Chicago Press, 1964); David P. Handlin, *The American Home: Architecture and Society, 1815–1915* (Boston: Little, Brown, 1979), 386–425; and Daniel E. Sutherland, *Americans and Their Servants: Domestic Service in the United States from 1800 to 1920* (Baton Rouge: Louisiana State University Press, 1981). Neither Wright nor his clients made many specific references to the kitchen, especially in comparison with the fireplace and the living room. The elimination of the parlor caused the most concern among his socially oriented clients. See Wright, *Autobiography*, 139–50, and Eaton, *Two Chicago Architects*, 65–134.

18. John Wynn, as quoted in Whetstone, "Historical Factors," 80, and ibid., 15–20, 73–86.

19. The origins of electrical service in the suburbs are surveyed in two related histories written in 1928 by employees of the PSCNI. The first, a manuscript without an author listed, formed the basis for the typescript

written by Imogene Whetstone. See "A History of the Public Service Company of Northern Illinois," ca. 1928, in CEC-HA, box 4025, and Whetstone, "Historical Factors," in CEC-HA, box 9003. The records of the predecessor companies of the PSCNI constitute a major resource of the company's historical archives. Reference to the materials of individual companies will be noted below as appropriate.

20. See Whetstone, "Historical Factors," 14–20, 73–78; *Important Papers of the North Shore Electric Company*, 3 vols. (Chicago: North Shore Electric Company, 1902–11), in CEC-HA, box 4025; and the records of the predecessor electric companies in CEC-HA, boxes 4220–24. For records of the Northwestern Gas Light and Coke Company between 1874 and 1913, see CEC-HA, boxes 4242–47.

21. Whetstone, "Historical Factors," 53–60, 75–78. On Walsh's tangled involvements in Chicago utilities and politics, see Joel Tarr, "John R. Walsh of Chicago: A Case Study of Banking and Politics, 1881–1905," *Business History Review* 40 (Winter 1966): 451–66. Walsh was also tied to Insull and became an original investor in Insull's suburban experiment in Highland Park. See Highland Park Electric Light Company, Corporate Records, 1890–1902, in CEC-HA, box 4207. In 1912 a special report by the United States Census Bureau underscored three major developments in the electric utility business. The trends noted were the consolidation of central stations "far beyond the original city limits" into unified networks, the spreading use of the holding company, and the increasing exploitation of waterpower sites to generate electricity. See Thomas C. Martin, "Part II: Technical," in U.S. Census, Special Report, Electric and Street Railways, 1912, 111–13.

22. Samuel Insull, "Memoirs of Samuel Insull," 76, in Samuel Insull papers, Loyola University of Chicago, box 17 for the quotation, 94–98 for Insull's account of his first encounters with Frank Baker. See Samuel Insull, "Massing of Energy Production an Economic Necessity," paper presented to the General Electric Company, Boston, 25 February 1910, reprinted in William E. Keily, ed., *Central Station Electrical Service: Its Commercial Development and Economic Significance as Set Forth in the Public Addresses (1897–1914) of Samuel Insull* (Chicago: privately printed, 1915), 136, for Insull's discussion with Emil Rathenau. For comparative perspectives on electrification of European cities, see P. Jundersfeld, "Electric Power Systems in Some European Cities," *Journal of the Western Society of Engineers* 9 (November–December 1906): 686–92, and Thomas P. Hughes, *Networks of Power: Electrification in Western Society, 1880–1930* (Baltimore: Johns Hopkins University Press, 1983). On Frank Baker (1864–1922), see Whetstone, "Historical Factors," 181–82, and his obituary in Public Service Company of Northern Illinois *News* 7 (January 1923), in the Insull Papers, box 74. Baker helped Frank J. Sprague prove the viability of electric traction in Richmond, Virginia. Arthur Lundy of the engineering firm of Sargent and Lundy was the only other Chicagoan to be involved in this historic effort.

23. Insull bought the Highland Park Electric Company for $92,350 and

sold it for $500,000 ($350,000 in cash and $150,000 in NSEC securities). See "History of the Public Service Company," 25–27; Highland Park Electric Light Company, Corporate Records, 1890–1902, in CEC-HA, box 4205.

24. Samuel Insull to Charles Coffin, letter of 21 May 1909, in the Insull Papers, box 1, for the first quotation; Coffin to Insull, letter of 19 May 1909, in ibid., for the second quotation; Insull to Coffin, letter of 24 May 1909, in ibid., for the third quotation. Also see Commonwealth Edison Company Board of Directors Meeting of 4 February 1908, in which Insull revealed that he was "anxious to tie them [the electric railway companies] up for, say, not less than ten years for the whole of their power," in CEC-HA, box 300. In December 1906 the NSEC signed up its first interurban line, the Chicago Milwaukee Electric Railroad Company. Copies of this contract and several others are found in CEC-HA, box 4023, file 871. Also see Central Electric Railfans' Association, *Interurban to Milwaukee* (Chicago: Central Electric Railfans' Association, 1962), 8–23, and William D. Middleton, *North Shore: America's Fastest Interurban* (San Marino, Calif.: Golden West Books, 1964), 9–61. On the purchase of the Cosmopolitan Electric Company, see the *Chicago Record-Herald*, 28 November, 3 December 1908, and Commonwealth Edison Company Board of Directors Meetings of 12 August, 25 September 1913, in CEC-HA, box 300. This central station electric company was set up by the political boss Roger Sullivan and a group of gas company investors. Insull's biographer, Forrest McDonald, agrees that 1907 marked a turning point in the utility executive's career. See Forest McDonald, *Insull* (Chicago: University of Chicago Press, 1962), 133.

25. John H. Volp, *The First Hundred Years, 1835–1935: Historical Review of Blue Island* (Blue Island, Ill.: Volp, 1935), 47–50; Louise Carroll Wade, *Chicago's Pride: The Stockyards, Packingtown, and Environs in the Nineteenth Century* (Urbana: University of Illinois Press, 1987); Mario Kijewski, David Brusch, and Robert Balanda, *The Historical Development of Three Chicago Millgates* (Chicago: Illinois Labor History Society, 1973); Monchow, *Seventy Years*, passim; William T. Huchinson, *Cyrus Hall McCormick*, 2 vols. (New York: Appleton-Century, 1935), 1:512–21; Stanley Buder, *Pullman: An Experiment in Industrial Order and Community Planning, 1880–1930* (New York: Oxford University Press, 1967).

26. Warner, *Urban Wilderness*, 85–112. On the one-story factory, see the suggestive ideas of John R. Stilgoe, *Metropolitan Corridor: Railroads and the American Scene* (New Haven: Yale University Press, 1983), 82–85. For an introduction to the reorganization of factory work, see Richard DuBoff, "The Introduction of Electric Power in American Manufacturing," *Economic History Review* 20 (December 1967): 509–18, and Warren D. Devine, "From Shafts to Wires: Historical Perspectives on Electrification," *Journal of Economic History* 43 (June 1983): 347–72. The effects of electricity on manufacturing will be the focus of the following chapter. Here industrial uses of energy are put in the context of the spatial and social effects of suburbanization.

27. Harvey Land Association, *The Town of Harvey: Manufacturing Suburb*

of Chicago (Chicago: Harvey Land Association, 1892), 22 for the first quotation, 36 for the second quotation; [Alec C. Kerr, ed.], *History: The City of Harvey* (Chicago: First National Bank in Harvey, 1962), 17, for the description "magic city." The early investors in the Harvey Land Association included the Northfield (Massachusetts) Institute, an organization affiliated with Chicago evangelist Dwight Moody, businessman-reformer Sidney Kent, sporting goods manufacturer A. G. Spalding, and the president of the Chicago Telephone Company, Henry B. Stone. See [Kerr], *History,* 15–18. The records of the land company's utility enterprises are contained in CEC-HA, box 4253. For perspectives on the industrial suburb in the wake of the failure of the Pullman experiment, see Gwendolyn Wright, *Building the Dream: A Social History of Housing in America* (Cambridge: MIT Press, 1983), 177–92. Harvey was not a lone exception. About six miles farther south, for example, Chicago Heights was developed after the annexation of 1889 by a similar group of investors who were led by Charles Wacker. See Whetstone, "Historical Factors," 91–93; CEC-HA, boxes 4252–53; Harold M. Mayer and Richard C. Wade, *Chicago: Growth of a Metropolis* (Chicago: University of Chicago Press, 1969), 186; and Monchow, *Seventy Years,* 98–118. For insight on subdivision developments, see Ann Durkin Keating, *Building Chicago: Suburban Developers and the Creation of a Divided Metropolis* (Columbus: Ohio State University Press, 1989). For a contemporary study of the planned industrial suburb with a focus on Gary, Indiana, see the book by one of Jane Addams's close associates, Taylor, *Satellite Cities.*

28. Volp, *Blue Island,* 163, 184–95.

29. Volp, *Blue Island,* 122–41, 163–66. On the aspirations of the working class for homeownership, see Michael J. Doucet and John C. Waver, "Material Culture and the North American House: The Era of the Common Man," *Journal of American History* 72 (December 1985): 560–87.

30. "Electric Service in Chicago Suburbs," *Electric World* 61 (7 June 1913): 1243–54; H. A. Seymour, "History of Commonwealth Edison," manuscript, 1934, CEC-HA, box 9001, Appendix D, 171. To compare the state of services in the suburbs and city, see "Electric Service in Chicago," *Electric World* 61 (31 May 1913): 1137–45. By 1909, a decade after Insull's presidential address to the National Electric Light Association, the very definition of the utility industry was undergoing change, refinement, and standardization as part of the institutionalization of Insull's gospel of consumption. For the new understanding of load factor, see E. W. Lloyd, "Compilation of Load Factors," *Proceedings of the National Electric Light Association,* 1909, 2:586–600 (hereafter cited as NELA, *Proceedings*).

31. "History of the Public Service Company," 49; "Station and Distribution Output," in CEC-HA, box 4227. Also see George H. Lukes, "Distribution in Suburban Districts," NELA, *Proceedings,* 1908, 67–87, and William Abbott, "Central Station Economies," *Journal of the Western Society of Engineers* 15 (January–February 1910): 41–42.

32. On residential rates, see Whetstone, "Historical Factors," 138; and

Seymour, "History of Commonwealth Edison," Appendix D, 83–84. On the output of the NSEC, see "History of the Public Service Company," 42. In comparison to the 21 million kwh sold in the suburbs by the NSEC in 1912, Commonwealth Edison generated 703 million kwh. See Seymour, "History of Commonwealth Edison," Appendix D, 176.

33. NSEC, *Annual Report,* 1907.

34. For the concept of unwitting metropolitan communities, see Sam Bass Warner, Jr., "Technology and Its Impacts on Urban Culture," paper presented at the Lowell Conference on Industrial History, Lowell, Massachusetts, 1983.

35. Mayer and Wade, *Growth of a Metropolis,* 186, 247–48; Carl Condit, *Chicago, 1910–1929: Building, Planning, and Urban Technology* (Chicago: University of Chicago Press, 1973), 42; Jerry A. Pinkepank, "The Belt Railway of Chicago: A Railroad's Railroad," *Trains* 26(September 1966): 36–46; Louis P. Cain, "The Creation of Chicago's Sanitary District and Construction of the Sanitary and Ship Canal," *Chicago History* 8 (Summer 1979): 98–111.

36. Monchow, *Seventy Years,* 88–98; Whetstone, "Historical Factors," 141–46; Public Service Company of Northern Illinois (PSCNI), *Annual Report,* 1912, in CEC-HA, box 4094.

37. The Chicago Suburban Light and Power Company (CSLPC) served Oak Park, River Forest, and the surrounding communities in Cicero township. With origins as a waterworks in the late 1880s, the CSLPC began supplying electricity in 1891, faced the competition from Yaryan's company at the end of the decade, and in 1906 started purchasing electricity wholesale from the NSEC. Nonetheless, in 1910 it provided residents with energy at an average rate of about 10¢ per kwh compared with a charge by the NSEC of almost 14¢. In addition, the CSLPC served over 25 percent of residents in its service territory, or more than twice the proportion of the NSEC customers. Even restricting the comparison of the NSEC to the most affluent North Shore district, Insull served about 13 percent of the residents. See "Study of the Properties Which Have Been Assembled to Form the Electrical Department of the Public Service Company of Northern Illinois with Special Reference to Rates and Hours of Service of the Former Plant," Illinois Commerce Commission Case 22353 (1933), Company Exhibit 2, in CEC-HA, box 4093. For a more complete analysis of this company, see CEC-HA, boxes 4228–34. It is beyond the purposes of this book to examine the history of suburban gas services in general and the NGLCC in particular. Commonwealth Edison's historical archives have an extensive collection of the company's records. See CEC-HA, boxes 4242–47. Unfortunately, historians have uncritically accepted the myth of the technological and economic superiority of the Insull system, a myth largely created by Insull himself. Until recently, the history of electrical energy in northern Illinois has been written by company employees; see the works cited here by Edkins, Seymour, and Whetstone. The most recent company history in this public relations genre is John Hogan, *A Spirit*

Capable: The Story of Commonwealth Edison (Chicago: Mobium, 1986). Hogan is the company's director of public communications. In 1962 a professional historian, Forrest McDonald, wrote the first account of Insull and his business leadership that was not sponsored by the company. However, *Insull* was hardly unbiased, since he worked closely with his subject's son, Samuel Insull, Jr. The result was an unbalanced business history that was heavily weighted in Insull's favor. Reviewers of the book were correct in highlighting McDonald's lack of objectivity. See, for example, the reviews by Thomas P. Hughes in *American Historical Review* 68(April 1963): 770–71, and by Wayne Andrews in the *Nation* 196(19 January 1963): 55–56. To a large extent, the best recent historical study of electrification also falls into the trap of the technological inevitability of regional power grids. See the award-winning work Hughes, *Networks of Power*, 1–17, passim. In contrast, the thesis presented here in the achievement of efficient, low-cost electricity lays first importance on the diversity of energy consumers, not the territorial extent of the distribution grid. As Insull himself maintained, the key to reducing the unit cost of electricity was managing the load to maximize the full-time use of the generating equipment. Whether farms and extractive industries outside the metropolitan area added significantly to the system's load factor is open to question. The case studies offered below suggest that several systems may have coexisted in the Chicago region.

38. For a survey of Joliet history, see Whetstone, "Historical Factors," 29–37. For a more detailed review, see "History of the Public Service Company," tables 14–17, and Economy Light and Power Company, "Corporate Record Book, 1890–1906," 5–179, in CEC-HA, box 4210 (hereafter cited as ELPC). Other statistics of the company are found in CEC-HA, boxes 4223–27, passim. On the history of the Illinois River valley, see Jones Piety, "The Illinois and Michigan Canal and the Early Historical Geography of Chicago," *Geographical Perspectives* 47(Spring 1981): 30–37, and Conzens, "Illinois and Michigan Canal."

39. For a contemporary description of the city's history, see *Joliet Daily News* 28 November 1911 [copy at the Chicago Historical Society]. For recent perspectives see Michael P. Conzen, ed., *Time and Place in Joliet: Essays on the Geographical Evolution of the City* (Chicago: Committee on Geographical Studies, University of Chicago, 1988). On the growth of industry and its use of electricity, see "Electric Service in Chicago Suburbs," 1249–51.

40. ELPC, "Corporate Record Book, 1890–1906," 179–219, in CEC-HA, box 4210. By the time the project was started in 1901, the price tag on the modernization program had risen to close to $1 million because of the addition of a new coal-fired station to the project. On Insull's network of financial and political allies, see McDonald, *Insull*, 82–95.

41. ELPC, "Corporate Record Book, 1890–1906," 197–211 for the 1901 power contract with the Joliet Railroad Company, 179–260 for the growth of the company to 1906; ELPC, "Corporate Record Book, 1906–1939," 1–43, for developments in the period 1906–11. Statistics are also provided in

"Study of the Properties," 9–13, 40. Insull, of course, was quick to appreciate the significance of Monroe's railway contract for building and diversifying the electrical load. Monroe became a member of Insull's inner circle of advisers, and he shared in the profits from the organization of the holding company on a regional scale. See McDonald, *Insull*, 95–101.

42. Whetstone, "Historical Factors," 21–25, 39–50, passim. For records of the early electric utilities in Kankakee and Streator, see CEC-HA, boxes 4235–38.

43. McDonald, *Insull*, 133–39, on Hawthorne Farm, which grew from an original 160 acres to 4,000 acres by the twenties. Also see Glenn Price, *Our Town: The Story of the Growth and Development of a Typical American Town, Libertyville* (Libertyville, Ill.: Libertyville Lions Club, 1942). A copy of this local history is contained in the Insull Papers, box 74. For perspectives on rural electrification, see D. Clayton Brown, *Electricity for Rural America: The Fight for the REA* (Westport, Conn.: Greenwood, 1980), and Richard A. Peace, *The Next Greatest Thing* (Washington, D.C.: National Rural Electric Cooperative Association, 1984).

44. For the report on the farm experiment, see Samuel Insull, "The Production and Distribution of Energy," lecture presented at the Franklin Institute, Philadelphia, 1913, in Keily, *Central-Station Service*, 357–91. The experiment attracted favorable notice by United States census takers. See Thomas C. Martin, "Part II: Technical," in U.S. Census, Special Report, Electric and Street Railways, 1912, 111–13, 154.

45. Whetstone, "Historical Factors," 141–46. For a more detailed survey of the Illinois Valley Gas and Electric Company, see the compilation of the holding company's franchises and records, *The Important Papers of the Illinois Valley Gas and Electric Company*, vol. 1, 1911, in CEC-HA, boxes 4045–46. Additional records of the short-lived company and its predecessors are found in CEC-HA, boxes 4235–37.

46. Whetstone, "Historical Factors," 150–64; "History of the Public Service Company," 55–70, passim; *Important Papers of the Public Service Company of Northern Illinois*, vol. 1, 1916, in CEC-HA, boxes 4045–46. On the number of customers, see PSCNI, "Reference Statistics" (photocopy in my possession). (Unfortunately, this important group of records was destroyed shortly after I copied a portion of them. The photocopy is all that remains.) At the same time, Insull set up a separate holding company, the Illinois Northern Utilities Company, to control the rural areas along the northern rim of the state. See *Important Papers of the Illinois Northern Utilities Company*, vol. 1, 1912, in CEC-HA, box 6003. In Du Page County, a mostly rural area before the 1950s, a holding company, the Western United Gas and Electric Company, operated independently of Insull. See CEC-HA, box 5001, passim.

47. PSCNI, "Reference Statistics"; "Electric Service in Chicago Suburbs," 1243–54.

48. John C. Learned, "10,211 Wiring Contracts," *Electrical Merchandise* 15(March 1916): 87–88. For perspectives on this type of sales campaign,

see F. H. Golding, "How to Get the Old Buildings Wired," NELA, *Proceedings,* 1907, 71–76, and "Bring the Old Home up-to-Date," *Electric City* 10(September 1912): 7–10.

49. For the governors' messages and the legislative history of the state utility commission, see Illinois, *Journal of the House of Representatives,* 1913, 120–21, 202–4, 1324–26, 2068–78, 2109, 2124. The act was upheld in *State Public Utilities Commission vs. Monarch Refrigerating Company* 267 Ill. 528 (1915). For Insull's role, see Forrest McDonald, "Sam Insull and the Movement for State Utility Commissions," *Business History Review* 32(Autumn 1958): 251–53. Also see Edward F. Dunne, *Illinois: The Heart of the Nation,* 5 vols. (Chicago: Lewis, 1933), 2:308–79, and Charles Maynard Kneier, *State Regulation of Public Utilities in Illinois,* University of Illinois Studies in the Social Sciences, 16, no. 1 (Urbana: University of Illinois, 1927).

50. City of Chicago, *Proceedings of the City Council,* 1913, 2646–47, 2802–4. The case of the gas company will be considered in the next chapter. For insight into the tangled politics of public transportation and Insull's role in the consolidation of the surface lines, see Barrett, *Automobile and Urban Transit,* 96–99.

51. "Electric Service in Chicago," 1137.

Chapter Eight

1. *Chicago Tribune,* 11 January 1918, for the quotation, 11 December 1917, 8 January 1918 for reports of people frozen to death, 2 January 1918 for a report on the coal riot in New York City. Also see James P. Johnson, "The Wilsonians as War Managers: Coal and the 1917–1918 Winter Crisis," *Prologue* 9(Winter 1977): 193–208, for a survey of the energy crisis. Johnson's negative commentary on the response of government officials to the crisis is based on an unfortunate use of historical hindsight.

2. *Chicago Tribune,* 7, 13, 19, 27 January 1918, for the situation in Chicago. Also see David M. Kennedy, *Over Here: The First World War and American Society* (New York: Oxford University Press, 1980), 123–28, and Robert C. Cuff, "Harry Garfield, the Fuel Administration, and the Search for a Cooperative Order during World War I," *American Quarterly* 30(Spring 1978): 39–53.

3. The most recent major work in this vein is Kennedy, *Over Here,* 93–142. Also consider similar interpretations in recent studies by Ellis Hawley, *The Great War and the Search for a Modern Order: A History of the American People and Their Institutions, 1917–1933* (New York: St. Martin's, 1979), and Jim Potter, *The American Economy between the World Wars* (New York: Wiley, 1974).

4. Gilman M. Ostrander, *American Civilization in the First Machine Age: 1890–1940* (New York: Harper and Row, 1970), 200–201, for the quotation; Richard Guy Wilson, "America and the Machine Age," in *The Machine Age,* ed. Richard Guy Wilson, Dianne H. Pilgrim, and Dickran Tashjian (New York: Brooklyn Museum and Harry N. Abrams, 1986), 23–42, for insight

on contemporary perceptions of the birth of a new age. For an analogous study of American literature during this period, see Henry F. May, *The End of American Innocence: A Study of the First Years of Our Own Time, 1912–1917* (Chicago: Quadrangle, 1959). For insight on the debate among intellectuals on the meaning of the war and the changes it produced, see Warren I. Susman, *Culture as History: The Transformation of American Society in the Twentieth Century* (New York: Pantheon, 1973), 105–21, and Cecelia Tichi, *Shifting Gears: Technology, Literature, Culture in Modernist America* (Chapel Hill: North Carolina University Press, 1987).

5. Neil A. Wynn, *From Progressivism to Prosperity: World War I and American Society* (New York: Holmes and Meier, 1986), 80 for the quotation, 71–82 for a brief survey of the war's impact on American industry. Wynn also points out that the study of war and society languishes in the United States while providing a topic of endless fascination in Europe. He suggests that American optimism and actual experience in the twentieth century have combined to create a mythical interpretation of war's impact on society. Americans believe that military engagements can be fought without major disruptions on the home front. This myth may account for the narrow focus of historical research on political life in Washington, D.C. According to Wynn, a tight focus on the effect of the war on the political economy has reinforced a false impression of World War I as merely a short-term interruption in long-term trends. Most studies of the home front have been preoccupied with a short list of rather disconcerting topics such as the rise of centralized bureaucracies, the government-inspired attack on civil liberties, and the red scare of 1919. Recent research on these subjects, in fact, has gone far to minimize even the growth of national power by recasting the War Industries Board and other wartime agencies in traditional terms of voluntarism, corporate commonwealth, and regulation without coercion. Cf. Wynn, *From Progressivism to Prosperity,* xiii–xxii; Kennedy, *Over Here,* 139–42; and Robert C. Cuff, *The War Industries Board: Business-Government Relations during World War I* (Baltimore: Johns Hopkins University Press, 1973).

6. Hawley, *Great War,* 80–81 for the first quotation, 81 for the second quotation. The anticonsumer bias of American intellectuals has a long history. For insight on the evolution of this strain of thought in the twentieth century, see T. J. Jackson Lears, *No Place of Grace: Antimodernism and the Transformation of American Culture, 1880–1920* (New York: Pantheon, 1981), and Daniel Horowitz, "Consumption and Its Discontents: Simon N. Patten, Thorstein Veblen, and George Gunton," *Journal of American History* 67(September 1980): 301–17.

7. For an introduction to the rise of a new culture of consumption, see Ostrander, *First Machine Age;* Roland Marchand, *Advertising the American Dream: Making Way for Modernity, 1920–1940* (Berkeley: University of California Press, 1986); and David A. Hounshell, *From the American System to Mass Production: The Development of Manufacturing Technology in the United States* (Baltimore: Johns Hopkins University Press, 1984), 303–9, for the concept of an "ethos of mass production."

8. John L'Brant, "Technological Change in American Manufacturing during the 1920s," *Journal of Economic History* 27(June 1927): 243, for the quotation. For the concept of "Fordism," see James J. Flink, *The Car Culture* (Cambridge: MIT Press, 1975), 67–112, and Hounshell, *Mass Production*, 1–15. For additional studies of the second industrial revolution, see Hounshell, *Mass Production*, 217–61; Richard DuBoff, "The Introduction of Electric Power in American Manufacturing," *Economic History Review* 20(December 1967): 508–18; and Warren D. Devine, "From Shafts to Wires: Historical Perspectives on Electrification," *Journal of Economic History* 43(June 1983): 347–72. An invaluable reference source for this study was one of President Herbert Hoover's carefully prepared reports. See Committee on Recent Economic Changes of President's Conference on Unemployment, *Recent Economic Changes in the United States*, 2 vols. (New York: McGraw-Hill, 1929).

9. Harvey Levenstein, *Revolution at the Table: The Transformation of the American Diet* (New York: Oxford University Press, 1988), 136, for the quotation. Also see *Recent Economic Changes*, 1:13–51. Unfortunately, Levenstein's pioneering study of the American diet became available only after a draft of this chapter had been written. In general, its findings complement and reinforce the interpretation presented here, but Levenstein deserves credit for highlighting the contrast between the effects of food reform on the middle and working classes. To a large extent his insights have been incorporated in revisions of the text.

10. Harry A. Garfield, quoted in *Chicago Tribune*, 18 January 1918, 1.

11. Johnson, "Coal Crisis," 194–201, for general perspectives; *Chicago Tribune*, 4, 17, 22 January 1918, for statistics.

12. Illinois State Council of Defense, *Final Report* (Springfield, Ill.: State of Illinois, 1920), v, 43–50. Also consider Insull's observations on the preparedness campaign. See Samuel Insull, "What Preparedness Meant in Wartime Utility Service," paper presented to the Commonwealth Edison Company Section of the National Electric Light Association (NELA), Chicago, October 1917, reprinted in William E. Keily, ed., *Public Utilities in Modern Life: Selected Speeches (1914–1923) by Samuel Insull* (Chicago: privately printed, 1924), 138–44. Insull's role in the war is described by Forrest McDonald, *Insull* (Chicago: University of Chicago Press, 1962), 162–76.

13. Kennedy, *Over Here*, 116–28; Johnson, "Coal Crisis," 200–202. For insight on Garfield's vision of a corporate commonwealth, see Cuff, "Harry Garfield," 39–53. For a sympathetic biography of Hoover, see Joan Hoff Wilson, *Herbert Hoover: Forgotten Progressive* (Boston: Little, Brown, 1975). For a recent careful study of government fuel policy during the war, see John G. Clark, *Energy and the Federal Government: Fossil Fuel Policies, 1900–1946* (Urbana: University of Illinois Press, 1987), 49–80.

14. *Chicago Tribune*, 1–10 December 1917; Samuel Insull, "The Electric Industry and the War," paper presented to the Electric Development League of San Francisco, San Francisco, March 1918, in Keily, *Public Utilities*, 149–50; and Samuel Insull, "Utilities in Wartime: A Note of Assur-

ance," paper presented to the annual meeting of the National Electric Light Association (NELA), Atlantic City, June 1918, in ibid., 154–55.

15. *Chicago Tribune*, 10–21 December 1918; Johnson, "Coal Crisis," 202–8.

16. *Chicago Tribune*, 7 January 1918, 1 for the first quotation, 2 for the second quotation, 29 December 1917–7 January 1918 for related stories.

17. *Chicago Tribune*, 13 January 1918 for the quotation, 12–16 January 1918 for full coverage of the crisis.

18. *Chicago Tribune*, 17–27 January 1918; Johnson, "Coal Crisis," 202–8; Clark, *Energy and the Federal Government*, 81–140, on the postwar fuel situation.

19. Sigfried Giedion, *Mechanization Takes Command: A Contribution to Anonymous History* (New York: Norton, 1948). Also see the pioneering work of Lewis Mumford, *Technics and Civilization* (New York: Harcourt, Brace, 1934). For perspectives on these important observers of technology and society, see William Kuhns, *The Post-Industrial Prophets: Interpretations of Technology* (New York: Harper and Row, 1971). For a contrary, radical interpretation of the second industrial revolution that shows how human choice, not technological determinism, shaped history, see David F. Noble, *American by Design: Science, Technology, and the Rise of Corporate Capitalism* (New York: Oxford University Press, 1979).

20. Statistics compiled from H. A. Seymour, "History of Commonwealth Edison," 3 vols., unpublished manuscript, 1934, Appendix, 174, 175, in CEC-HA, box 9001: "Net Increase in Contracted Business from 1889 to 1925 inclusive, in CEC-LF, F4-E5; E. W. Lloyd, "Compilation of Load Factors," National Electric Light Association, *Proceedings of the National Electric Light Association*, 1909, 590 (hereafter cited as NELA, *Proceedings*); U.S. Census, Manufacturers, 1919, 8:226.

21. John F. Gilchrist, "The Isolated Plant," in *Electric Service in Chicago: A Series of Advertisements by the Commonwealth Edison Company*, Chicago, 1916, no. 17 (in my possession); Devine, "From Shafts to Wires," 347–61; DuBoff, "Introduction of Electric Power," 509–14. For insight by Insull's top salesman on the difficulties of displacing the self-contained electric plant, see John Gilchrist, "The Competition of Isolated Plants," paper presented to the annual meeting of the Association of Edison Illuminating Companies (AEIC), 1901, reprinted in John Gilchrist, *Public Utility Subjects*, 2 vols. (Chicago: privately printed, 1940), 21–35, and John Gilchrist, "Commercial Problems," paper presented to the AEIC, 1909, reprinted in ibid., 47–58.

22. See above, chapter 5.

23. Devine, "From Shafts to Wires," 347–72; DuBoff, "Introduction of Electric Power," 509–18. In 1906 Commonwealth Edison had 110 industrial customers, including 56 who used group drive and 54 who used unit drive. For a complete listing of these customers, see E. W. Lloyd, "Notes on the Sale of Electric Power," pamphlet, Chicago, 1906, 32–36, in CEC-LF, F4-E2. Another important source of information about the technolog-

ical state of electric motors in the prewar period is NELA, *The Electrical Solicitors' Handbook* (New York: NELA, 1909).

24. Clearing Industrial District, "Locate Your Business Where It Will Grow," advertising pamphlet, 1915, for the quotation. In 1915 the district included the Argo Company's largest cornstarch plant in the world, the Smith Form-a-Truck Company, the Snyder electric furnace works, and the railroad car repair shop of Arthur J. O'Leary and Sons. For the development of industrial facilities in Chicago, see Carl W. Condit, *Chicago, 1910– 1929: Building, Planning, and Urban Technology* (Chicago: University of Chicago Press, 1973), 136–44. Also see Lloyd, "Notes on the Sale of Electric Power," 32–36, which lists the various uses of group and unit drive by different industries. In 1906 Lloyd calculated that about 80 percent of their connected load was group drive.

25. Hounshell, *Mass Production,* 217–61. For a biography of Ford, see Allan Nevins, in collaboration with Frank E. Hill, *Ford: The Times, the Man, the Company, 1865–1915* (New York: Scribner's Sons, 1954). For Ford's link to Edison, see Henry Ford, in collaboration with Samuel Crowther, *Edison as I Knew Him* (New York: Cosmopolitan, 1930).

26. On the coal strikes after the war, see William E. Leuchtenburg, *The Perils of Prosperity, 1914–1932* (Chicago: University of Chicago Press, 1958), 75–76; Robert K. Murray, *The Red Scare: A Study in National Hysteria, 1919– 1920* (New York: McGraw-Hill, 1955), 153–65; McDonald, *Insull,* 197–200; Edward Eyre Hunt et al., eds., *What the Coal Commission Found: An Authoritative Summary by the Staff* (Baltimore: Williams and Wilkins, 1925); and Commonwealth Edison Company, *Annual Report,* 1919–22, in CEC-HA, box 113. On mine workers, see Irving Bernstein, *The Lean Years: A History of the American Worker, 1920–1933* (Baltimore: Penguin, 1966), 117–36, 358– 90. Also consider the far-ranging symposium "The Price of Coal," *American Academy of Political and Social Science Annals* 111(1924) (hereafter cited as *Annals*). On the resulting transition in the use of energy fuels, see Sam Schurr and Bruce Netschert, *Energy in the American Economy, 1850–1975* (Baltimore: Johns Hopkins University Press/Resources for the Future, 1960), 31–43, 69–83, 180–89.

27. See George H. Jones, "Timely Power Loads," CEC pamphlet, 1919, 6–10, in CEC-LF, F4B-E4, for statistics on coal consumption; also see Samuel Insull, "Progress of Economic Power Generation and Distribution," paper presented to a meeting of the American Society of Mechanical Engineers, New Haven, Conn., 1916, in Keily, *Public Utilities,* 50–53, and Samuel Insull, "Production and Distribution of Electric Energy in the Central Portion of the Mississippi Valley," paper presented at the Cyrus Fogg Brackett Lectures in Applied Engineering Technology, Princeton University, 1921, in ibid., 272–82. On new directions in government policy, see Charles Keller, *The Power Situation during the War* (Washington, D.C.: Government Printing Office, 1921).

28. Chicago Department of Electricity, *Annual Report,* 1914–1920; Arthur George Woolf, "Energy and Technology in American Manufacturing,

1900–1929" (Ph.D. diss., University of Wisconsin–Madison, 1980), 10–15, 31–39; Seymour, "History of Commonwealth Edison," Appendix, 171; L'Brant, "Technological Change," 243–46. Although the trend toward fewer, larger generating units was cost effective for most manufacturers, there was no technological determinism involved. On the contrary, people made decisions that led to the abandonment of an important alternative source of industrial power, the self-contained electric plant. To be sure, larger turbogenerators are more cost effective than smaller units, making the decision for central station power the correct one for most manufacturing processing. However, when large amounts of heat are generated as a by-product of the manufacturing process, then the self-contained "cogenerator" might make the best economic sense. In the Chicago area the steel and metals mills, the gas company itself, food canning companies, bakeries, and breweries were some of the leading heat-producing industries. In Insull's time, for example, the steel mills were venting huge quantities of fuel-grade gas as a waste product. In rehabilitating the gas company, Insull was able to purchase the gas for an average price of 15¢ MCF and sell it to residential customers for 85¢ MCF (McDonald, *Insull,* 211). Had the steel mills used the gas themselves to run a self-contained electric plant, the savings in electric bills could have paid off the cost of the equipment and ultimately resulted in a net reduction in their energy costs. During the 1910s and 1920s, moreover, the navy and the shipping industry sponsored the technological development of a more efficient small-scale turbogenerator. Of course it is impossible to prove a historical hypothesis for a road not taken, yet the response of some of Chicago's steel mills to the energy crisis of the 1970s is instructive. The concept of "cogeneration" of electricity from by-product heat has been revived, especially as Commonwealth Edison's rates have soared to among the highest in the nation. See Harold Henderson, "Commonwealth Edison Has a Deal for You," *Reader* 16 (10 April 1987), and "Edison Rates Turning off Firms," *Chicago Tribune* 9 August 1987. For a more careful analysis of the fallacies of technological determinism, see Noble, *America by Design.*

29. "Contract Department Review Work Done during Year 1916," *Edison Round Table,* 20 January 1917; "Coal Conservation Result of Our Taking over Big Buildings," *Edison Round Table,* 17 April 1918; "Contract Department," unsigned manuscript, 1919; "Power Division Does Big Business," *Edison Round Table,* 31 January 1913, "Wholesale Power Load Shows Rapid Growth," ibid., 31 August 1915, and "Edison Service Displaces 27 Isolated Plants in 1926," ibid., 15 February 1927, all in CEC-LF, F4D(1)-E2; Commonwealth Edison Company, *Yearbook,* 1922, 19, 1927, 18. Also see "649 Private Plants Replaced by Edison Service from 1914 to 1936," *Edison Round Table,* 1 June 1937, 10, in CEC-LF, F4D(1)-E2, for a list of the number of plants replaced by type of industry and their power capacity, which amounted to 241,000 HP over the twenty-two-year period. On the emergence of youth culture, see Ostrander, *First Machine Age,* 237–74.

30. For the first electric steel furnace, see *Edison Round Table Weekly,* 20

January 1917, in CEC-LF, F4D(1)-E2, and "The Cry for Electric Steel and Why," *Electric City* 15(July 1917): 7–9, 20, in CEC-LF. For a review of the war's impact on the steel and iron industry, see Bernard M. Baruch, *American Industry in the War: A Report of the War Industries Board (March, 1921)* (New York: Prentice-Hall, 1941), 117–35; for the profits of the steel companies, see Wynn, *Progress to Prosperity*, 78–82. The war also led to the widespread adoption of electric arc welding techniques by metals fabricators, including the fast-growing automobile industry. For the role of the government in promoting electric arc welding, see Noble, *America by Design*, 165–66.

31. *Recent Economic Changes*, ix for the first quotation, 80 for the second quotation.

32. Seymour, "History of Commonwealth Edison," Appendix, 176.

33. Seymour, "History of Commonwealth Edison," Appendix, 176, for electrical consumption; U.S. Census, Population, 1910, 1920, 1930, for population statistics; James E. Clark, "The Impact of Transportation Technology on Suburbanization in the Chicago Region, 1830–1920" (Ph.D. diss., Northwestern University, 1977), 70, for figures on automobile ownership. For the decline of public transportation and the rise of the automobile in Chicago, see Paul Barrett, *The Automobile and Urban Transit: The Formation of Public Policy in Chicago, 1900–1930* (Philadelphia: Temple University Press, 1983), 129–83. Also consider Flink, *Car Culture*, 140–90.

34. U.S. Census, Manufacturing, 1919, 8:226, 1919, 3:141; DuBoff, "Introduction of Electric Power," 513; Woolf, "Energy and Technology, 39–48.

35. Carter Goodrich, "Power and Working Life," *Annals* 117(March 1925): 97, for the quotation; "Modern Manufacturing," *Annals* 85(1919); "The Second Industrial Revolution and Its Significance," *Annals* 149(1930), pt. 1; Stephen Meyer III, *The Five Dollar Day: Labor Management and Social Control in the Ford Motor Company, 1908–1921* (Albany: State University of New York Press, 1981); Hounshell, *Mass Production*, 11, 217–62.

36. Statistical Department, Commonwealth Edison Company, "Output . . . Customers . . . of 500,000 K.W.H. or More, Grouped as to Classes of Business," 1927, in CEC-LF, F23-E3; Statistical Department, Commonwealth Edison Company, "Output . . . Customers Having Annual Consumption of 1,000,000 K.W.H. or More—1924 and 1925," in CEC-LF, F23-E3. On government sponsorship of factory modernization during the war, see *Recent Economic Trends*, 79–83, and Noble, *America by Design*, 69–83. In 1929 the U.S. Census Bureau recorded the use of power equipment in manufacturing industries in Cook County, Illinois. The breakdown on the 1,644,000 HP in use in the county was 1,022,000 HP driven by central station service, 392,000 driven by self-contained electric plants, and 360,000 driven by steam engines. See U.S. Census, Manufacturing, 1929, 1:123.

37. J. Ogden Armour, "Diversify, Fertilize, Motorize, Specialize," *Saturday Evening Post*, 14 July 1917, reprinted in J. Ogden Armour, *Business Problems of the War* (Chicago: Donnelley, 1917), 5.

38. Wynn, *Progress to Prosperity*, 120–32; Levenstein, *Revolution at the Table*, 98–121; Alfred W. Crosby, Jr., *Epidemic and Peace, 1918* (Westport, Conn.: Greenwood, 1976), 5 for the quotation, 53–61, 203–7, for a description of the epidemic in Chicago. The number of lives lost from influenza in the armed services mounted to 43,000, which was equal to 80 percent of the war's combat deaths. Crosby estimates that 550,000 Americans died and 30 to 40 million people died worldwide; see *Epidemic and Peace*, 203–7.

39. *Recent Economic Trends*, 47, for the first quotation; Robert S. Lynd and Helen Merrell Lynd, *Middletown: A Study in Modern American Culture* (New York: Harcourt, Brace and World, 1929), 156, for the second quotation. The Lynds studied Muncie, Indiana. On the emergence of the modern fertilizer business as a by-product of the wartime research, see Arthur A. Noyes, "The Supply of Nitrogen Products for the Manufacture of Explosives," in *The New World of Science: Its Development during the War*, ed. Robert M. Yerkes (Freeport, N.Y.: Books for Libraries Press, 1969 [1920]), 123–47, and Preston J. Hubbard, *Origins of the TVA: The Muscle Shoals Controversy, 1920–1932* (Nashville, Tenn.: Vanderbilt University Press, 1961).

40. For the history of the ice business in Chicago up to the mid-eighties, see A[lfred] T[heodore] Andreas, *History of Chicago*, 3 vols. (New York: Arno, 1975 [1884–86]), 3:337–38. For a careful technological and economic study of the general subject, see Oscar E. Anderson, *Refrigeration in America: A History of a New Technology and Its Impact* (Princeton: Princeton University Press, 1953), 37–54. For the meat packers, see Mary Y. Kujovich, "The Refrigerator Car and the Growth of the American Dressed Beef Industry," *Business History Review* 44(Winter 1970): 460–82.

41. Anderson, *Refrigeration*, 54–70; Lynd and Lynd, *Middletown*, 153–58; Levenstein, *Revolution at the Table*, 10–29, 60–71.

42. Anderson, *Refrigeration*, 86–113. Air-conditioning remained a luxury in the period leading up to the Great Depression. The principles were worked out in the 1900s, but installations were confined to a few commercial and industrial establishments. In Chicago, a system was first placed in the Congress Hotel in 1906, followed six years later by the first office installation in Armour and Company's stockyard building. The main users, however, were candy manufacturers, printing plants, and bakeries. By 1932 the city had 130 industrial, 229 commercial, and 20 residential installations. See "Industrial Air-Conditioning Progress in Chicago," *Central Manufacturing District Magazine* 20(May 1936): 7–15, and Knight C. Porter, "Air-Conditioning Progress in Chicago," *Central Manufacturing District Magazine*, 21(May 1937): 9–11. Copies of both articles can be found in CEC-LF, F4D-E2.

43. Lloyd, "Compilation of Load Factors," 592–94; Samuel Insull, "Developments in Electric Utility Operating," address presented at the annual meeting of the American Institute of Electrical Engineers, Pittsfield, Massachusetts, 1917, in Keily, *Public Utilities*, 126; Commonwealth Edison Company, "Electric Refrigeration," pamphlet, 1913, in CEC-LF, F4D-E2; E. W. Lloyd to William J. Norton, memo, 29 August 1913, CEC-LF, F4D-

E2; Commonwealth Edison Company, "The Double Burden," advertising flier, 1915, CEC-LF, F4D-E2. In 1913 the company files listed three ice-making customers, a number that increased to twenty-one over the next three years.

44. Anderson, *Refrigeration*, 103–13, 207–11, for general perspectives; Commonwealth Edison Company, *Yearbook*, 1922, 18; Statistical Department, "Output . . . 1924–1925"; Statistical Department, "Output . . . Grouped as to Classes of Business"; and "Commercial Refrigeration in Chicago," *Electric World* 89(23 April 1927): 872, which reported that the utility company had captured 100 percent of the business.

45. Armour, "Diversity, Fertilize," 7, for the quotation; Anderson, *Refrigeration*, 127–50; Georg Borgstrom, "Food Processing and Packaging," in *Technology in Western Civilization*, ed. Melvin Kranzberg and Carroll W. Purcell, Jr., 2 vols. (New York: Oxford University Press, 1967), 2:386–402; Levenstein, *Revolution at the Table*, 30–43, 150–55. In the 1900s the mechanized assembly lines of canning companies in Chicago furnished Ford engineers with a model for speeding up the production process. See Hounshell, *Mass Production*, 241. Of course the "disassembly" line of the meat packers provided an even earlier prototype of modern techniques of mass production. See Giedion, *Mechanization*, 209–46.

46. Edward C. Hampe, Jr., and Merle Wittenberg, *The Lifeline of America: Development of the Food Industry* (New York: McGraw-Hill, 1964), 275–328; Wynn, *Progress to Prosperity*, 108–20.

47. J. Ogden Armour, "Lest Women Realize," *Ladies' Home Journal*, July 1917, reprinted in Armour, *Business Problems*, 42, for the quotation: J. Ogden Armour, "The Truth about the Price of Meat," *Collier's Weekly*, 15 September 1917, reprinted in ibid., 25–26. Wynn, *Progressivism to Prosperity*, 65–70; William Clinton Mullendore, *History of the United States Food Administration, 1917–1919*, Hoover Library on War, Revolution and Peace, Publication 18 (Stanford, Calif.: Stanford University Press, 1941); and James H. Shideler, *Farm Crisis* (Berkeley: University of California Press, 1957), 10–35, for the activities of the Food Administration; Illinois, State Council of Defense, *Report*, 96–102; Vernon Kellogg, "The Food Problem," in Yerkes, *New World of Science*, 265–76. See Levenstein, *Revolution at the Table*, 137–46, 175, for confirmation of Armour's estimate of the proportion of the family budget spent on food. Levenstein shows how food reform in the guise of patriotic sacrifice had a major impact on the diets of the upper and middle classes. However, the war seemed to have just the opposite effect on the working classes. As their income improved during the war, they tended to buy more beef and animal fat products. Levenstein does not evaluate how this change in diet affected people's overall health.

48. Armour, "Truth," 40 for the quotation, 24–41; Armour, "Lest Women Realize," 42–48.

49. Lynd and Lynd, *Middletown*, 158 for the quotation, 153–58; Levenstein, *Revolution at the Table*, 147–82; and Riva Apple, "'They Need It Now': Science, Advertising and Vitamins, 1925–1940," *Journal of Popular Culture*

22(Winter 1988): 65–84. Compare Mark H. Rose and John G. Clark, "Light, Heat, and Power: Energy Choices in Kansas City, Wichita and Denver, 1900–1935," *Journal of Urban History* 5(May 1979): 340–64. On the technological improvement of the home refrigerator, see Anderson, *Refrigeration*, 195–98, and Ruth Schwartz Cowan, *More Work for Mother: The Ironies of Household Technology from the Open Hearth to the Microwave* (New York: Basic, 1983), 128–43. Electric refrigerators were so uncommon in 1923 that a survey of household appliances by Commonwealth Edison did not list them. In contrast, a 1935 survey showed that 56 percent of the utility's customers now had them. Cf. H. B. Gear, "A Survey of Electrical Appliances Used in Chicago Residences," paper read at the annual meeting of the Illinois State Electric Association, Springfield, 1923, in CEC-LF, F4A-E2, and "Extent of Ownership of Electrical Appliances among Electricity Customers of the Commonwealth Edison Company," 1935 (photocopy in my possession). On a national level, the Hoover economic report listed 27,000 refrigerators in use in 1922 and 755,000 in 1927. In comparison, households had 3,850,000 vacuum cleaners in 1922 and 6,828,000 five years later. See *Recent Economic Changes,* 325.

50. Donald R. Richberg, *My Hero: The Indiscreet Memoirs of an Eventful but Unheroic Life* (New York: Putnam's Sons, 1954), 106, for the quotation. Also see McDonald, *Insull,* 180–82.

51. For the governors' messages and the legislative history of the state utility commission, see Illinois, *Journal of the House of Representatives,* 1913, 120–21, 202–4, 1324–26, 2068–78, 2109, 2124. The act was upheld in *State Public Utilities Commission vs. Monarch Refrigerating Company* 267 Ill. 528 (1915). Also see Charles Maynard Kneier, *State Regulation of Public Utilities in Illinois,* University of Illinois Studies in the Social Sciences, 16, no. 1 (Urbana: University of Illinois, 1927). On Insull's role in the influential report of the National Civic Federation on municipal public services, see above, chapter 5.

52. See Edward F. Dunne, *History of Illinois: The Heart of the Nation,* 5 vols. (Chicago: Lewis, 1933), 2:308–79, for the governor's version of utility reform. For Insull's role, see Forrest McDonald, "Sam Insull and the Movement for State Utility Regulatory Commissions," *Business History Review* 32 (Autumn 1958): 251–53.

53. Edward Bemis, Letter to the City Council, 25 June 1917, in *Proceedings of the City Council,* 1917–18, 353–54, for the oil statistics; W. G. C. Rice, *Seventy-five Years of Gas Service in Chicago* (Chicago: Peoples Gas Light and Coke Company, 1925), 43–45; McDonald, *Insull,* 206–13, for a general history of the gas company's relations with city hall. The legal maneuvering of the two sides became extremely convoluted, and they did not reach a settlement until 1927, when changed economic and technological conditions made the decisions irrelevant in terms of public policy. The leading case against the 1905 law was *Sutter vs. Peoples Gas Light and Coke Company,* 284 Ill. 634 (1918). This suit by a customer ended with the court's declaring the act unconstitutional. The legislative history of the act showed that it

was originally intended as an enabling act for municipal ownership. However, the United States Supreme Court decision of 1904 overruling Chicago's rate ordinance of 1900 prompted the legislators to add rate-making powers to the municipal ownership bill. This procedure, according the Illinois high bench, violated the state constitution's prohibition (article 4, section 13) against more than one subject's being contained in each law. By 1918, of course, the power to set utility rates had percolated to the state commission. Two other suits were initiated in the wake of the 1911 rate-fixing ordinance: *Chicago vs. Peoples Gas Light and Coke Company*, 170 Ill. App. 98 (1912) and *Mills vs. Peoples Gas Light and Coke Company* 327 Ill. 508 (1927) continued the battle until the company finally won a hollow victory.

54. On the problems of the gas company, see Rice, *Seventy-five Years*, 41–45; McDonald, *Insull*, 158–61, 205–11; Chicago, *Proceedings of the City Council* 1(1916–17): 1297–1305, 1(1917–18): 1902. For Mayor Thompson's role in the gas settlement, see ibid., 1(1915): 4–5, 2483–84, 1(1917–18): 1381–82. For a recent assessment of Thompson, see Douglas Bukowski, "Big Bill Thompson: The 'Model Politician,'" in *The Mayors: The Chicago Political Tradition*, ed. Paul M. Green and Melvin G. Holli (Carbondale: Southern Illinois University Press, 1987), 61–81. According to Forrest McDonald, Thompson was virtually Insull's man in city hall in matters concerning public utilities. McDonald points out that the mayor appointed Samuel Ettleson, a close associate of Insull, to be the city's corporation counsel. See McDonald, *Insull*, 178–82, 219.

55. Chicago, *Proceedings of the City Council* 1(1916–17): 1297–1305, 1(1917–18): 353–54, 731–38; Richberg, *My Hero*, 102–6; Insull, "Some Remarks on Diversity and Rate Making," paper delivered to the annual meeting of the National Commercial Gas Association, Atlantic City, 1916, in Keily, *Public Utilities*, 69–107.

56. McDonald, *Insull*, 207, for the quotation. See Chicago, *Proceedings of the City Council* 1(1917–18): 732–38, for the ordinance-contract. Also compare the responses of Insull's gas utilities outside Chicago, which included the transition to gas by-products technology and the purchase of additional coal mines. See Public Service Company of Northern Illinois (PSCNI), *Annual Report*, 1916–17, in CEC-HA, box 4094.

57. Chicago, *Proceedings of the City Council* 1(1917–18): 1242–43, 1309, 1467–72, 1902, 2026, 2380; Kneier, *State Utility Regulation*, 25–35; Dunne, *Illinois*, 2:424–54; Rice, *Seventy-five Years*, 44–47; McDonald, *Insull*, 177–81. The state commission held hearings on the gas company's request for a substantial rate increase in 1919 and granted the request in June of the following year. For Insull's testimony, see *Chicago Tribune*, 19–21 June 1919.

58. Rice, *Seventy-five Years*, 47–50; McDonald, *Insull*, 205–13; Insull, "Diversity and Rate Making," 69–107; Samuel Insull, "The Gas Industry's Biggest Task," paper presented to the annual meeting of the American Gas Association, Chicago, 1921, in Keily, *Public Utilities*, 252–62. Insull followed his usual method of experimenting with new ideas in controlled test markets before instituting them systemwide. In the case of differential

gas rates for industrial use, he used the suburban territory of the Public Service Company of Northern Illinois. On 13 April 1916, Insull gained the approval of the state regulatory commission to try a demand system of rates. Seven months later the suburban utility completed the first welded pipeline in its system, a fifty-mile, four-inch extension line linking the Evanston gas plant to Wauconda. The new line was welded to make it strong enough to withstand higher pressures, which eliminated the need for intermediate pumping stations between the gas plant and consumers. Gas from the new line was immediately put in the service of a creamery plant in Lake Zurich to replace coal-fired steam boilers. Other early industrial customers to switch from coal to gas on the demand system included the Buda Company in Harvey and the Argo Corn Products Company in the Clearing Industrial District. See PSCNI, *Public Service Lumen* 1(November 1916): 1, 4, and 1(January 1917): 4, in CEC-Library.

59. Commonwealth Edison Company, *Annual Report*, 1919; Commonwealth Edison Company, *Yearbook*, 1922, 12, and 1928, 32, in CEC-HA, box 113; McDonald, *Insull*, 182–87. Within five years, the number of investors in Commonwealth Edison increased from 6,500 to over 40,000. In 1923 Insull estimated that 90 percent of the stockholders were Chicago residents and another 5 percent were citizens of Illinois who lived outside the city. See Commonwealth Edison Company, *Yearbook*, 1923, 3. For insight on the propaganda and the stock purchase schemes, see the reports of his chief sales assistant, John Gilchrist. For example, see John Gilchrist, "The Development of Favorable Public Opinion towards Public Utilities," paper presented to the annual meeting of the Illinois State Electric Association, Chicago, 1919, in Gilchrist, *Public Utility Subjects*, 162–71; John Gilchrist, "Report on the Committee on Public Utility Information," paper presented to the annual meeting of the NELA, Pasadena, 1920, in ibid., 172–82; John Gilchrist, "Taking Your Customers into Partnership," paper presented at the annual meeting of the NELA, Chicago, 1921, in ibid., 187–213. By the late 1920s, Insull's propaganda efforts and his financial support of political candidates became the objects of federal investigation. See the muckraking books by Ernest Gruening, *The Public Pays* (New York: Vanguard, 1931), and M. L. Ramsay, *Pyramids of Power: The Story of Roosevelt, Insull and the Utility Wars* (Indianapolis: Bobbs-Merrill, 1937).

60. Insull, "Gas Industry's Biggest Problem," 261, for the quotation. Also see McDonald, *Insull*, 206–11; Rice, *Seventy-five Years*, 44–50.

61. By the early 1920s, the fallacy of setting differential rates in a community of energy consumers served by a mature electrical grid was evident to practitioners and theorists alike. See William J. Hausman and John L. Neufeld, "Engineers and Economists: Historical Perspectives on the Pricing of Electricity," *Technology and Culture* 30(January 1989): 83–104.

62. Baruch, *American Industry in War*, 298–304; Grosvenor B. Clarkson, *Industrial America in the World War: The Strategy behind the Line, 1917–1918* (Boston: Houghton Mifflin, 1923), 461–64. Clarkson was a New York advertising man who joined the War Industries Board and later became di-

rector of the Council of National Defense. See Cuff, *War Industries Board,* 20–40. For the wartime survey of electric utilities, see Keller, *Report on the Power Situation.* Also see Morris Llewellyn Cooke, "The Long Look Ahead," *Survey Graphic* 51(1 March 1924): 602–4, and Thomas P. Hughes, *Networks of Power: Electrification in Western Society, 1880–1930* (Baltimore: Johns Hopkins University Press, 1983), 285–323, on the new focus of attention on regional power systems in the United States and Europe during the war.

63. Samuel P. Hays, *Conservation and the Gospel of Efficiency: The Progressive Conservation Movement, 1890–1920* (Cambridge: Harvard University Press, 1959), 240 for the quotation, 81, 238–40; Hubbard, *Origins of TVA,* 1–27; Martin V. Melosi, *Coping with Abundance: Energy and Environment in Industrial America* (Philadelphia: Temple University Press, 1985), 11–25. Also see the issues of *Survey Graphic* 51(March 1924), and the *Annals* 117(March 1925), for a wide-ranging debate on regional power grids and their promise of social benefit.

64. Baruch, *American Industry in the War,* 301, for the quotation; Noble, *America by Design,* 147–66, for insight on the growth of government involvement in industrial research. These were the directors of the NRC's Industrial Relations Division: Raymond Bacon, director of the Mellon Institute; J. J. Carty, chief engineer of American Telephone and Telegraph Company; Frank Jewett, director of the Western Electric laboratories; Arthur D. Little, president of the nation's largest engineering consulting firm; C. E. Shimmer, director of research at the Westinghouse Company; and Willis Whitney, director of research at the General Electric Company. See Noble, *America by Design,* 155–56. For a historical introduction to energy planning in the twenties, see Hughes, *Networks of Power,* 324–459, and Melosi, *Coping with Abundance,* 121–37. For contemporary debate on the advent of mass production, see "The Second Industrial Revolution," *Annals* 149(1930), pt. 1.

65. Edward A. Filene, in collaboration with Charles W. Wood, *Successful Living in the Machine Age* (New York: Simon and Schuster, 1932), 1.

Chapter Nine

1. Commonwealth Edison Company, *Presentation of the Commonwealth Edison Company in Competition for the Charles A. Coffin Award, 1925* (Chicago: Commonwealth Edison Company, 1925), in CEC-HA, box 9006. The award was actually presented in May 1926. See Commonwealth Edison Company, *Annual Report,* 1926, 8–9, in CEC-HA, box 113, and Commonwealth Edison Company, *Yearbook,* 1926, 10, 22, 32, in CEC-HA, box 113. At the end of 1925, the company listed 679,497 residential customers. By extrapolating from the 1920 and 1930 census reports, the number of families in Chicago can be estimated at 733,245, meaning that 92.7 percent had service. Insull believed that 95 percent had service. Insull's figure may be

more accurate, because some families boarded with others. See U.S. Census, Population, 1920, 3:274, 1930, 6:388. Insull's claim that Chicago was the world's most energy-intensive city was based on the fact that in 1924 Commonwealth Edison carried the largest maximum load of 663,000 kw and generated the second largest amount of electricity, 2,573,287,000 kwh. In comparison, the New York Edison Company reached a maximum load of 497,000 kw, while the Niagara Falls Power Company generated 2,595,847,000 kwh. See *Yearbook*, 1924, 4. Compare similar figures for 1926, when the Chicago company retained its claim to the greatest maximum load but slipped to third place in the amount of energy generated, behind the Buffalo, Niagara and Eastern Power Corporation and the Hydro Electric Commission of Ontario, Canada. See Commonwealth Edison Company, Statistical Department, "Relative Position of Commonwealth Edison Company with Electricity Producers . . . ," manuscript, 10 May 1927, in CEC-LF, F23-E3.

2. U.S. Census, Population, 1920, 1:393, 3:274, 1930, 3, pt. 1:656, 676. For an introduction to the physical transformation of Chicago, see Carl W. Condit, *Chicago, 1910–1929: Building, Planning, and Urban Technology* (Chicago: University of Chicago Press, 1973), 1–79. Of course, not all Chicagoans shared in the prosperity of the twenties. Cf. Thomas Lee Philpott, *The Slum and the Ghetto: Neighborhood Deterioration and Middle Class Reform, Chicago, 1880–1930* (New York: Oxford University Press, 1978).

3. The classic analysis of the economics of the real estate industry in the city is Homer Hoyt, *One Hundred Years of Land Values in Chicago* (Chicago: University of Chicago Press, 1933), 232–78. Also see Condit, *Chicago, 1910–1929*, 79–301, and Frank A. Randall, *History of the Development of Building Construction in Chicago* (Urbana: University of Illinois Press, 1949). For perspectives on the 1920s, see Charles W. Eagles, "Urban-Rural Conflict in the 1920s: A Historiographical Assessment," *Historian* 49(November 1986): 26–48. The best insight on the changing attitudes of farmers toward the city and its modern technology is Don S. Kirschner, *City and Country: Rural Responses to Urbanization in the 1920s* (Westport, Conn.: Greenwood, 1970).

4. Bruce Barton, "How Long Should a Housewife Live?" in Commonwealth Edison Company, *Coffin Presentation*, fig. 5. For a perceptive analysis of Barton, see Warren Susman, *Culture as History: The Transformation of American Society in the Twentieth Century* (New York: Pantheon, 1984), 122–31.

5. On the 1920s cult of "technological filiarchy," see Gilman M. Ostrander, *American Civilization in the First Machine Age: 1890–1940* (New York: Harper and Row, 1970), 237–74. On the conservative use of technological innovation in domestic life, see Ruth Schwartz Cowan, *More Work for Mother: The Ironies of Household Technology from the Open Hearth to the Microwave* (New York: Basic, 1983).

6. Roland Marchand, *Advertising the American Dream: Making Way for Modernity, 1920–1940* (Berkeley: University of California Press, 1985), 1 for the

quotation, 1–51 for insight on the effects of the war on the advertising industry, including the formation in 1919 of Barton, Durstine and Osborn.

7. H. A. Seymour, "History of Commonwealth Edison Company," 3 vols., unpublished manuscript, 1934, Appendix, 9, 81, for statistics on Chicago's residential use and cost of electricity; Commonwealth Edison Company, Statistical Department, "Cost of Living," manuscript, 10 May 1928, in CEC-LF, F23-E3; and Arthur George Woolf, "Energy and Technology in American Manufacturing: 1900–1929" (Ph.D. diss., University of Wisconsin–Madison, 1980), 18, 63, for cost-of-living statistics. Also see Sam Schurr and Burce Netschert, *Energy in the American Economy, 1850–1975: An Economic Study of Its History and Prospects* (Baltimore: Resources for the Future, 1960), 31–99, 180–89, and Martin V. Melosi, *Coping with Abundance: Energy and Environment in Industrial America* (Philadelphia: Temple University Press, 1985), 103–16, for broad overviews of the shifting costs and uses of various fuels in the United States during the twenties. National energy policy is the subject of John G. Clark, *Energy and the Federal Government: Fossil Fuel Policies, 1900–1946* (Urbana: University of Illinois Press, 1987), 141–92.

8. Daniel E. Sutherland, *Americans and Their Servants: Domestic Service in the United States from 1800 to 1920* (Baton Rouge: Louisiana State University Press, 1981), 10; David M. Katzman, *Seven Days a Week: Women and Domestic Service in Industrializing America* (New York: Oxford University Press, 1978), 95–128; Bettina Berch, "Scientific Management in the Home: The Empress' New Clothes," *Journal of American Culture* 3(Fall 1980): 440–45. For a typical expression of contemporary middle-class attitudes toward the use of electrical technology in the home, see Mary Pattison, "The Abolition of Household Slavery," *American Academy of Political and Social Science Annals* 117(March 1925): 124–27 (hereafter cited as *Annals*). Of course the movement of women out of the home and into the work force was not restricted to factory work. For the growth of white-collar jobs for women, see Elyce J. Rotella, *From Home to Office: U.S. Women at Work, 1870–1930* (Ann Arbor, Mich.: UMI Research, 1977), and Margery W. Davis, *Women's Place Is at the Typewriter: Office Work and Office Workers, 1870–1930* (Philadelphia: Temple University Press, 1982).

9. William E. Keily, "History of the Electric Shop," manuscript, 29 July 1920, in CEC-LF, F4B-E4, for the quotation. Also see the same statement of the company stores' "oversold" condition in comments of E. L. Lloyd at the convention of the Illinois State Electric Association, Chicago, October 1919, in Commonwealth Edison Company, "Symposium on Merchandise and Service of the Commonwealth Edison Company," pamphlet, 1920, in CEC-LF, F4B-E4.

10. George H. Jones, "Timely Power Loads," in Commonwealth Edison Company, "Symposium," 6, for the first quotation; Edgar A. Edkins, "The Story of the Commonwealth Edison Company," unpublished manuscript, revised to 4 March 1922, 20, in CEC-LF, F4B-E4, for the second quotation; Committee on Recent Economic Changes of President's Conference on

Unemployment, *Recent Economic Changes in the United States,* 2 vols. (New York: McGraw-Hill, 1929), 1:321, for the third quotation.

11. *Recent Economic Changes,* 1:56 for the quotation, 1:56–58, 414–19.

12. Commonwealth Edison Company, "Catalogue of Commonwealth Edison Electric Shops," pamphlet, 1920, in CEC-LF, F4B-E4; H. B. Gear, "A Survey of Electrical Appliances Used in Chicago Residences," paper presented at the annual meeting of the Illinois State Electric Association, Chicago, 1923, in CEC-LF, F4A-E2; Commonwealth Edison Company, *Yearbook,* 1922, 24, and 1927, 22.

13. Samuel Insull, "Chicago Central Station Development," reprinted in William E. Keily, ed., *Central-Station Electric Service . . . as Set Forth in the Public Addresses (1897–1914) of Samuel Insull* (Chicago: privately printed, 1915), 383, for the quotation; Seymour, "History of Commonwealth Edison," Appendix, 174; Commonwealth Edison Company, *Yearbook,* 1923, 4, and 1927, 12, for statistics. By 1917 the company appliance sales techniques had reached a point of maturity and sophistication, including special appeals to housewives by saleswomen, and the use of college students as door-to-door solicitors. For example, see Clara H. Zillessen, "Advertising to Women," *Electrical Merchandise* 15(January 1916): 211–13; Clara H. Zillessen, "When the Woman Buys," *Electrical Merchandise* 16(August 1916): 62–63; Oliver R. Hogue, "Summer Appliance Campaigning with College Students," *Electrical Merchandise* 16(August 1916): 69–71; "10,211 Wiring Contracts," *Electrical Merchandise* 15(March 1916): 87–88; and Edgar A. Edkins, "An Idea-Journey through a Modern Electric Shop," *Electrical Merchandise* 16(July 1916): 9–13. In 1922, company statistician Ed Fowler prepared a report for Insull that predicted the saturation of the residential market in the next few years. See Commonwealth Edison Company, Statistical Department, "Comparison of the Growth of Chicago . . . ," manuscript, 8 February 1922, in CEC-LF, F23-E3. Four years later, a second report confirmed Fowler's predictions. See Commonwealth Edison Company, Statistical Department, "Growth of Residential Customers," manuscript, 10 February 1926, in CEC-LF, F4-E5.

14. See, for example, Dolores Hayden, *The Grand Domestic Revolution: A History of Feminist Designs for American Homes, Neighborhoods, and Cities* (Cambridge: MIT Press, 1981), and Stuart Ewen, *Captains of Consciousness: Advertising and the Social Roots of the Consumer Society* (New York: McGraw-Hill, 1976), for the conspiracy interpretation. For a contrary view, see Marchand, *Advertising the American Dream,* and Roland Marchand, "Contested Terrain: *Where* Do Consumers and Advertisers Meet?" paper presented to the annual meeting of the Social Science History Association, New Orleans, 1987.

15. Gear, "Survey of Electrical Appliances"; Committee on Residence Survey, "Survey of Residence Electrical Installations: Part II, Synopsis," report presented at the annual meeting of the Association of Edison Illuminating Companies, in *Minutes,* 1923, 609–37. On the slow adoption of the "convenience outlet," see Fred E. H. Schroeder, "More 'Small Things

Forgotten': Domestic Electric Plugs and Receptacles, 1881–1931," *Technology and Culture* 27(July 1986): 525–43.

16. Charles A. Thrall, "The Conservative Use of Modern Household Technology," *Technology and Culture* 23(April 1982): 175–76.

17. John B. Rae, *American Automobile Manufacturers: The First Forty Years* (Philadelphia: Chilton, 1959), 67–71; Walker Vehicle Company file, CEC-LF, F32-E3.

18. C. C. McMahon, "Problems of Transportation," manuscript, 16 May 1917, in CEC-LF, F3A-E4; "New Electric Truck Service Ready in Chicago," *Electric Review* 71 (29 December 1917): 1091–94; "Electric Truck Gaining Favor in Chicago," *Electric World* 77 (14 May 1921): 1095–97; "Chicago Vehicle Men Plan Co-operative Association," *Electric World* 77(25 June 1921): 1503. In Chicago, the electric vehicle achieved success only as a delivery truck for bakeries, department stores, and express companies. Insull's advertisements stressed the economy of the electric truck for this type of short-haul delivery service. See Commonwealth Edison Company, "Partial List of Chicago Companies Using Electric Truck Successfully," advertising flier, January 1918, in CEC-LF, F3A-E4, and Commonwealth Edison Company, "In a Nutshell: Facts about City Deliveries for the Busy Executive," advertising flier, October 1922, in CEC-LF, F3A-E4.

19. J. Fred McDonald, *Don't Touch That Dial: Radio Programming in American Life from 1920 to 1960* (Chicago: Nelson-Hall, 1979), 1–30; Samuel Insull, "Report to the Stockholders," in Commonwealth Edison Company, *Annual Report,* 1925, 9; U.S. Census, Population, 1930, 6:388; "Wiring and Appliance Survey in Chicago Homes Reveals Extensive Future Market," *Edison Round Table* 21(31 December 1929), in CEC-Library; "Extend Ownership of Electrical Appliances . . . ," in *Public Opinion and the Peoples Gas Light and Coke Company and the Commonwealth Edison Company* (Chicago: Lord and Thomas, 1935), 42, (photocopy in my possession). The census of 2,982 electric customers showed that 94.1 percent owned radios compared with 92.2 percent owning floor lamps, 88.8 percent irons, 85.5 percent table lamps, and 71.9 percent vacuum cleaners. In 1926 the cost of an RCA fully electric radio ranged from $115 for a five-tube, Radiola 20 table model to $575 for an eight-tube Radiola 50 cabinet model. Battery-operated models by Mohawk and by Atwater Kent fell into the $110 to $130 range. See *Chicago Daily News,* 14 October 1926, sec. 4.

20. Robert S. Lynd and Helen Merrell Lynd, *Middletown: A Study in Modern American Culture* (New York: Harcourt, Brace and World, 1929), 271 for the quotation, 251–71. The impact of the radio on the lives of farmers was even more sweeping and deep-seated than its effect on city dwellers. See James E. Evans, *Prairie Farmer and WLS: The Burridge D. Butler Years* (Urbana: University of Illinois Press, 1969). The plight of the farmer in the 1920s will be discussed later in this chapter.

21. Public Service Company of Northern Illinois (PSCNI), *Presentation of PSCNI in Competition for the Charles A. Coffin Award: April 1st, 1924* (Chicago: PSCNI, 1924), in CEC-LF, F1, for the quotations and the description

of the model house. Two other models were built in Waukegan and in Joliet. See Gwendolyn Wright, *Building the Dream: A Social History of Housing in America* (New York: Pantheon, 1981), 208–10, for insight on the relation between the appeal of the suburbs and contemporary anxieties about child rearing.

22. Wright, *Building the Dream* 193–214; Ann Durkin Keating, "Real Estate Developers and the Emergence of Suburban Government: The Case of Nineteenth Century Chicago," paper presented at the annual meeting of the American Historical Association, Chicago, 1986; Philpott, *Slum and Ghetto,* 189–98; Allan H. Spear, *Black Chicago: The Making of a Negro Ghetto* (Chicago: University of Chicago Press, 1967), 212–29. For perspectives on working-class housing patterns, see Michael J. Doucet and John C. Weaver, "Material Culture and the North American House: The Era of the Common Man," *Journal of American History* 72(December 1985): 560–87. World War I had a powerful effect on the building industry and the perceptions of housing reformers. The war experience helped spawn concepts of community building and regional planning. For many reformers, the regional power grid of the electrical utilities was a compelling image, reorienting visions of the natural social unit toward a regional scale. See Roy Lubove, "Homes and 'a Few Well Placed Fruit Trees': An Object Lesson in Federal Housing," *Social Research* 27(Winter 1960): 469–86, and Roy Lubove, *Community Planning in the 1920s: The Contributions of the Regional Planning Association of America* (Pittsburgh: University of Pittsburgh Press, 1963), 17–27.

23. *Recent Economic Changes,* 1:219 for the first quotation, 226 for the second quotation. The real estate industry in Chicago during the postwar decade is the subject of Hoyt, *Land Values,* 232–78. See Helen Lefkowitz Horowitz, *Culture and the City: Cultural Philanthropy in Chicago from the 1880s to 1917* (Lexington: University of Kentucky Press, 1976), 93–125, for the growth of highbrow institutions, and Mark H. Haller, "Organized Crime in Urban Society: Chicago in the Twentieth Century," *Journal of Social History* 5(Winter 1971): 210–34, for an analysis of the origins of organized crime in the city.

24. Condit, *Chicago, 1910–1929,* 145, for the quotation. Also see George D. Bushnell, "Chicago's Magnificent Movie Palaces," *Chicago History* 6(Summer 1977): 99–106, for an introduction to the topic.

25. John Eberson, "A Description of the Capitol Theater, Chicago," *Architectural Forum* 42(June 1925): 376, for the first quotation; D. D. Kimball, "Ventilating and Cooling of Motion Picture Theaters," *Architectural Forum* 42(June 1925): 397, for the second quotation.

26. See Condit, *Chicago, 1910–1929,* 98–99, on the Wrigley Building; Hoyt, *Land Values,* 240–42, on the construction of skyscrapers; *Recent Economic Changes,* 143–46, on rising standards of lighting in the workplace, including the Post Office; William Henry Leffingwell, *Office Management: Principles and Practice* (Chicago: Shaw, 1925), 300–308, for contemporary ideas about office work. Also see Rotella, *From Home to Office,* 151–59, and

Davis, *Women's Place,* 97–128, for insight on the relationship between women in the labor force and the mechanization of office work.

27. Celia Hilliard, "'Rent Reasonable to Right Parties': Gold Coast Apartment Buildings, 1906–1929," *Chicago History* 8(Summer 1979): 66–77; Condit, *Chicago, 1910–1929,* 155–64; Hoyt, *Land Values,* 242–45. The classic study of urban society in this area of Chicago is Harvey Warren Zorbaugh, *The Gold Coast and the Slum: A Sociological Study of Chicago's Near North Side* (Chicago: University of Chicago Press, 1929). For a more graphic description of the Near North area, see Frank Norris, *The Pit* (New York: Doubleday, 1928), 55–58.

28. Condit, *Chicago, 1910–1919,* 161–62; Hoyt, *Land Values,* 245; *Recent Economic Changes,* 1:224–30; Wim De Wit, "Apartment Houses and Bungalows: Building the Flat City," *Chicago History* 12(Winter 1983–84): 18–29.

29. *Recent Economic Changes,* 1:232 for the first quotation, 1:230–33; Condit, *Chicago, 1910–1929,* 164, for the second quotation; Schroeder, "Electric Plugs," 539–43; Wright, *Building the Dream,* 168–72.

30. De Wit, "Apartment Houses and Bungalows," 18–29; Zorbaugh, *Gold Coast and the Slum;* Philpott, *Slum and the Ghetto;* Hoyt, *Land Values,* 238, 245–48; Condit, *Chicago, 1910–1929,* 164–66.

31. Daniel Presser, "Chicago and the Bungalow Boom of the 1920s," *Chicago History* 10(Summer 1981): 86–95; De Wit, "Apartment Houses and Bungalows," 18–29; Hoyt, *Land Values,* 245–46, 255–65; "Wiring and Appliance Survey," 10.

32. *Recent Economic Changes,* 1:236, for the first quotation; Harlan Paul Douglass, *The Suburban Trend* (New York: Century, 1925), 225, for the second quotation. See Richard Sennett, *Families against the City: Middle Class Homes of Industrial Chicago, 1872–1890* (Cambridge: Harvard University Press, 1970), and chapter 7 above for interpretations of the roots of the suburban trend. Also see Kenneth T. Jackson, *The Crabgrass Frontier: The Suburbanization of the United States* (New York: Oxford University Press, 1985), 157–89, for an introduction to suburbanization during the interwar years.

33. See Robert M. Fogelson, *The Fragmented Metropolis: Los Angeles, 1850–1930* (Cambridge: Harvard University Press, 1967); and Fred W. Viehe, "Black Gold Suburb: The Influence of the Extractive Industry on the Suburbanization of Los Angeles, 1890–1930," *Journal of Urban History* 8 (November 1981): 3–26.

34. *Recent Economic Change,* 1:52 for the quotation, 228–30 on auto-related construction; Sinclair Lewis, *Babbitt* (New York: Harcourt, Brace, Jovanovich, 1922), 57–63; Lynd and Lynd, *Middletown,* 261, 251–63.

35. Paul Barrett, *The Automobile and Urban Transit: The Formation of Public Policy in Chicago, 1900–1930* (Philadelphia: Temple University Press, 1983), 139–53; James E. Clark "The Impact of Transportation Technology on Suburbanization in the Chicago Region, 1830–1920" (Ph.D. diss., Northwestern University, 1977), 70; James J. Flink, *The Car Culture* (Cambridge: MIT Press, 1975), 140–90.

36. Central Area Railfans Association, *Route of the Electroliners*, Bulletin 107 (Chicago: Central Area Railfans Association, 1963), 6–7, 23–34; William D. Middleton, *North Shore: America's Fastest Interurban* (San Marino, Calif.: Golden West, 1964), 9–61; Commonwealth Edison Company, *Yearbook*, 1927, 4; Condit, *Chicago, 1910–1929*, 50–52, 235–90; *Chicago Daily News*, 7–9 August 1926; Helen Corbin Monchow, *Seventy Years of Real Estate Subdividing in the Region of Chicago* (Evanston, Ill.: Northwestern University, 1939), 96–98, 103–13; Hoyt, *Land Values*, 255. For Insull's ideas about railroad electrification and a detailed proposal for its achievement, see Samuel Insull, "The Relation of Central-Station Generation to Railroad Electrification," paper presented to the annual meeting of the American Institute of Electrical Engineers, New York, 1912, in Keily, *Central-Station Electric Service*, 255–315.

37. PSCNI newspaper advertisements, as reproduced in PSCNI, *Coffin Award*. Also see copies of the originals in PSCNI, "Secretary's Scrapbook, 1911–1928" in CEC-HA, box 4182.

38. "Average Number of Electric Customers by Years," and "Kilowatt Hours Sold—Districts 'All,' " in PSCNI, "Reference Statistics [1912–35]," CEC-HA, box 4145 (photocopies in my possession); Rate History files in CEC-HA, boxes 4047–49; and additional material on rates in "Study of Properties," Illinois Commerce Commission Case 22353 (1933), Company Exhibit 2, 63, in CEC-HA, box 4093. Also see the useful statistical information in PSCNI, *Yearbook*, 1924–31, in CEC-HA, box 4094.

39. PSCNI, *Yearbook*, 1924, 8, 1929, 4, PSCNI, *Annual Report*, 1923, all in CEC-HA, box 4094; Commonwealth Edison Company, *Yearbook*, 1924, 5; Imogene Whetstone, "Historical Factors in the Development of Northern Illinois and Its Utilities," typescript, Chicago, PSCNI, 1928, 167–69, in CEC-Library.

40. "Average Number of Gas Customers by Year," in PSCNI, "Reference Statistics": PSCNI, *Annual Report*, 1923; PSCNI, *Yearbook*, 1929, 4, 6, 10–11, in CEC-HA, box 4094.

41. Douglass, *Suburban Trend*, 149, for the quotation; "Kilowatt Hours Accounted For," and "Kilowatt Hours Sold—Districts 'All,' " in PSCNI, "Reference Statistics." For comparative perspectives, see Joel Tarr and Gabriel Dupuy, eds., *Technology and the Rise of the Networked City in Europe and America* (Philadelphia: Temple University Press, 1988).

42. Monchow, *Real Estate Subdividing*, 118, for the quotation. Also see Alan Gowans, *The Comfortable House: North American Suburban Architecture, 1890–1930* (Cambridge: MIT Press, 1986), for a catalog of popular housing styles.

43. See Monchow, *Real Estate Subdividing*, chap. 3, for an analysis of the oversupply of building lots; Marc A. Weiss, *The Rise of the Community Builders: The American Real Estate Industry and Urban Land Planning* (New York: Columbia University Press, 1987), 40–78, for insight on the evolving trends in home building; and above, chapter 4, for Cochran's Edgewater. For a case study of one of the most influential community builders in the

United States, see Mark H. Rose, "'There Is Less Smoke in the District': J. C. Nichols, Urban Change, and Technological Systems," *Journal of the West* 25(January–February 1986): 44–54.

44. PSCNI, Statistical Department, "Average Annual Kilowatt Hours per Customer by Towns" (photocopy in my possession; unfortunately, these records were inadvertently[?] destroyed by the company after I photocopied a portion of them); U.S. Census, Population, 1920, 1:393–405, 1930, 3, 1:656–72. The townships included in each statistical district are as follows: mixed suburb—Evanston; residential suburb—Deerfield and Shields; industrial suburb—Waukegan; industrial suburb—Bloom; satellite city—Joliet and Troy.

45. Douglass, *Suburban Trend*, 84, 85, for the first and second quotations.

46. See Monchow, *Real Estate Subdividing*, 108–10, for the growth of shopping centers in the suburbs; Hoyt, *Land Values*, 248–55, for a more complete analysis of these commercial corners in the city, including lists and maps of their locations.

47. *Chicago Daily News*, 15 October 1926, 5 for the Edison quotation, the phrase "Daylight Way," and the plans of the Randolph Street merchants, 14 October 1926 for details of the State Street lights. As the newspaper remarked, similar systems had been installed in twenty cities, including London, Paris, San Francisco, and Salt Lake City. Also see Commonwealth Edison Company, *Yearbook*, 1927, 11, for additional technical information about the street lighting system.

48. Hoyt, *Land Values*, 246, for the phrase "blighted area."

49. Richard A. Pence, ed., *The Next Greatest Thing* (Washington, D.C.: National Rural Electric Cooperative Association, 1984), 2, for the quotation. The text of this picture book was taken from a chapter titled "The Sad Irons," in Robert Caro, *The Lyndon Johnson Years: The Path to Power* (New York: Knopf, 1982), 502–15. Caro's description of Johnson's district in Texas during the late 1920s and early 1930s is the best narrative account I have read on the hardships of rural life.

50. Kirschner, *City and Country*, 1–21; James H. Shideler, *Farm Crisis, 1919–1923* (Berkeley: University of California Press, 1957); Theodore Saloutos and John D. Hicks, *Agricultural Discontent in the Middle West, 1900–1939* (Madison: University of Wisconsin Press, 1951), 100–110; George M. Marsden, *Fundamentalism and American Culture: The Shaping of Twentieth-Century Evangelicalism, 1870–1925* (New York: Oxford University Press, 1980), 141–64; U.S. Department of Commerce, Bureau of Statistics, *Historical Statistics of the United States: Colonial Times to 1970*, 2 vols. (Washington, D.C.: Government Printing Office, 1975), 1:27, 458, 462.

51. Ann Dennis Bursch, "Electric Power and the Public Welfare," paper presented to the annual convention of the National League of Women Voters, Saint Louis, 1926, 4, for the quotation. For a similar statement, also see Martha Bensley Bruere, "What Is Giant Power For?" *Annals* 117(March 1925): 120–23. For rural conditions, see A. M. Daniels, "Electric Light and Power in the Farm Home," in U.S. Department of Agriculture, *Yearbook*,

(Washington, D.C.: Government Printing Office, 1919), 223, and D. Clayton Brown, *Electricity for Rural America: The Fight for the REA* (Westport, Conn.: Greenwood, 1980), 7–12.

52. Harvey Levenstein, *Revolution at the Table: The Transformation of the American Diet* (New York: Oxford University Press, 1988), 173–82.

53. See Eagles, "Urban-Rural Conflict," 26–48, and James H. Shideler, "Flappers and Philosophers and Farmers: Rural-Urban Tensions of the Twenties," *Agricultural History* 47(October 1973): 283–99.

54. Kirschner, *City and Country*, 249, 250, for the first and second quotations; Marsden, *Fundamentalism*, 176–95.

55. Evans, *Prairie Farmer and WLS*, 153–75; Kirschner, *City and Country*, 70–73.

56. Gifford Pinchot, as quoted in Brown, *Electricity for Rural America*, xvi; Samuel P. Hays, *Conservation and the Gospel of Efficiency: The Progressive Conservation Movement, 1890–1920* (Cambridge: Harvard University Press, 1959); Preston J. Hubbard, *Origins of the TVA: The Muscle Shoals Controversy, 1920–1932* (Nashville, Tenn.: Vanderbilt University Press, 1961). Also consider two sources on the Regional Planning Association of America, Lubove, *Community Planning*, and Carl Sussman, ed., *Planning the Fourth Migration: The Neglected Vision of the Regional Planning Association of America* (Cambridge: MIT Press, 1976).

57. Morris L. Cooke, "The Long Look Ahead," *Survey Graphic* 51(1 March 1924): 601, for the quotation; Pence, *Next Greatest Thing*, 40–43; Brown, *Electricity for Rural America*, 22–34; Lubove, *Community Planning*, 100–102, for background on Cooke and the Giant Power Survey. In 1931 Governor Franklin D. Roosevelt would appoint Cooke to the new Power Authority of the State of New York, setting the stage for the coming of the TVA.

58. See Morris L. Cooke, "A Note on Rural Electric Rates," *Annals* 117(March 1925): 52–59, and Brown, *Electricity for Rural America*, 3–12. For reflections on the history of rural electrification from the perspective of the early 1940s, compare Morris L. Cooke, "The Early Days of the Rural Electrification Idea," *American Political Science Review* 42(June 1948): 431–37; Royden Stewart, "Rural Electrification in the United States: The Pioneer Period, 1906–1923," *Edison Electric Institute Bulletin* 9(September 1941): 381–86; Royden Stewart, "Rural Electrification in the United States: National Development, 1924–1935: The Committee on the Relation of Electricity to Agriculture," *Edison Electric Institute Bulletin* 9(October 1941): 409–15.

59. U.S. Census, Agriculture, 1930, 2, 1:570–75, 584–89; PSCNI, *Thirty-seven Years of Farm Electric Service*, pamphlet, 1950, in Chicago Historical Society Library. The calculations used here are estimates based on the number of farm families in the following counties: Lake, Cook, Will, Kendall, Grundy, Kankakee, Livingston, and Marshall. During the 1920s, the number of farm families in these counties declined from 21,063 to 17,337, while the value of farms fell significantly in all the counties except Lake and Cook, which were undergoing rapid suburbanization.

60. See above, note 44.

61. Hubbard, *Origins of the TVA;* Brown, *Electricity for Rural America,* 11–33; Richard Lowitt, *George W. Norris: The Persistence of a Progressive, 1913–1933* (Urbana: University of Illinois Press, 1971), 149–216, 260–71, 330–64.

Chapter Ten

1. See Commonwealth Edison Company, *Yearbook,* 1928, 23, in CEC-HA, box 113, for the quotation; Chicago Arc Light and Power Company files, CEC-LF, F6-E3, for additional materials on the building's use as a central station; above, chapter 3, for Bullock's role as an agent for Charles Brush's arc light system; and Forrest McDonald, *Insull* (Chicago: University of Chicago Press, 1962), 242–44, 283–84, for Insull's patronage of the opera. Also see Carl Condit, *Chicago, 1910–1929: Building, Planning, and Urban Technology* (Chicago: University of Chicago Press, 1973), 125–29, for a description of the impressive new structure.

2. Commonwealth Edison Company, *Yearbook,* 1928, 12, in CEC-HA, box 113.

3. Condit, *Chicago, 1910–1929,* 121–25; Commonwealth Edison Company, *Annual Report,* 1929, 8, in CEC-HA, box 113.

4. Jeffrey L. Meikle, *Twentieth Century Limited: Industrial Design in America, 1925–1939* (Philadelphia: Temple University Press, 1979), 27 for the quotation, 7–38 for the emergence of industrial design. Although Meikle does not place a precise date on the appearance of the first products designed with machine age aesthetics, he makes frequent reference to the period 1925–28 as formative. Also see Richard Guy Wilson, Dianne H. Pilgrim, and Dickran Tashjian, eds., *The Machine Age* (New York: Brooklyn Museum and Harry N. Abrams, 1986).

5. See Meikle, *Twentieth Century Limited,* 103–4, and Ruth Schwartz Cowan, *More Work for Mother: The Ironies of Household Technology from the Open Hearth to the Microwave* (New York: Basic, 1983), 128–43, on the monitor-top refrigerator. See Commonwealth Edison Company, *Yearbook,* 1930, 8, and 1931, 14, in CEC-HA, box 113; "Industrial Air-Conditioning Progress in Chicago, *Central Manufacturing Magazine* 20(May 1936): 7–15; and Knight C. Porter, "Air Condition Progress in Chicago," *Central Manufacturing Magazine* 21(May 1937): 9–11 (copies of the two articles are found in CEC-LF, F4D-E2), for the spread of air-conditioning in the city. The importance of air-conditioning as a new source of "load building" is underscored by its mention in virtually every annual report of the utility company during the depression years. See Commonwealth Edison Company, *Annual Report,* 1930–39, in CEC-HA, boxes 113–14.

6. Public Service Company of Northern Illinois (PSCNI), *Yearbook,* 1928, 21–22, and 1929, 12–13, in CEC-HA, box 4094; PSCNI, *Thirty-seven Years of Farm Electric Service* (Chicago: PSCNI, 1950); "model farm" file, in CEC-LF, F35B; and Samuel Insull, "Address of Mr. Samuel Insull at the Opening of the Model Farm," Lake County, Ill., PSCNI, 1928, pamphlet (copy at the Chicago Historical Society). An intriguing memo submitted as an exhibit

in a 1933–34 rate case suggests that rural electrification during the first half of the depression was limited largely to the country estates of the rich just beyond the suburbs in Lake, Cook, and Du Page counties. The 1934 note reveals that two-thirds of the 8 million kwh of electricity sold to farm customers in the PSCNI service territory came from these collar counties. See PSCNI, "Kilowatt Hours Sold to Farm Customers—Year 1932," 25 January 1934, Illinois Commerce Commission, Case 22353 (1933), Company Exhibit 31 in CEC-HA, box 4093.

7. PSCNI, *Yearbook*, 1928, 19, for the quotation and statistics.

8. Commonwealth Edison Company, *Yearbook*, 1928, 12, 15–18, PSCNI, *Yearbook*, 1928, 9–12, 18–19, 21–22. On 16 October 1931, a twenty-four-inch pipeline of natural gas opened between Texas and Joliet, Illinois, where the Chicago District Pipeline Company then distributed the energy fuel to a variety of local utility companies. See PSCNI, *Annual Report*, 1931, in CEC-HA, box 4094. For rich detail on the coming of natural gas to Chicago, consider the gas pipeline case, Illinois Commerce Commission, Case 12225 (1932), in CEC-HA, boxes 4087–92. The pioneering nature of this venture receives notice in Martin V. Melosi, *Coping with Abundance: Energy and Environment in Industrial America* (New York: Knopf, 1985), 156.

9. Paul Barrett, *The Automobile and Urban Transit: The Formation of Public Policy in Chicago, 1900–1930* (Philadelphia: Temple University Press, 1983), 182–83, 197–208; Linda Jane Lears, "The Aggressive Progressive: The Political Career of Harold L. Ickes, 1874–1933" (Ph.D. diss., George Washington University, 1974), 310–45; McDonald, *Insull*, 252–70.

10. Carroll H. Wooddy, *The Case of Frank L. Smith: A Study in Representative Government* (Chicago: University of Chicago Press, 1931), 55–58, 241–43.

11. Ernest Greuning, *The Public Pays* (New York: Vanguard, 1931); M. L. Ramsay, *Pyramids of Power* (Indianapolis: Bobbs-Merrill, 1937); George W. Norris, "The Power Trust in the Public Schools," *Nation* 127(18 September 1929): 296–97; Franklin D. Roosevelt, "Acceptance of the Nomination for the Governorship," 16 October 1928, in *The Public Papers and Addresses of Franklin D. Roosevelt*, 13 vols., comp. Samuel I. Rosenman (New York: Random House, 1938–50), 1:13–16; Franklin D. Roosevelt, "Campaign Address," 23 October 1928, in ibid., 44–51; Franklin D. Roosevelt, "Inaugural Address," 1 January 1929, in ibid., 75–86; Melosi, *Coping with Abundance*, 118–26; Philip J. Funigiello, *Toward a National Power Policy: The New Deal and the Electric Utility Industry, 1933–1941* (Pittsburgh: University of Pittsburgh Press, 1973), 3–31; McDonald, *Insull*, 255–73.

12. McDonald, *Insull*, 284 for the quotation, 274–87 for a description of his financial empire. Also see the organizational chart of the complete Insull group of utility companies in Ramsay, *Pyramids of Power*, 244.

13. Franklin Roosevelt, "New Conditions Impose New Requirements upon Government and Those Who Conduct Government," campaign address at the Commonwealth Club of San Francisco, 23 September 1932, as quoted in Rosenman, *Public Papers of Roosevelt*, 1:755. Also see Roosevelt's

statement two days earlier of his policy position on public utilities, which attacked the "Insull Monstrosity," in Franklin Roosevelt, "A National Yardstick to Prevent Extortion . . . ," campaign address at Portland, Oregon, 21 September 1932, in ibid., 727–42. See McDonald, *Insull,* 286–304, for a description of the collapse of Insull's financial empire.

14. McDonald, *Insull,* 305–39; Francis Xavier Busch, *Guilty or Not Guilty?* (Indianapolis: Bobbs-Merrill, 1952), 127–94. For the documents of the trial, see the Samuel Insull Papers, Chicago, Loyola University of Chicago Archives, boxes 56–63. Also see his correspondence with his son, Samuel Insull, Jr., which reveals a desperate effort to keep the financial empire from collapse, in Samuel Insull Papers, boxes 9–11.

15. See Funigiello, *Toward a National Power Policy,* 32–121; Ellis Hawley, *The New Deal and the Problem of Monopoly* (Princeton: Princeton University Press, 1966), 304–43; and Melosi, *Coping with Abundance,* 126–34, for an introduction to the reform legislation of the New Deal. Compare Richard Rudolph and Scott Ridley, *Power Struggle: The Hundred-Year War over Electricity* (New York: Harper and Row, 1986), which emphasizes the loopholes in the Holding Company Act that have allowed financial concentration of the utilities industry to continue mostly without effective restraint. Although written in a muckraking style by journalists, the book deserves serious consideration because of the authors' impressive research effort to document their allegations.

16. See Thomas K. McGraw, *TVA and the Power Fight, 1933–1939* (Philadelphia: Lippincott, 1971); Funigiello, *Toward a National Power Policy,* 122–25; and D. Clayton Brown, *Electricity for Rural America: The Fight for REA* (Westport, Conn.: Greenwood, 1980). For the ascendancy of a regional concept of planning, see Mel Scott, *American City Planning since 1890* (Berkeley: University of California Press, 1969), 170–82, 190–227; Roy Lubove, *Community Planning in the 1920s: The Contributions of the Regional Planning Association of America* (Pittsburgh: University of Pittsburgh Press, 1963); and Carl Sussman, ed., *Planning the Fourth Migration: The Neglected Vision of the Regional Planning Association of America* (Cambridge: MIT Press, 1976).

17. For the development of Insull's superpower system, see Commonwealth Edison Company, *Annual Report,* 1924–31, in CEC-HA, box 113, and PSCNI, *Annual Report,* 1924–31, in CEC-HA, box 4094. The records of the Superpower Company are contained in CEC-HA, boxes 3017–33. For Insull's plans for integrating the power networks of the Midwest, see National Electric Light Association (NELA), Power Survey Committee, *Electric Power Survey: Report on Area A—Great Lakes Division* (Chicago: NELA, 1924–28).

18. See above, note 16, and Melosi, *Coping with Abundance,* 134–37. But also see the scathing critique of the TVA in Jane Jacobs, *Cities and the Wealth of Nations* (New York: Random House, 1984), 105–23.

19. See Commonwealth Edison Company, *Annual Report,* 1928–88, in CEC-HA, boxes 113–16, for electric rates in Chicago; Melosi, *Coping with*

Abundance, 277–332, for an introduction to the energy transition of the seventies: H. K. Vietor, *Energy Policy in America since 1945* (New York: Cambridge University Press, 1984), for perspectives on the energy-fuels policies of the federal government. Also see Richard Hirsh, *Technology and Transformation in the American Electric Utility Industry* (New York: Cambridge University Press, 1989).

20. H. A. Seymour, "History of Commonwealth Edison," 3 vols., manuscript, 1934, vol. 3, Appendix, 81, 174, in CEC-HA, box 9001; Commonwealth Edison Company, *Annual Report,* 1933–40, in CEC-HA, box 114.

21. James Simpson, address at the annual meeting of the stockholders of the Commonwealth Edison Company, 26 February 1934, 9, in CEC-HA, box 114; Commonwealth Edison Company, *Annual Report,* 1933, 12, in CEC-HA, box 114; Commonwealth Edison Company, *Annual Report,* 1940, 14, in CEC-HA, box 114. On the World's Fair, see the materials in the "World's Fair" file in CEC-LF, F45B; Lenox Riley Lohr, *Fair Management, the Story of Century of Progress Exposition: A Guide for Future Fairs* (Chicago: Cuneo, 1952); and Century of Progress International Exposition, 1933–34, *Official Book of the Fair* (Chicago: Century of Progress International Exposition, 1932). Although the Chicago exposition has been largely ignored by cultural historians, the New York World's Fair of 1939 has received more attention. For a collection of excellent articles on the New York experience, including an essay by Warren Susman, see Helen A. Harrison, ed., *Dawn of a New Day: The New York World's Fair, 1939/40* (New York: Queens Museum and New York University Press, 1980).

22. Seymour, "History of Commonwealth Edison," Appendix, 174, 176; Commonwealth Edison Company, *Annual Report,* 1933–40.

23. Mina Edison, quoted in David Nye, "Edison's Reification at Light's Golden Jubilee," manuscript, Union College, [1983?], 11. Also see Wyn Wachhorst, *Thomas Alva Edison: An American Myth* (Cambridge: MIT Press, 1981), 131–68, for a complementary analysis of Edison's symbolism in American popular culture. For a brief account of the celebration of the jubilee in Chicago, see Commonwealth Edison Company, *Yearbook,* 1928, 21.

24. Joel A. Tarr and Gabriel Dupuy, eds., *Technology and the Rise of the Networked City in Europe and America* (Philadelphia: Temple University Press, 1989), for the phrase "networked city" and an excellent summary of recent research on the history of the urban infrastructure.

25. Tarr and Dupuy, *Technology,* for infrastructure; Sam Bass Warner, Jr., *The Urban Wilderness: A History of the American City* (New York: Harper and Row, 1972), for the best statement on the culture of the American city; Perry Duis, "Whose City? Public and Private Places in Nineteenth Century Chicago," *Chicago History* 7(Spring 1983): 2–27, and 7(Summer 1983): 2–23, for a taxonomy of urban space; Kenneth T. Jackson, *The Crabgrass Frontier: The Suburbanization of the United States* (New York: Oxford University Press, 1985); Margaret Marsh, "From Separation to Togetherness: The Social Construction of Domestic Space in American Suburbs, 1840–1915," *Journal*

of American History 76(September 1989): 506–27, for patterns of social segregation; Siefried Giedion, *Mechanization Takes Command* (New York: Oxford University Press, 1948), 659–93, and Nelson Manfred Blake, *Water for the Cities: A History of the Urban Water Supply Problem in the United States* (Syracuse: Syracuse University Press, 1956), for the spread of water service into the home.

26. Giedion, *Mechanization Takes Command,* 556 for the quotation, 41, 93, 548–96.

27. George Rodgers Taylor, "The Beginnings of Mass Transportation in Urban America," *Smithsonian Journal of History* 1(Summer 1966): 35–50, and 1(Autumn 1966): 39–52; and Charles W. Cheape, *Moving the Masses: Urban Public Transit in New York, Boston, and Philadelphia, 1880–1912* (Cambridge: Harvard University Press, 1980).

28. Robert H. Wiebe, *The Search for Order, 1877–1920* (New York: Hill and Wang, 1967), for the phrase "island communities."

29. Lewis Mumford, *Technics and Civilization* (New York: Harcourt Brace and World, 1934). Also see John B. Rae, *The Road and the Car in American Life* (Cambridge: MIT Press, 1971); and Joel A. Tarr and Josef W. Konvitz, "Patterns in the Development of the Urban Infrastructure," in *American Urbanism: A Historiographical Review,* ed. Howard Gillette, Jr., and Zane L. Miller (Westport, Conn.: Greenwood, 1987), 195–226.

30. Patrick Geddes, *Cities in Evolution* (London: Williams and Norgate, 1915); Ebenezer Howard, *Garden Cities of Tomorrow* (London: Sonnenschein, 1902); for the idea of the green-belt city; Richard Ingersoll, Elizabeth Hollander, et al., "The Re-Creation of the Street," papers read at a symposium at the Chicago Historical Society, Chicago, 1989, for new directions in city planning.

31. Marshall McLuhan, *Understanding Media: The Extensions of Man* (New York: McGraw-Hill, 1964); Marshall McLuhan and Quentin Fiore, *War and Peace in the Global Village* (New York: Bantam, 1968).

32. McLuhan, *Understanding Media,* vi for the quotation, 3–21.

33. Liz Cohen, "Encountering Mass Culture at the Grassroots: The Experience of Chicago Workers in the 1920s," *American Quarterly* 41(March 1989): 6–33.

34. For insight on this transformation, see Warren Susman, "Culture and Civilization: The Nineteen-Twenties," in *Culture as History: The Transformation of American Society in the Twentieth Century,* ed. Warren Susman (New York: Pantheon, 1984), 105–21.

35. Eugene A. Santomasso, "The Design of Reason: Architecture and Planning at the 1939/40 New York World's Fair," in Harrison, *Dawn of a New Day,* 29 for the quotation, 29–41. Also see the other excellent essays on this theme by Joseph P. Cusker, Warren I. Susman, and Helen A. Harrison, in ibid., 3–56, passim, and the sources listed above in note 21 for the Century of Progress.

Index

Illinois and Michigan canal. *See* Illinois River valley
Illinois Central Railroad Company, 177, 207, 252
Illinois Council of Defense, 206–7, 225, 231. *See also* World War I
Illinois Maintenance Company, 102
Illinois River valley, 8; and regional integration, 166, 183, 185–90, 192; and transportation of coal, 11. *See also* Regional integration; *and names of individual communities*
Illinois Valley Gas and Electric Company, 190. *See also* Public Service Company of Northern Illinois
Incandescent lights: alternative systems of, 71–74; in Chicago, introduction of, 33–39; in homes, early use of, 34, 153–54, 170; improved bulbs, 145–46; and Light's Golden Jubilee, 279; in Loop, early use of, 53–55; new levels of intensity of, 247, 260, 278, 286. *See also* Arc lights; Chicago Edison Company; Electric lighting
Industrialization: and coal, 9–10; and conversion to electric power, 176, 203–4, 208–20; effect of World War I on, 201–4, 208–10, 211–20, 232–34, 237–39; and steam power, 208–10, 268; and suburbanization, 174–78, 186–87, 190–95, 210–11, 254, 256–61, 280–85. *See also* Central station electric systems; Consumerism; Electric power; Food, processing of; Self-contained electric systems; Technology and culture, and industrial design
Insull, Samuel, 66–67, 152, 189, 273; and Chicago Edison Company, 67–68, 74–78, 82; and Edison, 66–67, 83, 279; and gas business, 229–32, 270–73; and gospel of consumption, 111–12, 115–19, 122–24, 137–38, 173, 204, 229, 269, 275–79; and government-business relations, 125–37, 196–97, 227, 229–32, 270–73; and Great Depression, 271–73; and ice business, 223; legacy of, 273–79; and regional integration,

183, 187–88, 188–90, 269–70, 274; and sales campaign, 156–59, 231, 235, 237, 239–40, 268–69; and street railways, 173–74, 270–71; and suburban utilities, 181. *See also* Chicago Edison Company; Commonwealth Edison Company; Public Service Company of Northern Illinois
Interurban railroads, 141–42, 149; and suburbanization, 164, 173, 183, 186, 187, 245, 252–53. *See also* Regional integration; Street railways; Suburbanization
"Isolated" electric systems. *See* Self-contained electric systems

Joliet, Ill., 71, 166, 185–87, 256–61

Kankakee, Ill., 166, 188
Kerosene lights, 14, 263
Kingsland, P. S., 22, 29

La Follette, Robert M., Jr., 266
Lake Forest, Ill. *See* Highland Park, Ill.; North Shore
Lake Street Elevated Company, 120, 126. *See also* Street railways
La Salle, Ill., 166
Lemont, Ill., 187
Lever Food and Fuel Act (1917), 206. *See also* United States Food Administration; United States Fuel Administration
Libertyville, Ill., 189–90
Lighting. *See* Electric lighting; Gas services
Light's Golden Jubilee (1929), 279
Lloyd, Henry Demarest, 169
Lockport, Ill., 166
Lowden, Frank O., 206
Lundy, Ayres D. *See* Sargent and Lundy
Lynd, Robert and Helen, 221, 222, 226, 244, 252

McAdoo, William, 207
Machine age. *See* Technology and culture
McLuhan, Marshall, 285–88